Coastal Construction Manual

Principles and Practices of Planning, Siting, Designing, Constructing, and Maintaining Residential Buildings in Coastal Areas
(Fourth Edition)

FEMA P-55 / Volume II / August 2011

FEMA

Preface

The 2011 *Coastal Construction Manual*, Fourth Edition (FEMA P-55), is a two-volume publication that provides a comprehensive approach to planning, siting, designing, constructing, and maintaining homes in the coastal environment. Volume I of the *Coastal Construction Manual* provides information about hazard identification, siting decisions, regulatory requirements, economic implications, and risk management. The primary audience for Volume I is design professionals, officials, and those involved in the decision-making process.

Volume II contains in-depth descriptions of design, construction, and maintenance practices that, when followed, will increase the durability of residential buildings in the harsh coastal environment and reduce economic losses associated with coastal natural disasters. The primary audience for Volume II is the design professional who is familiar with building codes and standards and has a basic understanding of engineering principles.

Volume II is not a standalone reference for designing homes in the coastal environment. The designer should have access to and be familiar with the building codes and standards that are discussed in Volume II and listed in the reference section at the end of each chapter. The designer should also have access to the building codes and standards that have been adopted by the local jurisdiction if they differ from the standards and codes that are cited in Volume II. If the local jurisdiction having authority has not adopted a building code, the most recent code should be used. Engineering judgment is sometimes necessary, but designers should not make decisions that will result in a design that does not meet locally adopted building codes.

The topics that are covered in Volume II are as follows:

- **Chapter 7 –** Introduction to the design process, minimum design requirements, losses from natural hazards in coastal areas, cost and insurance implications of design and construction decisions, sustainable design, and inspections.

■ **Chapter 8 –** Site-specific loads, including from snow, flooding, tsunamis, high winds, tornadoes, seismic events, and combinations of loads. Example problems are provided to illustrate the application of design load provisions of ASCE 7-10, *Minimum Design Loads for Buildings and Other Structures*.

■ **Chapter 9 –** Load paths, structural connections, structural failure modes, breakaway walls, building materials, and appurtenances.

■ **Chapter 10 –** Foundations, including design criteria, requirements and recommendations, style selection (e.g., open, closed), pile capacity in soil, and installation.

■ **Chapter 11 –** Building envelope, including floors in elevated buildings, exterior doors, windows and skylights, non-loading-bearing walls, exterior wall coverings, soffits, roof systems, and attic vents.

■ **Chapter 12 –** Installing mechanical equipment and utilities.

■ **Chapter 13 –** Construction, including the foundation, structural frame, and building envelope. Common construction mistakes, material selection and durability, and techniques for improving resistance to decay and corrosion are also discussed.

■ **Chapter 14 –** Maintenance of new and existing buildings, including preventing damage from corrosion, moisture, weathering, and termites; building elements that require frequent maintenance; and hazard-specific maintenance techniques.

■ **Chapter 15 –** Evaluating existing buildings for the need for and feasibility of retrofitting for wildfire, seismic, flood, and wind hazards and implementing the retrofitting. Wind retrofit packages that can be implemented during routine maintenance are also discussed (e.g., replacing roof shingles).

For additional information on residential coastal construction, see the FEMA Residential Coastal Construction Web site at http://www.fema.gov/rebuild/mat/fema55.shtm.

Acknowledgments

Fourth Edition Authors and Key Contributors
William Coulbourne, Applied Technology Council
Christopher P. Jones, Durham, NC
Omar Kapur, URS Group, Inc.
Vasso Koumoudis, URS Group, Inc.
Philip Line, URS Group, Inc.
David K. Low, DK Low and Associates
Glenn Overcash, URS Group, Inc.
Samantha Passman, URS Group, Inc.
Adam Reeder, Atkins
Laura Seitz, URS Group, Inc.
Thomas Smith, TLSmith Consulting
Scott Tezak, URS Group, Inc. – Consultant Project Manager

Fourth Edition Volume II Reviewers and Contributors
Katy Goolsby-Brown, FEMA Region IV
John Ingargiola, FEMA Headquarters – Technical Assistance and Research Contracts Program Manager
John Plisich, FEMA Region IV
Paul Tertell, FEMA Headquarters – Project Manager
Ronald Wanhanen, FEMA Region VI
Gregory P. Wilson, FEMA Headquarters
Brad Douglas, American Forest and Paper Association
Gary Ehrlich, National Association of Home Builders
Dennis Graber, National Concrete Masonry Association
David Kriebel, United States Naval Academy
Marc Levitan, National Institute of Standards and Technology
Tim Mays, The Military College of South Carolina
Sam Nelson, Texas Department of Insurance
Janice Olshesky, Olshesky Design Group, LLC
Michael Powell, Delaware Department of Natural Resources and Environmental Control
David Prevatt, University of Florida
Timothy Reinhold, Insurance Institute for Business & Home Safety
Tom Reynolds, URS Group, Inc.
Michael Rimoldi, Federal Alliance for Safe Homes
Randy Shackelford, Simpson Strong-Tie
John Squerciati, Dewberry
Keqi Zhang, Florida International University

Fourth Edition Technical Editing, Layout, and Illustration
Diana Burke, URS Group, Inc.
Lee-Ann Lyons, URS Group, Inc.
Susan Ide Patton, URS Group, Inc.
Billy Ruppert, URS Group, Inc.

Contents

Chapter 8. Determining Site-Specific Loads

List of Figures

Chapter 9

Chapter 10

Chapter 11

Chapter 12

Chapter 13

Chapter 14

Chapter 15

List of Tables

Chapter 7

Chapter 8

Chapter 9

Chapter 10

List of Equations

Chapter 10

Chapter 13

List of Examples

Chapter 8

Chapter 9

Chapter 10

List of Worksheets

Chapter 8

Pre-Design Considerations

This chapter provides an overview of the issues that should be considered before the building is designed.

Coastal development has increased in recent years, and some of the sites that are chosen for development have higher risks of impact from natural hazards than in the past. Examples of sites with higher risks are those that are close to the ocean, on high bluffs that are subject to erosion, and on artificial fill deposits. In addition, many of the residential buildings constructed today are larger and more costly than before, leading to the potential for larger economic losses if disaster strikes. However, studies conducted by the Federal Emergency Management Agency (FEMA) and others after major coastal disasters have consistently shown that coastal residential buildings that are properly sited, designed, and constructed have generally performed well during natural hazard events.

Important decisions need to be made prior to designing the building. The decisions should be based on an understanding of regulatory requirements, the natural hazard and other risks associated with constructing a building on a particular site (see Chapter 4), and the financial implications of the decisions. The financial implications of siting decisions include the cost of hazard insurance, degree of hazard resistance and sustainability in the design, and permits and inspections.

CROSS REFERENCE

For resources that augment the guidance and other information in this Manual, see the Residential Coastal Construction Web site (http://www.fema.gov/rebuild/mat/fema55.shtm).

Once a site has been selected, decisions must be made concerning building placement, orientation, and design. These decisions are driven primarily by the following:

- Owner, designer, and builder awareness of natural hazards

- Risk tolerance of the owner

- Aesthetic considerations (e.g., building appearance, proximity to the water, views from within the building, size and number of windows)

- Building use (e.g., full-time residence, part-time residence, rental property)

- Requirements of Federal, State, and local regulations and codes

- Initial and long-term costs

The interrelationships among aesthetics, building use, regulatory and code requirements, and initial cost become apparent during siting and design, and decisions are made according to the individual needs or goals of the property owner, designer, or builder. However, an understanding of the effect of these decisions on long-term and operational costs is often lacking. The consequences of the decisions can range from increased maintenance and utility costs to the ultimate loss of the building. The goal of this Manual is to provide the reader with an understanding of these natural hazards and provide guidance on concepts for designing a more hazard-resistant residential building.

7.1 Design Process

The design process includes a consideration of the types of natural hazards that occur in the area where the building site is located and the design elements that allow a building to effectively withstand the potential damaging effects of the natural hazards (see Figure 7-1). The intent of this Manual is to provide sufficient technical information, including relevant examples, to help the designer effectively design a coastal residential building.

This Manual does not describe all combinations of loads, types of material, building shapes and functions, hazard zones, and elevations applicable to building design in the coastal environment. The designer must apply engineering judgment to a range of problems. In addition, good design by itself is not enough to guarantee a high-quality structure. Although designing building components to withstand site-specific loads is important, a holistic approach that also includes good construction, inspection, and maintenance practices can lead to a more resilient structure.

Before designing a building and to optimize the usefulness of Volume II of this Manual, the designer should obtain the codes and standards, such as ASCE 7 and ASCE 24, that are listed in the reference section of each chapter and other relevant information such as locally adopted building codes and appropriate testing protocols.

Although codes and standards provide minimums, the designer may pursue a higher standard. Many decisions require the designer's judgment, but it is never appropriate to use a value or detail that will result in a building that is not constructed to code.

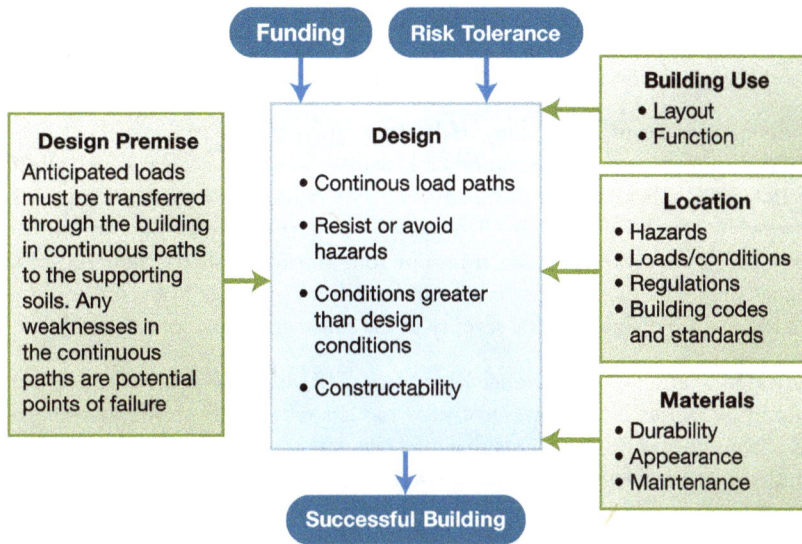

Figure 7-1.
Design framework for
a successful building,
incorporating cost, risk
tolerance, use, location,
materials, and hazard
resistance

Volume II contains many design equations, but they do not cover all of the design calculations that are necessary and are provided only as examples.

7.2   Design Requirements

The minimum design requirements for loads, materials, and material resistances for a given building design are normally specified in the locally adopted building code. Nothing in this Manual is intended to recommend the use of materials or systems outside the uses permitted in building code requirements. The loads used in this Manual are based on ASCE 7-10, which is the reference load standard in model building codes. Material and material resistance requirements cited in this Manual are based on the minimum requirements of applicable building codes. However, designers are encouraged throughout the Manual to seek out information on loads and materials that exceed the minimum requirements of the building code. Other sources of information for loads and materials are also provided.

7.3   Determining the Natural Hazard Risk

Assessing risk to coastal buildings and building sites requires identifying or delineating hazardous areas and considering the following factors:

- Types of hazards known to affect a region

- Geographic variations in hazard occurrence and severity

- Methods and assumptions underlying existing hazard identification maps or products

- "Acceptable" level of risk

- Consequences of using (or not using) recommended siting, design, and construction practices

Geographic variations in coastal hazards occur, both along and relative (perpendicular) to the coastline. Hazards affecting one region of the country may not affect another. Hazards such as wave loads, which affect construction close to the shoreline, usually have a lesser or no effect farther inland. For example, Figure 7-2 shows how building damage caused by Hurricane Eloise in 1975 was greatest at the shoreline but diminished rapidly in the inland direction. The figure represents data from only one storm but shows the trend of a typical storm surge event on coastlines (i.e., damage decreases significantly as wave height decreases). The level of damage and distance landward are dictated by the severity of the storm and geographic location.

Through Flood Insurance Studies (FISs) and Flood Insurance Rate Maps (FIRMs), FEMA provides detailed coastal flood hazard information (see Section 3.5). However, these products reflect only flood hazards and do not include a consideration of a number of other hazards that affect coastal areas. Other Federal agencies and some states and communities have completed additional coastal hazard studies and delineations. The Residential Coastal Construction Web site (http://www.fema.gov/rebuild/mat/fema55.shtm) provides introductory information concerning more than 25 hazard zone delineations developed by or for individual communities or states (see "Web Sites for Information about Storms, Big Waves, and Water Level"). Some delineations have been incorporated into mandatory siting and/or construction requirements.

When reviewing the hazard maps and delineations that are provided on the Residential Coastal Construction Web site, designers should be aware that coastal hazards are often mapped using different levels of risk or recurrence intervals. Thus, the **consistent** and **acceptable level of risk** (the level of risk judged by the designer to be appropriate for a particular building) should be considered early in the planning and design process (see Chapter 6). The hazard maps and delineations are provided as a historical reference only. The most up-to-date information can be obtained by contacting local officials.

Figure 7-2.
Average damage per structure (in thousands of 1975 dollars) versus distance from the Florida Coastal Construction Control Line for Bay County, FL, Hurricane Eloise (Florida, 1975)
SOURCE: ADAPTED FROM SHOWS 1978

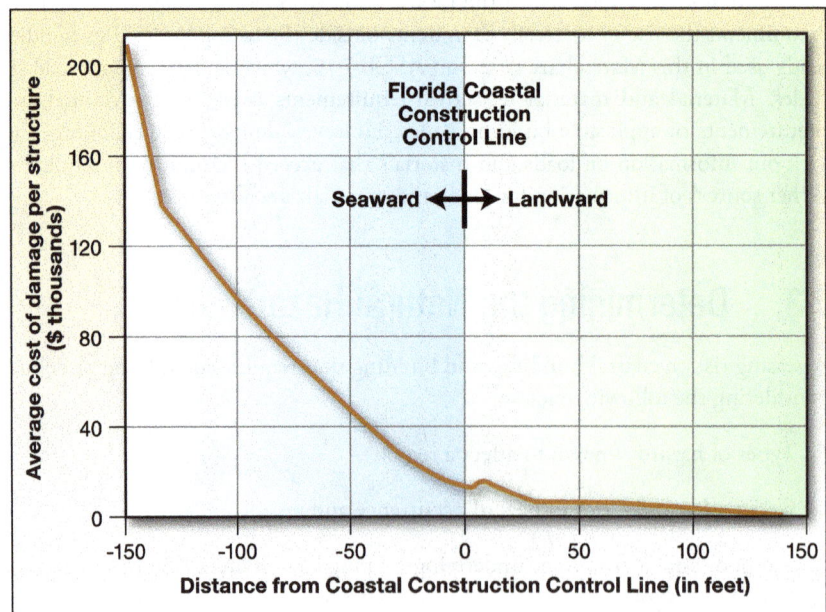

7.4   Losses Due to Natural Hazards in Coastal Areas

It is easy for property owners to become complacent about the potential for a natural disaster to affect their properties. Hurricanes and earthquakes are generally infrequent events. A geographic area may escape a major hazard event for 20 or more years. Or, if an area has recently been affected, residents may believe the chances of a recurrence in the near future are remote. These perceptions are based on inaccurate assumptions and/or a lack of understanding of natural hazards and the risk of damage.

The population and property values along the U.S. coast are both rapidly increasing. Although better warning systems have reduced the number of fatalities and injuries associated with natural disasters, increases in the number and value of structures along the coast have dramatically increased potential property losses.

From 2000 through 2009, there were 13 presidentially declared disasters resulting from hurricanes and tropical systems, each causing more than $1 billion in losses. Hurricane Katrina in 2005 was the most expensive natural disaster in U.S. history, causing estimated economic losses of more than $125 billion and insured losses of $35 billion, surpassing Hurricane Andrews's $26.5 billion in losses in 1992. Other recent memorable storms are Tropical Storm Allison (2001), Hurricane Rita (2005), Hurricane Wilma (2005), Hurricane Ike (2008), and the 2004 hurricane season in which four storms (Charley, Frances, Ivan, and Jeanne) affected much of the East Coast in both coastal and inland areas.

Following Hurricane Andrew, which ravaged south Florida in 1992, studies were conducted to determine whether the damage suffered was attributable more to the intensity of the storm or to the location and type of development. According to the Insurance Institute for Business and Home Safety (IBHS):

> Conservative estimates from claim studies reveal that approximately 25 percent of Andrew-caused insurance losses (about $4 billion) were attributable to construction that failed to meet the code due to poor enforcement, as well as shoddy workmanship. At the same time, concentrations of population and of property exposed to hurricane winds in southern Florida grew many-fold (IBHS 1999).

After Hurricane Andrew, codes and regulations were enacted that support stronger building practices and wind protection. IBHS conducted a study in 2004 following Hurricane Charley that found:

> … homes built after the adoption of these new standards resulted in a decrease in the frequency and severity of damage to various building components. Furthermore, based on the analysis of additional living expense records, it is concluded that the new building code requirements allowed homeowners to return to their home more quickly and likely reduced the disruption of their day to day lives (IBHS 2004, p. 5).

NOTE

According to the Mortgage Bankers Association (2006), from 1985 to 2005, hurricanes and tropical storms accounted for the major share of all catastrophic insurance losses. The percentages of property damage caused by various catastrophic events during this period were:

- 43.7 percent from hurricane/tropical storms
- 23.3 percent from wind/thunderstorms
- 5.1 percent from earthquakes

Approximately 94.4 percent of all catastrophic events occurring during this period were attributed to natural disasters.

The past several decades have not resulted in major losses along the Pacific coast or Great Lakes, but periodic reminders support the need for maintaining a vigilant approach to hazard-resistant design for coastal structures in other parts of the country. In February 2009, the Hawaiian Islands and portions of California were under a tsunami watch. This type of watch occurs periodically in sections of northern California, Oregon, Washington, and Alaska and supports the need to construct buildings on elevated foundations. Although tsunamis on the Pacific coast may be less frequent than coastal hazard events on the Atlantic coast, ignoring the threat can result in devastating losses.

Hazard events on the coastlines of the Great Lakes have resulted in damage to coastal structures and losses that are consistent with nor'easters on the Atlantic coast. Surge levels and high winds can occur every year on the Great Lakes, and it is important for designers to ensure that homeowners and builders understand the nature of storms on the Great Lakes. As in other regions, storm-related losses can result in the need to live in a house during lengthy repairs or be displaced for extended periods while the house is being repaired. The loss of irreplaceable possessions or property not covered by flood or homeowners insurance policies are issues a homeowner should be warned of and are incentives to taking a more hazard-resistant design approach.

Chapters 2 and 3 contain more information about the hazards and risks associated with building in coastal areas.

7.5   Initial, Long-Term, and Operational Costs

Like all buildings, coastal residential buildings have initial, long-term, and operational costs.

- *Initial costs* include property evaluation, acquisition, permitting, design, and construction.

- *Long-term costs* include preventive maintenance and repair and replacement of deteriorated or damaged building components. A hazard-resistant design can result in lower long-term costs by preventing or reducing losses from natural hazard events.

- *Operational costs* include costs associated with the use of the building, such as the cost of utilities and insurance. Optimizing energy efficiency may result in a higher initial cost but save in operational costs.

In general, the decision to build in any area subject to significant natural hazards—especially coastal areas—increases the initial, long-term, and operational costs of building ownership. Initial costs are higher because the natural hazards must be identified, the associated risks assessed, and the building designed and constructed to resist damage from the natural hazard forces. Long-term costs are likely to be higher because a building in a high-risk area usually requires more frequent and more extensive maintenance and repairs than a building sited elsewhere. Operational costs are often higher because of higher insurance costs and, in some instances, higher utility costs. Although these costs may seem higher, benefits such as potential reductions in insurance premiums and reduced repair time following a natural disaster may offset the higher costs.

7.5.1   Cost Implications of Siting Decisions

The cost implications of siting decisions are as follows:

- The closer buildings are sited to the water, the more likely they are to be affected by flooding, wave action, erosion, scour, debris impact, overwash, and corrosion. In addition, wind speeds are typically higher along coastlines, particularly within the first several hundred feet inland. *Repeated exposure to these hazards, even when buildings are designed to resist their effects, can lead to increased long-term costs for maintenance and damage repair.*

- Erosion—especially long-term erosion—poses a serious threat to buildings near the water and on high bluffs above the floodplain. Wind-induced erosion can lower ground elevations around coastal buildings, exposing Zone V buildings to higher-than-anticipated forces, and exposing Zone A buildings to Zone V flood hazards. *Maintenance and repair costs are high for buildings in erosion hazard areas, not only because of damage to the building, but also because of the need for remedial measures (e.g., building relocation or erosion protection projects, such as seawalls, revetments, and beach nourishment, where permitted).*

COST CONSIDERATION

Designers and homeowners should recognize that erosion control measures can be expensive, both initially and over the lifetime of a building. In some instances, erosion control costs can equal or exceed the cost of the property or building being protected.

CROSS REFERENCE

For information on siting coastal residential buildings, see Chapter 4.

- Sites nearest the water are likely to be in Zone V where building foundations, access stairs, parking slabs, and other components below the building are especially vulnerable to flood, erosion, and scour effects. As a result, *the potential for repeated damage and repair is greater for Zone V buildings than buildings in other zones, and the buildings have higher flood insurance rates and increased operational costs*. In addition, although elevating a building can protect the superstructure from flood damage, it may make the entire building more vulnerable to earthquake and wind damage.

7.5.2   Cost Implications of Design Decisions

The cost implications of design decisions are as follows:

- For aesthetic reasons, the walls of coastal buildings often include a large number of openings for windows and doors, especially in the walls that face the water. *Designs of this type lead to greater initial costs to strengthen the walls and to protect the windows and doors from wind and wind-borne debris (missiles).* If adequate protection in the form of shutter systems or impact-resistant glazing is not provided, long-term costs are greater because of (1) the need to repair damage to glazing and secondary damage by the penetration of wind-driven rain and sea spray and/or (2) the need to install retrofit protection devices at a later date.

NOTE

Over the long term, poor siting decisions are rarely overcome by building design.

■ As explained in Chapter 5, National Flood Insurance Program (NFIP) regulations allow buildings in Coastal A Zones to be constructed on perimeter wall (e.g., crawlspace) foundations or on earth fill. Open (pile, pier, or column) foundations are required only for Zone V buildings. Although a Coastal A Zone building on a perimeter wall foundation or fill may have a lower initial construction cost than a similar building on an open foundation, it may be subject to damaging waves, velocity flows, and/or erosion and scour over its useful life. As a result, ***the long-term costs for a building on a perimeter wall foundation or fill may actually be higher because of the increased potential for damage***.

■ In an effort to reduce initial construction costs, designers may select building materials that require high levels of maintenance. Unfortunately, the initial savings are often offset because (1) coastal buildings, particularly those near bodies of saltwater, are especially prone to the effects of corrosion, and (2) owners of coastal buildings frequently fail to sustain the continuing and time-consuming levels of maintenance required. ***The net effect is often increased building deterioration and sometimes a reduced capacity of structural and non-structural components to resist the effects of future natural hazard events.***

Table 7-1 provides examples of design elements and the cost considerations associated with implementing them. Although these elements may have increased costs when implementing them on a single building, developers may find that incorporating them into speculative houses with large-scale implementation can provide some savings.

Table 7-1. Examples of Flood and Wind Mitigation Measures

Mitigation Measure	Cross References[a]	Benefits/Advantages	Costs/Other Considerations
Adding 1 to 2 feet to the required elevation of the lowest floor or lowest horizontal structural member of the building	5.4.2 6.2.1.3	Reduces the potential for the structure to be damaged by waves and/or floodwaters; reduces flood insurance premiums	May conflict with community building height restrictions; may require additional seismic design considerations; longer pilings may cost more
Increasing embedment depth of pile foundations	10.2.3 13.1.2	Adds protection against scour and erosion	Longer pilings may cost more
Improving flashing and weather-stripping around windows and doors	11.4.1.2	Reduces water and wind infiltration into building	Increases the number of important tasks for a contractor to monitor
Installing fewer breakaway walls or more openings in continuous foundation walls than currently noted on the building plans	5.4.2	Decreases potential for damage to understory of structure; reduces amount of debris during storm event	Reduces the ability to use understory structure for storage for open foundations
Elevating a building in a Coastal Zone A on an open foundation or using only breakaway walls for enclosures below the lowest floor	5.4.2 10.3.1	Reduces the potential for the structure to be damaged by waves, erosion, and floodwaters	Breakaway walls still require flood openings in Zone A

Table 7-1. Examples of Flood and Wind Mitigation Measures (concluded)

Mitigation Measure	Cross References[a]	Benefits/Advantages	Costs/Other Considerations
Adding shutters for glazing protection	11.3.1.2	Reduces the potential for damage from wind-borne debris impact during a storm event; reduces potential for wind-driven rain water infiltration	Shutters require installation or activation before a storm event
Using asphalt roof shingles with high bond strength	11.5.1	Reduces shingle blowoff during high winds	High bond strength shingles are slightly more expensive
Instead of vinyl siding, installing cladding systems that have passed a test protocol that simulates design-level fluctuating wind pressures (on a realistic installed wall specimen)	11.4.1.1 14.2.2	Tested cladding systems reduce blowoff on walls during high winds	These systems may cost more than other materials and may require additional maintenance
Using metal connectors or fasteners with a thicker galvanized coating or connectors made of stainless steel	14.1.1 14.2.6	Increases useful life of connectors and fasteners	Thicker galvanized or stainless steel coatings are more costly
Installing roof sheathing using a high-wind prescriptive approach for improved fasteners, installing additional underlayments, or improving roof covering details as required	11.5 15.3.1	Reduces wind and water damage to roof covering and interior from a severe event	Minimal increased cost when these tasks are done during a reroofing project

(a) Sections in this Manual

DESIGNING FOR FLOOD LEVELS ABOVE THE BASE FLOOD ELEVATION (BFE)

Designers and owners should consider designing buildings for flood levels above the BFE for the following reasons:

- Floods more severe than the base flood can and do occur, and the consequences of flood levels above the BFE can be devastating.

- Older FIRMs may not reflect current base flood hazards.

- FIRMs do *not* account for the effect of future conditions flood hazards; future flood hazards may exceed present-day flood hazards because of sea level rise, coastal erosion, and other factors.

- Buildings elevated above the BFE will sustain less flood damage and will be damaged less often than buildings constructed at the BFE.

- For a given coastal foundation type, the costs of building higher than the BFE are nominal when compared to reduced future costs to the owner.

- Flood damage increases rapidly with flood elevation above the lowest floor, especially when waves are present. Lateral and vertical wave forces against elevated buildings ("wave slam") can be large and destructive. Waves as small as 1.5 feet high can destroy many residential walls.

- Elevated buildings whose floor systems and walls are submerged during a flood may enhance foundation scour by constricting flow between the elevated building and the ground.

- Over a 50-year lifetime, the chance of a base flood occurring is about 40 percent. For most coastal areas, the chance of a flood approximately 3 feet higher than the BFE occurring over 50 years will only be about 10 percent. Designing and constructing to an elevation of BFE + 3 feet is not normally difficult.

- Owners whose buildings are elevated above the BFE can save significant amounts of money through reduced flood insurance premiums. Premiums can be reduced by up to 50 to 70 percent, and savings can reach several thousands of dollars per year in Zone V.

7.5.3   Benefits and Cost Implications of Siting, Design, and Construction Decisions

This Manual is designed to help property owners manage some of the risk associated with constructing a residential building in a coastal area. As noted in Chapter 2, studies of the effects of natural disasters on buildings demonstrate that sound siting, design, engineering, construction, and maintenance practices are important factors in the ability of a building to survive a hazard event with little or no damage. This chapter and the remainder of Volume II provide detailed information about how to site, design, construct, and maintain a building to help manage risks.

CROSS REFERENCE

For more information on designing coastal residential buildings, see Chapter 9.

Constructing to a model building code and complying with regulatory siting requirements provides a building with a certain level of protection against damage from natural hazards. However, compliance with minimum code and regulatory requirements does not guarantee that a building is not at risk from a natural hazard. Exceeding code and minimum regulatory requirements provides an added measure of safety but also adds to the cost of construction, which must be weighed against the benefit gained.

The often minimal initial cost of mitigation measures offers long-term benefits that provide a cost savings from damage avoided over the life of the building. Incorporating mitigation measures can reduce a homeowner's insurance premiums and better protect the building, its contents, and occupants during a natural hazard event, thus decreasing potential losses. Similar to cost reductions provided by the U.S. Green Building Council *LEED [Leadership in Energy and Environmental Design]) for Homes Reference Guide* (USGBC 2009) and ICC 700-2008, incorporating hazard mitigation measures into a building may pay for themselves over a few years based on insurance premium savings and the improved energy efficiency that some of the techniques provide.

Table 7-1 lists examples of flood and wind mitigation measures that can be taken to help a structure withstand natural hazard events. The need for and benefit of some mitigation measures are difficult to predict. For example, elevating a building above the design flood elevation (DFE) could add to the cost of the building. This additional cost must be weighed against the probability of a flood or storm surge exceeding the DFE. Figure 7-3 illustrates the comparative relationship between damage, project costs, and benefits associated with a hazard mitigation project on a present-value[1] basis over the life of the project.

CROSS REFERENCE

Unless both questions presented in Section 4.8 of this Manual (regarding the acceptable level of residual risk at a site) can be answered affirmatively, the property owner should reconsider purchasing the property.

---

1 Present value is the current worth of future sums of money. For example, the present value of $100 to be received 10 years from now is about $38.55, using a discount rate equal to 10 percent interest compounded annually.

Figure 7-3.
Basic benefit-cost model

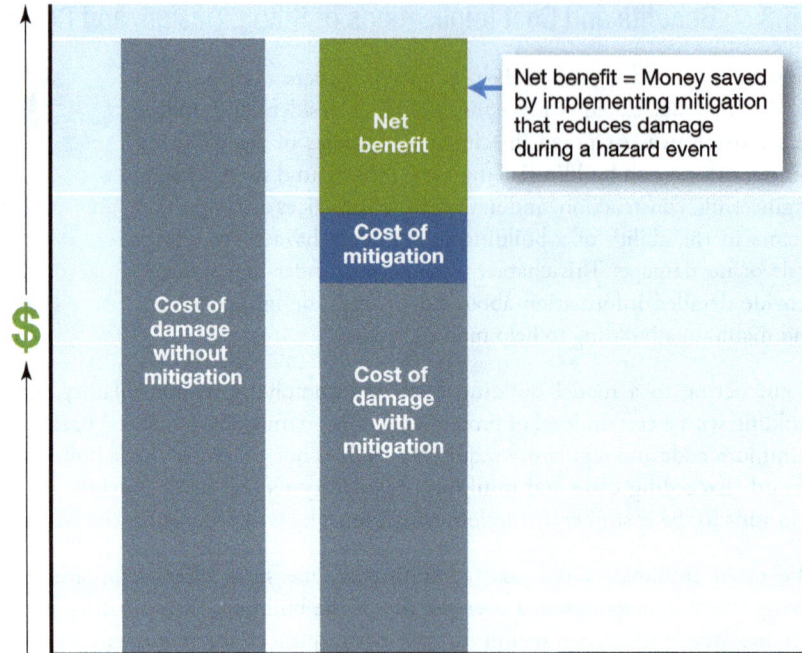

Net benefit = Money saved by implementing mitigation that reduces damage during a hazard event

7.6   Hazard Insurance

Insurance should never be viewed as an alternative to damage prevention. However, despite best efforts to manage risk, structures in coastal areas are always subject to potential damage during a natural hazard event. Hazard insurance to offset potential financial exposure is an important consideration for homeowners in coastal areas. Insurance companies base hazard insurance rates on the potential for a building to be damaged by various hazards and the predicted ability of the building to withstand the hazards. Hazard insurance rates include the following considerations:

- Type of building

- Area of building footprint

- Type of construction

- Location of building

- Date of construction

- Age of the building

NOTE

A single-family home is covered by homeowners insurance, and a multi-family building is covered by a dwelling policy. A homeowner policy is different from a dwelling policy. A homeowner policy is a multi-peril package policy that automatically includes fire and allied lines, theft, and liability coverage. For a dwelling policy, peril coverages are purchased separately. In addition to Federal and private flood insurance, this chapter focuses on homeowners insurance.

- Existence and effectiveness of a fire department and fire hydrants (or other dependable, year-round sources of water)

- Effectiveness of the building code and local building department at the time of construction

Although designers and builders may not be able to control the rates and availability of insurance, they should understand the implications of siting and construction decisions on insurance costs and should make homeowners aware of the risk and potential expense associated with owning a house in a high-hazard area. Insurance considerations can and do affect the decisions about the placement and height of coastal buildings and the materials used in their construction. Input from an insurance industry representative during the design process, rather than after the completion of the building, can positively influence important decisions in addition to potentially saving homeowners money on insurance premiums.

Standard homeowners insurance policies cover multiple perils, including fire, lightning, hail, explosion, riot, smoke, vandalism, theft, volcanic eruption, falling objects, weight of snow, and freezing. Wind is usually (but not always) covered, and an endorsement can often be added for earthquake coverage. Homeowners insurance also includes liability coverage. A separate policy is normally required for flooding.

7.6.1   Flood Insurance

As described in Chapter 5, flood insurance is offered through the NFIP (see Section 6.2.2.1) in communities that participate in the program (e.g., incorporated cities, towns, villages; unincorporated areas of counties, parishes, and federally recognized Indian tribal governments). This flood insurance is required as a condition of receiving federally backed, regulated, or insured financial assistance for the acquisition of buildings in Special Flood Hazard Areas (SFHAs). This includes almost all mortgages secured by property in an SFHA. NFIP flood insurance is not available in communities that do not participate in the NFIP. Most coastal communities participate in the program because they recognize the risk of flood hazard events and the need for flood insurance.

> **NOTE**
> Standard homeowners insurance policies do not normally cover damage from flood or earth movement (e.g., earthquakes, mudslides).

The following sections summarize how coastal buildings are rated for flood insurance and how premiums are established.

7.6.1.1   Rating Factors

The insurance rate is a factor that is used to determine the amount to be charged for a certain amount of insurance coverage, called the premium. Premiums are discussed in Section 7.6.1.3. The following seven rating factors are used for flood insurance coverage for buildings (not including contents):

- Building occupancy
- Building type
- Flood insurance zone

> **NOTE**
> NFIP regulations define basement as any area of a building with the floor subgrade (i.e., below ground level) on all sides.

- Date of construction
- Elevation of lowest floor or bottom or the lowest horizontal structural member of the lowest floor
- Enclosures below the lowest floor
- Location of utilities and service equipment

CROSS REFERENCE

For additional information about enclosures, the use of space below elevated buildings, and flood insurance, see Chapter 5.

Building Occupancy

The NFIP bases rates for flood insurance in part on four types of building occupancy:

- Single-family
- Two- to four-family
- Other residential
- Non-residential

Only slight differences exist among the rates for the three types of residential buildings.

Building Type

The NFIP bases rates for flood insurance in part on the following building-type factors:

- Number of floors (one floor or multiple floors)
- Presence of a basement
- First floor elevation (whether the building is elevated and/or whether there is an enclosure below the lowest elevated floor)
- Manufactured home affixed to a permanent foundation

NFIP flood insurance is generally more expensive for buildings with basements and for buildings with enclosures below BFE.

Flood Insurance Zone

The NFIP bases rates for flood insurance in part on flood insurance zones. The zones are grouped as follows for rating purposes:

- **Zone V (V, VE, and V1–V30).** The zone closest to the water, subject to "coastal high hazard flooding" (i.e., flooding with wave heights greater than 3 feet). Flood insurance is most expensive in Zone V because of the severity of the hazard. However, the zone is often not very wide. Zones V1–V30 were used on FIRMs until 1986. FIRMs published since then show Zone VE.

- **Zone A (A, AE, AR, AO, and A1–A30).** Coastal flood hazard areas where the wave heights are less than 3 feet. Zones A1–A30 were used on FIRMs until 1986. FIRMs published since then show Zone AE.

- **Zones B, C, and X.** The zones outside the 100-year floodplain or SFHA. Flood insurance is least expensive in these zones and generally not required by mortgage lenders. Zone B and Zone C were used on FIRMs until 1986. FIRMs published since then show Zone X.

> **NOTE**
>
> Because Zones B, C, and X designate areas outside the SFHA, construction in these zones is not subject to NFIP floodplain regulations. Homeowners in these areas, however, can purchase Preferred Risk Policies of flood insurance. The rates in these areas are significantly lower than those in Zone V and Zone A.

FIRMs show areas designated as being in the Coastal Barrier Resource System (CBRS) or "otherwise protected areas." These areas (known as "CBRA zones") are identified in the Coastal Barrier Resources Act (CBRA) and amendments. Flood insurance is available for buildings in these zones only if the buildings were walled and roofed before the CBRA designation date shown in the FIRM legend and only if the community participates in the NFIP.

> **CROSS REFERENCE**
>
> For more information about the CBRA and CBRS, see Chapter 5.

Date of Construction

In communities participating in the NFIP, buildings constructed on or before the date of the first FIRM for that community or on or before December 31, 1974, whichever is later, have flood insurance rates that are "grandfathered" or "subsidized." These buildings are referred to as pre-FIRM. They are charged a flat rate based on building occupancy, building type, and flood insurance zone.

> **NOTE**
>
> Flood insurance is available through the NFIP for the following types of buildings: single-family, 2- to 4-family, other residential, and non-residential buildings. Condominium policies are also available. Designers may wish to consult knowledgeable insurance agents and the *Flood Insurance Manual* (FEMA 2011) for policy details and exclusions that affect building design and use. Additional information is available in FEMA FIA-2, *Answers to Questions about the National Flood Insurance Program* (2004).

The rates for buildings constructed after the date of the first FIRM (post-FIRM buildings) are based on building occupancy, building type, flood insurance zone, and two additional factors: (1) elevation of the top of the lowest floor (in Zone A) or bottom of the lowest horizontal structural member of the lowest floor (in Zone V), and (2) enclosed areas below the lowest floor in an elevated building.

If a pre-FIRM building is substantially improved (i.e., the value of the improvement exceeds 50 percent of the market value of the building before the improvement was made), it is rated as a post-FIRM building. If a pre-FIRM building is substantially damaged for any reason (i.e., the true cost of repairing the building to its pre-damaged condition exceeds 50 percent of the value of the building before it was damaged), it is also rated as a post-FIRM building regardless of the amount of repairs actually undertaken. The local building official or floodplain administrator, not the insurance agent, determines whether a building is substantially improved or substantially damaged. If a building is determined to be substantially improved or substantially damaged, the entire structure must be brought into compliance with the current FIRM requirements.

An additional insurance rate table is applied to buildings constructed in Zone V on or after October 1, 1981. The table differentiates between buildings with an obstruction below the elevated lowest floor and those without such an obstruction.

Elevation of Lowest Floor or Bottom or Lowest Horizontal Structural Member of the Lowest Floor

In Zone A, the rating for post-FIRM buildings is based on the elevation of the lowest floor in relation to the BFE. In Zone V, the rating for post-FIRM buildings is based on the elevation of the bottom of the lowest floor's lowest horizontal structural member in relation to the BFE. Flood insurance rates are lower for buildings elevated above the BFE. Rates are significantly higher for buildings rated at 1 foot or more below the BFE.

Ductwork or electrical, plumbing, or mechanical components under the lowest floor must either be designed to prevent water infiltration or elevated above the BFE. Additional elevation of the lowest floor may be required.

In Zone A, a building on a crawlspace must have openings in the crawlspace walls that allow for the unimpeded flow of floodwaters more than 1-foot deep. If the crawlspace walls do not have enough properly sized openings, the crawlspace is considered an enclosed floor, and the building may be rated as having its lowest floor at the elevation of the grade inside the crawlspace. Similarly, if furnaces and other equipment serving the building are below the BFE, the insurance agent must submit more information on the structure to the NFIP underwriting department before the policy's premium can be determined.

WARNING

Differences exist between what is permitted under floodplain management regulations and what is covered by NFIP flood insurance. Some building design considerations should be guided by floodplain management requirements and by knowledge of the design's impact on flood insurance policy premiums. Although allowable, some designs that meet NFIP requirements will result in higher premiums.

Enclosures Below the Lowest Floor

In Zone V, buildings built on or after October 31, 1981, are rated in one of three ways:

COST CONSIDERATION

Significant financial penalties may be associated with the improper design, construction, conversion, or use of areas below the lowest floor.

1. A building is rated as "free of obstruction" if there is no enclosure below the lowest floor other than insect screening or open wood latticework. "Open" means that at least 50 percent of the lattice construction is open.

2. A building is subject to a more expensive "with obstruction" rate if service equipment or utilities are located below the lowest floor or if breakaway walls enclose an area of less than 300 square feet below the lowest floor.

3. If the area below the lowest floor has more than 300 square feet enclosed by breakaway walls, has non-breakaway walls, or is finished, the floor of the enclosed area is the building's lowest floor and

the insurance agent must submit more information on the structure to the NFIP before the policy's premium can be determined.

Although the NFIP allows enclosures below the lowest floor, enclosures affect the flood insurance premiums. The addition of a floor system above the ground, but below the lowest floor of the living space, can result in additional impacts to flood insurance premiums.

7.6.1.2   Coverage

The flood insurance that is available under the NFIP is called a Standard Flood Insurance Policy (SFIP). See FEMA F-122, *National Flood Insurance Program Dwelling Form: Standard Flood Insurance Policy* (FEMA 2009a) for more information about NFIP coverage.

To be insurable under the NFIP, a building must be walled and roofed with two or more rigid exterior walls and must be more than 50 percent above grade. Examples of structures that are *not* insurable because they do not meet this definition are gazebos, pavilions, docks, campers, underground storage

> **NOTE**
>
> The amount of building and contents coverage should be based on replacement value, not market value. Replacement value is the actual cost of rebuilding the building or replacing the contents. This may be higher or lower than market value.

tanks, swimming pools, fences, retaining walls, seawalls, bulkheads, septic tanks, and tents. Buildings constructed entirely over water or seaward of mean high tide after October 1, 1982, are not eligible for flood insurance coverage. Certain parts of boathouses located partially over water (e.g., ceiling, roof over the area where boats are floated) are not eligible for coverage.

Coverage does not include contents. Contents of insurable walled and roofed buildings can be insured under separate coverage within the same policy. Finishing materials and contents in basements or in enclosures below the lowest elevated floor in post-FIRM buildings are not covered with some exceptions. Certain building components and contents in areas below the elevated floors of elevated buildings are covered. Coverage can even include some items *prohibited* by FEMA/local floodplain management regulations if the NFIP deems the items essential to the habitability of the building. *Designers and building owners should not confuse insurability with proper design and construction. Moreover, significant financial penalties (e.g., increased flood insurance rates, increased uninsured losses) may result from improper design or use of enclosed areas below the BFE.*

With the above caveats in mind, buildings insured under the NFIP include coverage (up to specified policy limits) for the following items below the BFE:

- Minimum-code-required utility connections, electrical outlets, switches, and circuit breaker boxes
- Footings, foundation, posts, pilings, piers, or other foundation walls and anchorage system(s) as required for the support of the building
- Drywall for walls and ceilings and nonflammable insulation (in basements only)
- Stairways and staircases attached to the building that are not separated from the building by an elevated walkway

■ Elevators, dumbwaiters, and relevant equipment, except for such relevant equipment installed below the BFE on or after October 1, 1987

■ Building and personal property items—necessary for the habitability of the building—connected to a power source and installed in their functioning location as long as building and personal property coverage has been purchased. Examples of building and personal property items are air conditioners, cisterns, fuel tanks, furnaces, hot water heaters, solar energy equipment, well water tanks and pumps, sump pumps, and clothes washers and dryers.

■ Debris removal for debris that is generated during a flood

An SFIP does *not* provide coverage for the following building components and contents in areas below the elevated floors of elevated residential buildings:

■ Breakaway walls and enclosures that do not provide support to the building

■ Drywall for walls and ceilings

■ Non-structural slabs beneath an elevated building

■ Walks, decks, driveways, and patios outside the perimeter of the exterior walls of the building

■ Underground structures and equipment, including wells, septic tanks, and septic systems

■ Equipment, machinery, appliances, and fixtures not deemed necessary for the habitability of the building

■ Fences, retaining walls, seawalls, and revetments

■ Indoor and outdoor swimming pools

■ Structures over water, including piers, docks, and boat houses

■ Personal property

■ Land and landscaping

7.6.1.3   Premiums

Premiums are based on the seven rating factors discussed in Section 7.6.1.1, plus the following:

■ An expense constant

■ A Federal policy fee

■ The cost of Increased Cost of Compliance coverage

■ The amount of deductible the insured chooses

If a community elects to exceed the minimum NFIP requirements, it may apply for a classification under the NFIP Community Rating System (CRS). Based on its floodplain management program, the community could receive a CRS classification that provides up to a 45 percent premium discount for property owners within the community. At the time of this publication, nearly 1,250 communities were participating in the CRS, representing more than 69 percent of all flood insurance policies. For more information on the CRS, see Section 5.2.4.

Tables 7-2, 7-3, and 7-4 list sample NFIP premiums for a post-FIRM, one-story, single-family residence without a basement located in various flood zones. For buildings in Zone V, premiums are somewhat higher for structures with breakaway obstructions, and premiums are dramatically higher for structures with obstructions (e.g., service equipment, utilities, non-breakaway walls) below the lowest floor.

Reductions in flood insurance premiums can quickly offset the increased costs associated with building above the BFE.

For buildings in Zone A, premiums are higher when proper flood openings are not provided in enclosed areas or when service equipment or utilities are located below the BFE.

Table 7-2. Sample NFIP Flood Insurance Premiums for Buildings in Zone A; $250,000 Building/$100,000 Contents Coverage

Floor Elevation above BFE	Reduction in Annual Flood Premium	Annual Premium	Savings
0	0%	$ 1,622	$ 0
1 foot	45%	$ 897	$ 725
2 feet	61%	$ 638	$ 984
3 feet	66%	$ 548	$ 1,074
4 feet	67%	$ 530	$ 1,092

Rates as of May 2011 per the National Flood Insurance Program Flood Insurance Manual (FEMA 2011) for a Zone V structure free of obstruction. Rates include building ($250,000), contents ($100,000), and associated fees, including increased cost of compliance.

Table 7-3. Sample NFIP Flood Insurance Premiums for Buildings in Zone V Free of Obstruction Below the Lowest Floor; $250,000 Building/$100,000 Contents Coverage

Floor Elevation above BFE	Reduction in Annual Flood Premium	Annual Premium	Savings
0	0%	$ 7,821	$ 0
1 foot	33%	$ 5,256	$ 2,565
2 feet	55%	$ 3,511	$ 4,310
3 feet	65%	$ 2,764	$ 5,057
4 feet	71%	$ 2,286	$ 5,535

Rates as of May 2011 per the National Flood Insurance Program Flood Insurance Manual (FEMA 2011) for a Zone V structure free of obstruction. Rates include building ($250,000), contents ($100,000), and associated fees, including increased cost of compliance; premium to be determined by NFIP underwriting.

Table 7-4. Sample NFIP Flood Insurance Premiums for Buildings in Zone V with Obstruction Below the Lowest Floor; $250,000 Building/$100,000 Contents Coverage

Floor Elevation above BFE	Reduction in Annual Flood Premium	Annual Premium	Savings
0	0%	$ 10,071	$ 0
1 foot	22%	$ 7,901	$ 2,170
2 feet	40%	$ 6,056	$ 4,015
3 feet	50%	$ 5,076	$ 4,995
4 feet	54%	$ 4,591	$ 5,480

Rates as of May 2011 per the National Flood Insurance Program Flood Insurance Manual (FEMA 2011) for a Zone V structure free of obstruction. Rates include building ($250,000), contents ($100,000), and associated fees, including increased cost of compliance; premium to be determined by NFIP underwriting.

7.6.1.4 Designing to Achieve Lower Flood Insurance Premiums

Tables 7-2, 7-3, and 7-4 demonstrate that considerable savings can be achieved on flood insurance premiums by elevating a building above the BFE and by constructing it to be free of obstruction. Other siting, design and construction decisions can also lower premiums. Designers should refer to FEMA's V Zone Risk Factor Rating Form to estimate flood insurance premium discounts and as a planning tool to use with building owners. The form is in Chapter 5 of the Flood Insurance Manual (FEMA 2011), available at http://www.fema.gov/business/nfip/manual.shtm. Discount points, which translate into reduced premiums, are awarded for:

- Lowest floor elevation
- Siting and environmental considerations
- Building support systems and design details
- Obstruction-free and enclosure construction considerations

In addition to lowest floor elevation and free-of-obstruction discounts illustrated in Tables 7-2 and 7-3, flood insurance premium discounts also can be obtained for:

- Distance from shoreline to building
- Presence of large dune seaward of the building
- Presence of certified erosion control device or ongoing beach nourishment project
- Foundation design based on eroded grade elevation and local scour
- Foundation design based on this Manual and ASCE 7-10 loads and load combinations
- Minimizing foundation bracing

- Spacing of piles/columns/piers

- Size and depth of piles and pier footings

- Superior connections between piles/columns/piers and girders

Some poor practices reduce discount points. Negative discount points, which result in higher flood insurance premiums, are given for:

- Shallow pile embedment

- Certain methods of pile installation

- Small-diameter piles or columns

- Non-bolted connections between piles/columns/piers and girders

- Over-notching of wood piles

- Small pier footings

- Presence of elevators, equipment, ductwork and obstructions below the BFE

- Presence of solid breakaway walls

- Presence of finished breakaway walls

Table 10 (V Zone Risk Relativities) in the *Flood Insurance Manual* (FEMA 2011) provides an indication of how building discount points translate into flood premium discounts. Designers and owners should review this table and consult with a knowledgeable flood insurance agent regarding the flood insurance premium implications of using or avoiding certain design construction practices

7.6.2   Wind Insurance

Wind insurance coverage is generally part of a homeowners insurance policy. At the time this Manual was published, underwriting associations (or "pools") provided last resort insurance to homeowners in coastal areas who could not obtain coverage from private companies. The following seven states had beach and windstorm insurance plans at the time this Manual was released: Alabama, Florida, Louisiana, Mississippi, North Carolina, South Carolina, and Texas. Georgia and New York provide this kind of coverage for windstorm and hail in certain coastal communities through other property pools. In addition, New Jersey operates the Windstorm Market Assistance Program (Wind-MAP) to help residents in coastal communities find homeowners insurance on the voluntary market. When Wind-MAP does not identify an insurance carrier for a homeowner, the New Jersey Insurance Underwriting Association, known as the FAIR Plan, may provide a policy for windstorm, hail, fire, and other perils but does not cover liability.

Many insurance companies encourage their policyholders to retrofit their homes to resist wind-related damage, and some companies have established discount programs to reduce premiums, and other types of financial incentives, to reflect the risk reduction for homes that have been properly retrofitted. Some State insurance departments also have put in place insurance discount programs for properly retrofitted homes. The IBHS FORTIFIED *for Existing Homes* Program has been designed with the support of IBHS member insurance companies, although each individual company makes its own decisions about how it is implemented.

Wind is only one part of the rating system for multi-peril insurance policies such as a homeowners insurance policy. Most companies rely on the Homeowner's Multistate General Rules and State-specific exceptions Manual of the Insurance Services Office (ISO) as the benchmark for developing their own manuals. ISO stresses that the rules in the manual are advisory only and that each company decides what to use and charge. The ISO publishes a homeowner's manual in every state except Hawaii, North Carolina, and Washington (where State-mandated insurance bureaus operate).

The seven basic factors in rating a homeowners insurance policy are:

- Form (determines type of coverage)

- Age of the structure

- Territory

- Fire protection class

- Building code effectiveness

- Construction type

- Protective devices

The last five factors are discussed below. Premiums can also vary because of factors such as amount of coverage and deductible, but these additional factors are not related to building construction. Some companies, however, adjust their higher optional deductible credit according to construction type, giving more credit to more fire-resistant concrete and masonry buildings.

7.6.2.1   Territory

Wind coverage credit varies by territory. An entire state may be one territory, but some states, such as Florida, are divided into county and sub-county territories. In Florida, the Intracoastal Waterway is often used as the boundary line.

7.6.2.2   Fire Protection Class

ISO publishes a public protection classification for each municipality or fire district based on an analysis of the local fire department, water system, and fire alarm system. This classification does not affect wind coverage but is an important part of the rate.

7.6.2.3   Building Code Effectiveness Grading Schedule

The adoption and enforcement of building codes by local jurisdictions are routinely assessed through the Building Code Effectiveness Grading Schedule (BCEGS) program, developed by the ISO. Participation in BCEGS is voluntary and may be declined by local governments if they do not wish to have their local building codes evaluated. The results of BCEGS assessments are routinely provided to ISO's member private insurance companies, which in turn may offer rating credits for new buildings constructed in communities with strong BCEGS classifications. Conceptually, communities with well-enforced, up-to-date codes should experience fewer disaster-related losses and as a result, should have lower insurance rates.

In conducting the assessment, ISO collects information related to personnel qualification and continuing education, as well as number of inspections performed per day. This type of information combined with local building codes is used to determine a grade for the jurisdiction. The grades range from 1 to 10, with a BCEGS grade of 1 representing exemplary commitment to building code enforcement, and a grade of 10 indicating less than minimum recognized protection. Most participating communities fall in the 3 to 5 grade range.

7.6.2.4   Construction Type

To simplify insurance underwriting procedures, buildings are identified as being in only one of four categories:

- **Frame:** exterior walls of wood or other combustible construction, including stucco and aluminum siding

- **Masonry veneer:** exterior walls of combustible material, veneered with brick or stone

- **Masonry:** exterior walls of masonry materials; floor and roof of combustible materials

- **Superior:** non-combustible, masonry non-combustible, or fire resistive

Masonry veneer and masonry are often difficult to differentiate and are therefore often given the same rating.

Not many single-family homes qualify for the superior category, which results in a 15 percent credit off rates for the masonry categories. A home in the superior category may also qualify for a wind credit because some insurers believe that buildings with walls, floors, and roofs made of concrete products offer good resistance to windstorms and Category 1 hurricanes. Therefore, a fire-resistive home may get a wind-resistive credit.

ISO's dwelling insurance program allows companies to collect data from the owner, the local building department, or their own inspectors to determine whether a house can be classified as wind-resistive or semi-wind-resistive for premium credit purposes.

7.6.2.5   Protective Devices

Protective devices are not considered basic factors but items that may deserve some credits. This approach is more common for fire and theft coverage than for wind. Fire and theft coverage credits sprinklers and fire and/or burglar alarms tied to the local fire or police stations. ISO's rules do not address wind-protective devices except in Florida. In Florida, a premium credit is given if exterior walls and roof openings (not including roof ridge and soffit vents) are fully protected with storm shutters of any style and material that are designed and properly installed to meet the latest ASCE 7-10 engineering standard. This standard has been adopted by Dade County. Shutters must be able to withstand impact from wind-borne debris in accordance with the standards set by the municipality, or if there are no local standards, by Dade County. The rules also provide specifications for alternatives to storm shutters, such as windstorm protective glazing material.

7.6.3   Earthquake Insurance

Earthquake insurance is an addition to a regular homeowners insurance policy. Earthquake insurance carries a very high deductible—usually 10 or 15 percent of the value of the house. In most states, ISO has developed advisory earthquake loss costs based on a seismic model used to estimate potential damage to individual properties in the event of an earthquake. The model is based on seismic data, soil types, damage information from previous earthquakes, and structural analysis of various types of buildings. Based on this model, postal Zip codes have been assigned to rating bands and loss costs developed for each band. The number of bands varies within each state and, at times, within a county.

In California, the California Earthquake Authority (CEA), a State-chartered insurance company, writes most earthquake policies for homeowners. These policies cover the dwelling and its contents and are subject to a 15-percent deductible. CEA rates are also based on a seismic model used to estimate potential damage to individual properties in the event of an earthquake.

7.7   Sustainable Design Considerations

Sustainability concepts are increasingly being incorporated into residential building design and construction. The voluntary green building rating systems of the past decade are being replaced with adoption by local and State jurisdictions of mandatory minimum levels of compliance with rating systems such as the U.S. Green Building Council *LEED for Homes Reference Guide* (USGBC 2009) or consensus-based standards such as ICC 700-2008. These programs and standards use a system in which credits are accumulated as points assigned to favorable green building attributes pertaining to lot design, resource efficiency, energy efficiency, water efficiency, and indoor environmental quality.

Although green building programs are implemented as above-minimum building code practices, many aspects of green construction and its impact on structural performance and durability are not readily apparent upon initial consideration. Green building programs such as National Association of Home Builders (NAHB) Green, EarthAdvantage, and other State and local programs may incorporate LEED for Homes, ICC 700-2008, EnergyStar and other rating systems or product certifications as part of their offerings. For example, a homeowner may decide to add a rooftop solar panel system after the home is built. Depending on its configuration, this system could act as a "sail" in high winds, adding significant uplift loads to the roof and possibly triggering localized structural failure. To maintain expected structural performance in a high-wind event, these additional loads not only must be accommodated by the roof framing, but the complete load path for these additional loads must be traced and connections or framing enhanced as needed.

Examples of other green attributes that may require additional design consideration for resistance to natural hazards are large roof overhangs for shading (due to the potential for increased wind loads), vegetative green roofs (due to the presence of added weight and moisture to sustain the roof), and optimized or advanced framing systems that reduce overall material usage and construction waste (due to larger spacing between framing members and smaller header and framing member sizes).

It is important to verify that the design wind speed for an area is not in excess of the recommended maximum wind speeds for these systems. Building for resilience should not work against other green practices by unreasonably increasing the material resources needed to construct the building. However, buildings constructed to survive natural hazards reduce the need to be rebuilt and thus provide a more sustainable design approach.

When new green building attributes introduce new technology or new building materials into the building design, new interactions may affect the building's structural integrity and durability. Examples of interactions between green building attributes and resistance to natural hazards (e.g., resilience of the building) are described in FEMA P-798, *Natural Hazard Sustainability of Residential Construction* (FEMA 2010c). When implementing green attributes into a design, the designer should consider that building for resilience is possibly the most important green building practice. A green building fails to provide benefits associated with green building practices if it is more susceptible to heavy damage from natural hazard events that result in lost building function and increased cost of repair.

7.8   Inspection Considerations

After the completion of building permits and construction plans, good inspection and enforcement procedures are crucial. For coastal construction, building inspectors, code officers, designers, and floodplain managers must understand the flood-resistant design and construction requirements for which they need to check. The earlier a deviation is found, the easier it is to take corrective action working with the homeowner and builder.

A plan review and inspection checklist tailored to flood-related requirements should be used. Some of the inspections that can be performed to meet compliance directives with the local community's flood-resistant provisions are listed below. For a community that does not have a DFE, a BFE is applicable.

- Stake-out or site inspection to verify the location of a building; distances from the flood source or body of water can also be checked

- Fill inspection to check compaction and final elevation when fills are allowed in SFHAs

- Footing or foundation inspection to check for flood-opening specifics for closed foundations, lowest floor inspection for slab-on-grade buildings, and embedment depth and pile plumbness for pile-supported structures

- Lowest floor inspection (floodplain inspection) per Section 109.3.3 of 2012 IBC and Section R109.1.3 of 2012 IRC. This is also a good time to verify that the mechanical and electrical utilities are above the BFE or DFE for additional protection.

- Final inspection points for flood-prone buildings can include:

  - Enclosures below elevated buildings for placement of flood vents and construction of breakaway walls, where applicable

  - Use of enclosures for consistency with the use in the permit

  - Placement of exterior fill, where permitted, according to plans and specifications

- Materials below the DFE for flood-resistance; see NFIP Technical Bulletin 2, *Flood Damage-Resistant Materials Requirements* (FEMA 2008)

- Building utilities to determine whether they have been elevated or, when instructions are provided, installed to resist flood damage

- Existence of as-built documentation of elevations

- If a plan review and inspection checklist have been used, verification that have been signed off and placed in the permit file with all other inspection documentation

More information regarding inspections is available in FEMA P-762, *Local Officials Guide for Coastal Construction* (FEMA 2009b).

7.9   References

ASCE (American Society of Civil Engineers). 2010. *Minimum Design Loads for Buildings and Other Structures*. ASCE Standard ASCE 7-10.

ASCE. *Flood Resistant Design and Construction*. ASCE Standard ASCE 24.

FEMA (Federal Emergency Management Agency). 2004. *Answers to Questions about the National Flood Insurance Program*. FEMA FIA-2.

FEMA. 2008. *Flood Damage-Resistant Materials Requirements*. Technical Bulletin 2.

FEMA. 2009a. *National Flood Insurance Program Dwelling Form: Standard Flood Insurance Policy*. FEMA F-122.

FEMA. 2009b. *Local Officials Guide for Coastal Construction*. FEMA P-762.

FEMA. 2010c. *Natural Hazard Sustainability for Residential Construction*. FEMA P-798.

FEMA. 2011. *Flood Insurance Manual*. Available at http://www.fema.gov/business/nfip/manual.shtm. Accessed May 2011.

IBHS (Insurance Institute for Business and Home Safety). 1999. *Coastal Exposure and Community Protection: Hurricane Andrew's Legacy*. Tampa.

IBHS and Insurance Research Council. 2004. *The Benefits of Modern Wind Resistant Building Codes on Hurricane Claim Frequency and Severity – A Summary Report*.

ICC (International Code Council). 2008. *National Green Building Standard*. ICC 700-2008. Country Club Hills, IL: ICC.

ICC. 2011a. *International Building Code* (2012 IBC). Country Club Hills, IL: ICC.

ICC. 2011b. *International Residential Code for One-and Two-Family Dwellings* (2012 IRC). Country Club Hills, IL: ICC.

Mortgage Bankers Association. 2006. *Natural Disaster Catastrophic Insurance The Commercial Real Estate Finance Perspective.* Washington, D.C.

Shows, E.W. 1978. "Florida's Setback Line—An Effort to Regulate Beachfront Development." *Coastal Zone Management Journal 4* (1,2): 151–164.

USGBC (U.S. Green Building Council). 2009. *LEED for Homes Reference Guide.*

8 Determining Site-Specific Loads

This chapter provides guidance on determining site-specific loads from high winds, flooding, and seismic events. The loads determined in accordance with this guidance are applied to the design of building elements described in Chapters 9 through 15.

The guidance is intended to illustrate important concepts and best practices in accordance with building codes and standards and does not represent an exhaustive collection of load calculation methods. Examples of problems are provided to illustrate the application of design load provisions of ASCE 7-10. For more detailed guidance, see the applicable building codes or standards.

Figure 8-1 shows the process of determining site-specific loads for three natural hazards (flood, wind, and seismic events). The process includes identifying the applicable building codes and standards for the selected site, identifying building characteristics that affect loads, and determining factored design loads using applicable load combinations. Model building codes and standards may not provide

CROSS REFERENCE

For resources that augment the guidance and other information in this Manual, see the Residential Coastal Construction Web site (http://www.fema.gov/rebuild/mat/fema55.shtm).

NOTE

All coastal residential buildings must be designed and constructed to prevent flotation, collapse, and lateral movement due to the effects of wind and water loads acting simultaneously.

load determination and design guidance for the hazards that are listed in the figure. In such instances, supplemental guidance should be sought.

The loads and load combinations used in this Manual are required by ASCE 7-10 unless otherwise noted. Although the design concepts that are presented in this Manual are applicable to both Allowable Stress Design (ASD) and Load and Resistance Factor Design (LRFD), all calculations, analyses, and load combinations are based on ASD. Extension of the design concepts presented in this Manual to the LRFD format can be achieved by modifying the calculations to use strength-level loads and resistances.

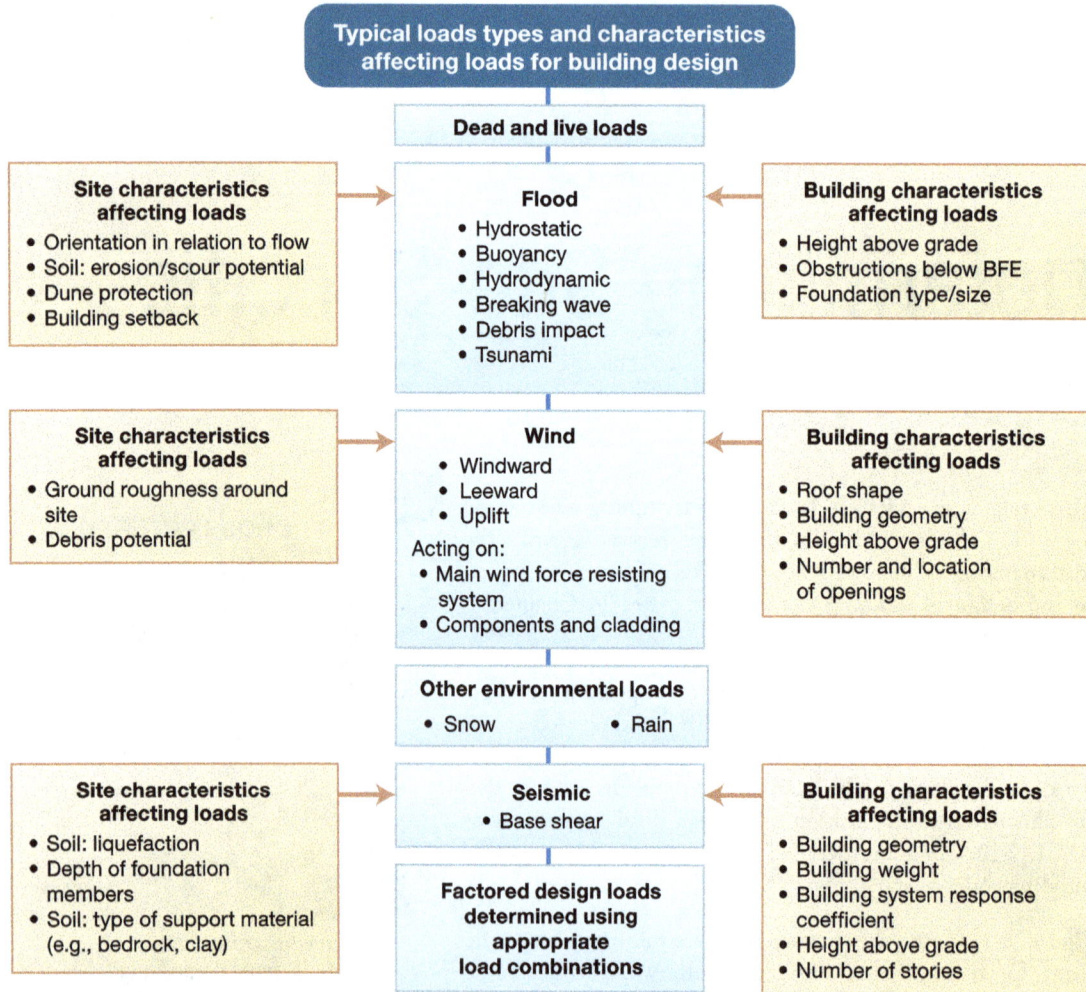

Typical loads types and characteristics affecting loads for building design

Dead and live loads

Site characteristics affecting loads
- Orientation in relation to flow
- Soil: erosion/scour potential
- Dune protection
- Building setback

Flood
- Hydrostatic
- Buoyancy
- Hydrodynamic
- Breaking wave
- Debris impact
- Tsunami

Building characteristics affecting loads
- Height above grade
- Obstructions below BFE
- Foundation type/size

Site characteristics affecting loads
- Ground roughness around site
- Debris potential

Wind
- Windward
- Leeward
- Uplift

Acting on:
- Main wind force resisting system
- Components and cladding

Building characteristics affecting loads
- Roof shape
- Building geometry
- Height above grade
- Number and location of openings

Other environmental loads
- Snow
- Rain

Site characteristics affecting loads
- Soil: liquefaction
- Depth of foundation members
- Soil: type of support material (e.g., bedrock, clay)

Seismic
- Base shear

Factored design loads determined using appropriate load combinations

Building characteristics affecting loads
- Building geometry
- Building weight
- Building system response coefficient
- Height above grade
- Number of stories

Figure 8-1.
Summary of typical loads and characteristics affecting determination of design load

8.1   Dead Loads

Dead load is defined in ASCE 7-10 as "… the weight of all materials of construction incorporated into the building including, but not limited to, walls, floors, roofs, ceilings, stairways, built-in partitions, finishes, cladding, and other similarly incorporated architectural and structural items, and fixed service equipment including the weight of cranes." The sum of the dead loads of all the individual elements equals the unoccupied weight of a building.

The total weight of a building is usually determined by multiplying the unit weight of the various building materials—expressed in pounds per unit area—by the surface area of the materials. Unit weights of building elements, such as exterior walls, floors and roofs, are commonly used to simplify the calculation of building weight. Minimum design dead loads are contained in ASCE 7-10, *Commentary*. Additional information about material weights can be found in *Architectural Graphic Standards* (The American Institute of Architects 2007) and other similar texts.

Determining the dead load is important for several reasons:

■ The dead load determines in part the required size of the foundation (e.g., footing width, pile embedment depth, number of piles).

■ Dead load counterbalances uplift forces from buoyancy when materials are below the stillwater depth (see Section 8.5.7) and from wind (see Example 8.9).

■ Dead load counterbalances wind and earthquake overturning moments.

■ Dead load changes the response of a building to impacts from floodborne debris and seismic forces.

■ Prescriptive design in the following code references and other code references is dependent on the dead load of the building. For example, wind uplift strap capacity, joist spans, and length of wall bracing required to resist seismic forces are dependent on dead load assumptions used to tabulate the prescriptive requirements in the following examples of codes and prescriptive standards:

   ▪ 2012 IRC, *International Residential Code for One-and Two-Family Dwellings* (ICC 2011b)

   ▪ 2012 IBC, *International Building Code* (ICC 2011a)

   ▪ ICC 600-2008, *Standard for Residential Construction in High-Wind Regions* (ICC 2008)

   ▪ WFCM-12, *Wood Frame Construction Manual for One- and Two-Family Dwellings* (AF&PA 2012)

   ▪ AISI S230-07, *Standard for Cold-formed Steel Framing-prescriptive Method for One- and Two-family Dwellings* (AISI 2007)

8.2   Live Loads

Live loads are defined in ASCE 7-10 as "… loads produced by the use and occupancy of the building … and do not include construction or environmental loads such as wind load, snow load, rain load, earthquake load, flood load, or dead load." Live loads are usually taken as a uniform load spread across the surface being designed. For residential one- and two-family buildings, the uniformly distributed live load for habitable areas (except sleeping and attic areas) in ASCE 7-10 is 40 pounds/square foot. For balconies and decks on one- and two-family buildings, live load is 1.5 times the live load of the occupancy served but not to exceed 100 pounds/square foot. This requirement typically translates to a live load of 60 pounds/square foot for a deck or balcony accessed from a living room or den, or a live load of 45 pounds/square foot for a deck or balcony accessed from a bedroom. ASCE 7-10 contains no requirements for supporting a concentrated load in a residential building.

NOTE

The live loads in the 2012 IBC and 2012 IRC for balconies and decks attached to one- and two-family dwellings differ from those in ASCE 7-10. Under the 2012 IBC, the live load for balconies and decks is the same as the occupancy served. Under the 2012 IRC, a minimum 40 pounds/square foot live load is specified for balconies and decks. Strict adherence to the ASCE 7-10 live loads for a residential deck requires a complete engineering design and does not permit use of the prescriptive deck ledger table in the 2012 IRC or the prescriptive provisions in AWC DCA6, which are based on a 40 pounds/square foot live load.

8.3 Concept of Tributary or Effective Area and Application of Loads to a Building

The tributary area of an element is the area of the floor, wall, roof, or other surface that is supported by that element. The tributary area is generally a rectangle formed by one-half the distance to the adjacent element in each applicable direction.

The tributary area concept is used to distribute loads to various building elements. Figure 8-2 illustrates tributary areas for roof loads, lateral wall loads, and column or pile loads. The tributary area is a factor in calculating wind pressure coefficients, as described in Examples 8.7 and 8.8.

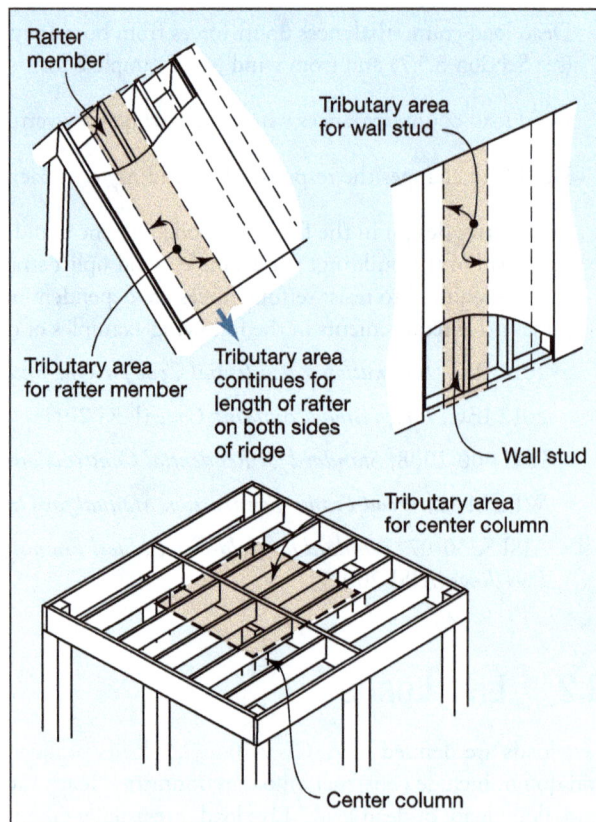

Figure 8-2.
Examples of tributary areas for different structural elements

8.4   Snow Loads

Snow loads are applied as a vertical load on the roof or other exposed surfaces such as porches or decks. Ground snow loads are normally specified by the local building code or building official. In the absence of local snow load information, ASCE 7-10 contains recommended snow loads shown on a map of the United States.

When the flat roof snow load exceeds 30 pounds/square foot, a portion of the weight of snow is added to the building weight when the seismic force is determined.

8.5   Flood Loads

Floodwaters can exert a variety of load types on building elements. Both hydrostatic and depth-limited breaking wave loads depend on flood depth.

Flood loads that must be considered in design include:

- Hydrostatic load – buoyancy (flotation) effects, lateral loads from standing water, slowly moving water, and nonbreaking waves

- Breaking wave load

- Hydrodynamic load – from rapidly moving water, including broken waves

- Debris impact load – from waterborne objects

> **NOTE**
>
> - Flood load calculation procedures cited in this Manual are conservative, given the uncertain conditions of a severe coastal event.
>
> - Background information and procedures for calculating coastal flood loads are presented in a number of publications, including ASCE 7-10 and the *Coastal Engineering Manual* (USACE 2008).

The effects of flood loads on buildings can be exacerbated by storm-induced erosion and localized scour and by long-term erosion, all of which can lower the ground surface around foundation elements and cause the loss of load-bearing capacity and loss of resistance to lateral and uplift loads. As discussed in Section 8.5.3, the lower the ground surface elevation, the deeper the water, and because the wave theory used in this Manual is based on depth-limited waves, deeper water creates larger waves and thus greater loads.

8.5.1   Design Flood

In this Manual, "design flood" refers to the locally adopted regulatory flood. If a community regulates to minimum NFIP requirements, the design flood is identical to the base flood (the 1-percent-annual-chance flood or 100-year flood). If a community has chosen to exceed minimum NFIP building elevation requirements, the design flood can exceed the base flood. The design flood is always equal to or greater than the base flood.

> **TERMINOLOGY: FREEBOARD**
>
> Freeboard is additional height incorporated into the DFE to account for uncertainties in determining flood elevations and to provide a greater level of flood protection. Freeboard may be required by State or local regulations or be desired by a property owner.

8.5.2   Design Flood Elevation

Many communities have chosen to exceed minimum NFIP building elevation requirements, usually by requiring freeboard above the base flood elevation (BFE) but sometimes by regulating to a more severe flood than the base flood. In this Manual, "design flood elevation" (DFE) refers to the locally adopted regulatory flood elevation.

In ASCE 24-05, the DFE is defined as the "elevation of the design flood, including wave height, relative to the datum specified on the community's flood hazard map." The design flood is the "greater of the following two flood events: (1) the base flood, affecting those areas identified as SFHAs on the community's FIRM or (2) the flood corresponding to the area designated as a flood hazard area on a community's flood hazard map or otherwise legally designated." The DFE is often taken as the BFE plus any freeboard required by a community, even if the community has not adopted a design flood more severe than the 100-year flood.

Coastal floods can and do exceed BFEs shown on FIRMs and minimum required DFEs established by local and State governments. When there are differences between the minimum required DFE and the recommended elevation based on consideration of other sources, the designer, in consultation with the owner, must decide whether elevating above the DFE provides benefits relative to the added costs of elevating higher than the minimum requirement. For example, substantially higher elevations require more stairs to access the main floor and may require revised designs to meet the community's height restriction. Benefits include reduced flood damage, reduced flood insurance premiums, and the ability to reoccupy homes faster than owners of homes constructed at the minimum allowable elevation. In both Hurricanes Katrina and Ike, high water marks after the storms indicated that if the building elevations had been set to the storm surge elevation, the buildings may have survived. See FEMA 549, *Hurricane Katrina in the Gulf Coast* (FEMA 2006), and FEMA P-757, *Hurricane Ike in Texas and Louisiana* (FEMA 2009), for more information.

In addition to considering the DFE per community regulations, designers should consider the following before deciding on an appropriate lowest floor elevation:

- **The 500-year flood elevation as specified in the Flood Insurance Study (FIS) or similar study.** The 500-year flood elevation (including wave effects) represents a larger but less frequent event than the typical basis for the DFE (e.g., the 100-year event). In order to compare the DFE to the 500-year flood elevation, the designer must obtain the 500-year wave crest elevation from the FIS or convert the 500-year stillwater level to a wave crest elevation if the latter is not included in the FIS report.

- **The elevation of the expected maximum storm surge as specified by hurricane evacuation maps.** Storm surge evacuation maps provide a maximum storm surge elevation for various hurricane categories. Depending on location, maps may include all hurricane categories (1 to 5), or elevations for selected storm categories only. Most storm surge evacuation maps are prepared by the U.S. Army Corps of Engineers (USACE) and are usually available from the USACE District Office or State/local emergency management agencies. Storm surge elevations are stillwater levels and do not include wave heights, so the designer must convert storm surge elevations to wave crest elevations.

  When storm surge evacuation maps are based on landmark boundaries (e.g., roads or other boundaries of convenience) rather than storm surge depths, the designer needs to obtain the surge elevations for a building site from the evacuation study (if available). The topographic map of the region may also provide information about the storm surge depths because the physical boundary elevation should establish the most landward extent of the storm surge.

■ **Historical information and advisory flood elevations.** Historical information showing flood levels and flood conditions during past flood events, if available, is an important consideration for comparison to the DFE. For areas subject to a recent coastal flood event, advisory flood elevations may be available based on the most recent flooding information unique to the site.

Community FIRMs do not account for the effects of long-term erosion, subsidence, or sea level rise, all of which could be considered when establishing lowest floor elevations in excess of the DFE. Erosion can increase future flood hazards by removing dunes and lowering ground levels (allowing larger waves to reach a building site). Sea level rise can increase future flood hazards by allowing smaller and more frequently occurring storms to inundate coastal areas and by increasing storm surge elevations.

Section 3.6 discusses the process a designer could follow to determine whether a FIRM represents flood hazards associated with the site under present-day and future-based flood conditions.

This section provides more information on translating erosion and sea level rise data into d_s (design flood depth) calculations. Figure 8-3 illustrates a procedure that designers can follow to determine d_s under a variety of future conditions. In essence, designers should determine the lowest expected ground elevation at the base of a building during its life and the highest expected stillwater elevation at the building during its life.

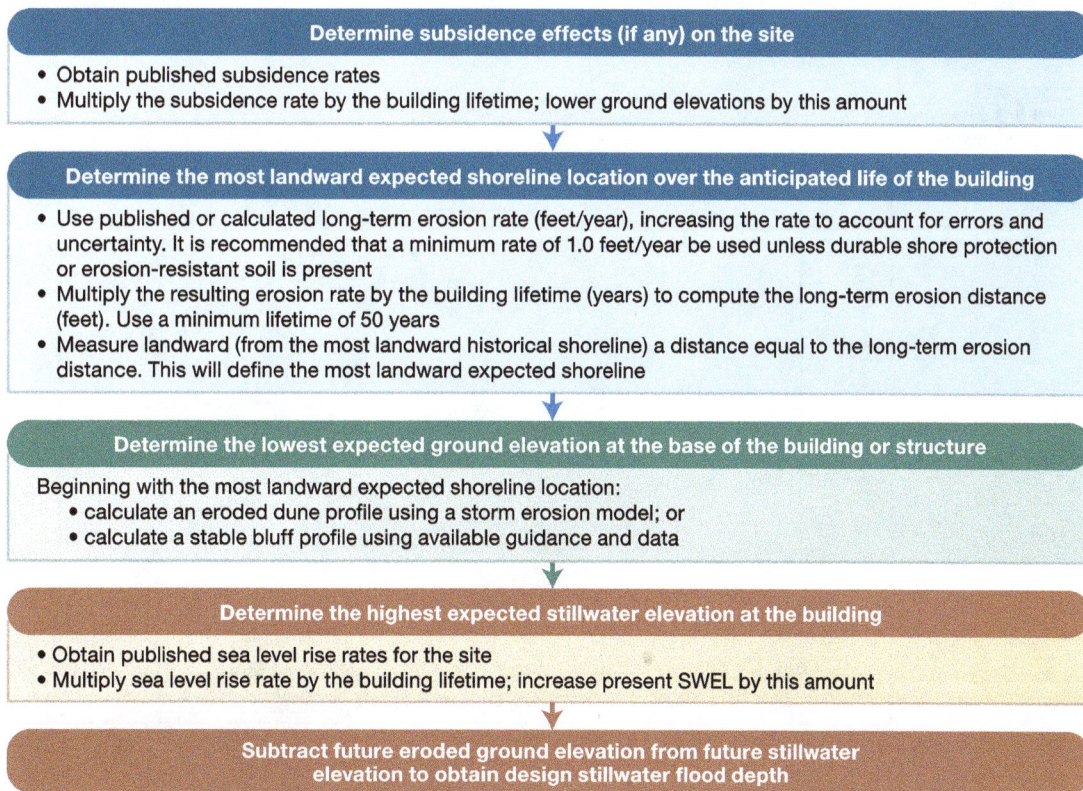

Determine subsidence effects (if any) on the site

- Obtain published subsidence rates
- Multiply the subsidence rate by the building lifetime; lower ground elevations by this amount

Determine the most landward expected shoreline location over the anticipated life of the building

- Use published or calculated long-term erosion rate (feet/year), increasing the rate to account for errors and uncertainty. It is recommended that a minimum rate of 1.0 feet/year be used unless durable shore protection or erosion-resistant soil is present
- Multiply the resulting erosion rate by the building lifetime (years) to compute the long-term erosion distance (feet). Use a minimum lifetime of 50 years
- Measure landward (from the most landward historical shoreline) a distance equal to the long-term erosion distance. This will define the most landward expected shoreline

Determine the lowest expected ground elevation at the base of the building or structure

Beginning with the most landward expected shoreline location:
- calculate an eroded dune profile using a storm erosion model; or
- calculate a stable bluff profile using available guidance and data

Determine the highest expected stillwater elevation at the building

- Obtain published sea level rise rates for the site
- Multiply sea level rise rate by the building lifetime; increase present SWEL by this amount

Subtract future eroded ground elevation from future stillwater elevation to obtain design stillwater flood depth

Figure 8-3. Flowchart for estimating maximum likely design stillwater flood depth at the site

- The lowest expected ground elevation is determined by considering subsidence, long-term erosion, and erosion during the base flood.

  - Subsidence effects can be estimated by lowering all existing ground elevations at the site by the product of the subsidence rate and the building lifetime. For example, if subsidence occurs at a rate of 0.005 foot/year and the building lifetime is 50 years, the profile should be lowered 0.25 foot.

  - Figure 8-4 illustrates a simple way to estimate long-term effects on ground elevations at the building. Translate the beach and dune portion of the profile landward by an amount equal to the product of the long-term erosion rate and the building lifetime. If the erosion rate is 3 feet/year and the building lifetime is 50 years, shift the profile back 150 feet.

  - Figure 8-4 also shows the next step in the process, which is to assess dune erosion (see Section 3.5.1) to determine whether the dune will be removed during a base flood event.

  - The lowest expected grade will be evident once the subsidence, long-term erosion, and dune erosion calculations are made.

- The stillwater level is calculated by adding the expected sea level rise element to the base flood stillwater elevation. For example, if the FIS states the 100-year stillwater elevation is 12.2 feet NAVD, and if sea level is rising at 0.01 foot/year, and if the building lifetime is 50 years, the future conditions stillwater level will be 12.7 feet NAVD (12.2 + [(50)(0.01)]).

- The design stillwater flood depth, d_s, is then calculated by subtracting the future conditions eroded grade elevation from the future conditions stillwater elevation.

Figure 8-4. Erosion's effects on ground elevation

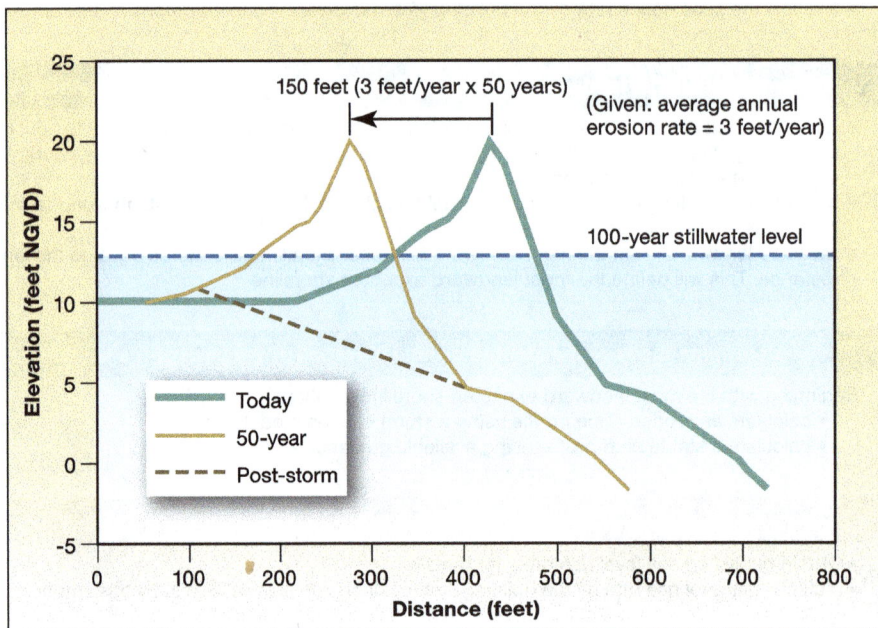

8.5.3   Design Stillwater Flood Depth

In a general sense, flood depth can refer to two different depths (see Figure 8-5):

- **Stillwater flood depth.** The vertical distance between the eroded ground elevation and the stillwater elevation associated with the design flood. This depth is referred to as the design stillwater flood depth (d_s).

> **NOTE**
>
> The design stillwater flood depth (d_s) (including wave setup; see Section 8.5.4) should be used for calculating wave heights and flood loads.

- **Design flood protection depth.** The vertical distance between the eroded ground elevation and the DFE. This depth is referred to as the design flood protection depth (d_{fp}) but is not used extensively in this Manual. This Manual emphasizes the use of the DFE as the minimum elevation to which flood-resistant design and construction efforts should be directed.

Determining the maximum design stillwater flood depth over the life of a building is the most important flood load calculation. Nearly every other coastal flood load parameter or calculation (e.g., hydrostatic load, design flood velocity, hydrodynamic load, design wave height, DFE, debris impact load, local scour depth) depends directly or indirectly on the design stillwater flood depth.

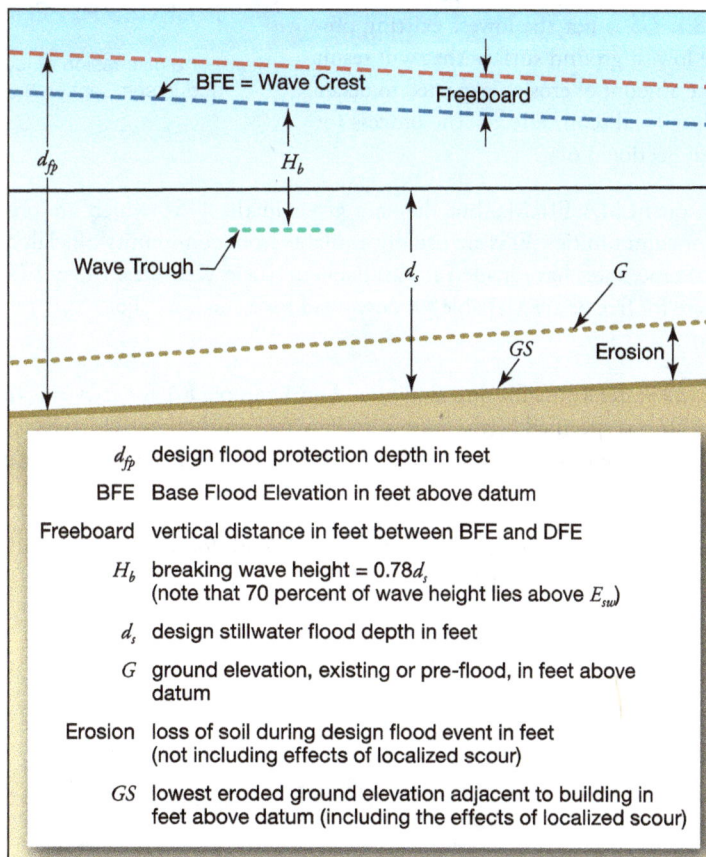

Figure 8-5.
Parameters that are determined or affected by flood depth

Legend within figure:

d_{fp}	design flood protection depth in feet
BFE	Base Flood Elevation in feet above datum
Freeboard	vertical distance in feet between BFE and DFE
H_b	breaking wave height = $0.78d_s$ (note that 70 percent of wave height lies above E_{sw})
d_s	design stillwater flood depth in feet
G	ground elevation, existing or pre-flood, in feet above datum
Erosion	loss of soil during design flood event in feet (not including effects of localized scour)
GS	lowest eroded ground elevation adjacent to building in feet above datum (including the effects of localized scour)

In this Manual, the design stillwater flood depth (d_s) is defined as the difference between the design stillwater flood elevation (E_{sw}) and the lowest eroded ground surface elevation (GS) adjacent to the building (see Equation 8.1) where wave setup is included in the stillwater flood elevation.

Σ EQUATION 8.1. DESIGN STILLWATER FLOOD DEPTH

$$d_s = E_{sw} - GS \qquad\qquad (Eq.\ 8.1)$$

where:

d_s = design stillwater flood depth (ft)

E_{sw} = design stillwater flood elevation in ft above datum (e.g., NGVD, NAVD)

GS = lowest eroded ground elevation, in ft above datum, adjacent to a building, excluding effects of localized scour around the foundation

Figure 8-5 illustrates the relationships among the various flood parameters that determine or are affected by flood depth. Note that in Figure 8-5 and Equation 8.1, GS is not the lowest existing pre-flood ground surface; it is the lowest ground surface that will result from long-term erosion and the amount of erosion expected to occur during a design flood, excluding local scour effects. The process for determining GS is described in Section 3.6.4.

CROSS REFERENCE

For a discussion of localized scour, see Section 8.5.10.

Values for E_{sw} are not shown on FEMA FIRMs, but they are given in the FISs, which are produced in conjunction with the FIRM for communities. FISs are usually available from community officials and NFIP State Coordinating Agencies. Some states have made FISs available on their Web sites. Many FISs are also available on the FEMA Web site for free or are available for download for a small fee. For more information, go to http://www.msc.fema.gov.

Design stillwater flood depth (d_s) is determined using Equation A in Example 8.1 for scenarios in which a non-100-year frequency-based DFE is specified by the Authority Having Jurisdiction (AHJ). Freeboard tied to the 100-year flood should not be used to increase d_s since load factors in ASCE 7 were developed for the 100-year nominal flood load.

Example 8.1 demonstrates the calculation of the design stillwater flood depth for five scenarios. All solutions to example problems are in bold text in this Manual.

EXAMPLE 8.1. DESIGN STILLWATER FLOOD DEPTH CALCULATIONS

Given:

- Oceanfront building site on landward side of a primary frontal dune (see Illustration A)

- Topography along transect perpendicular to shoreline is shown in Illustration B; existing ground elevation at seaward row of pilings = 7.0 ft NGVD

- Soil is dense sand; no terminating stratum above −25 ft NGVD

- Data from FIRM is as follows: flood hazard zone at site is Zone VE; BFE = 14.0 ft NGVD

- Data from FIS is as follows: 10-year stillwater elevation = 5.0 ft NGVD; 50-year stillwater elevation = 8.7 ft NGVD; 100-year stillwater elevation = 10.1 ft NGVD; 500-year stillwater elevation = 12.2 ft NGVD

- 500-year wave crest elevation (DFE) specified by AHJ = 18.0 ft NGVD

- Local government requires 1.0 ft freeboard; therefore DFE = 14.0 ft NGVD (BFE) + 1.0 ft = 15.0 ft NGVD

- Direction of wave and flow approach during design event is perpendicular to shoreline

- The eroded ground elevation (base flood conditions) at the seaward row of pilings = 5.5 ft NGVD

- Assume sea level rise is 0.01 ft/yr

- Assume long-term average annual erosion rate is 2.0 ft/yr, no beach nourishment or shoreline stabilization

- Assume building life = 50 years

Illustration A. Plan view of site and building location with flood hazard zones

EXAMPLE 8.1. DESIGN STILLWATER FLOOD DEPTH CALCULATIONS (continued)

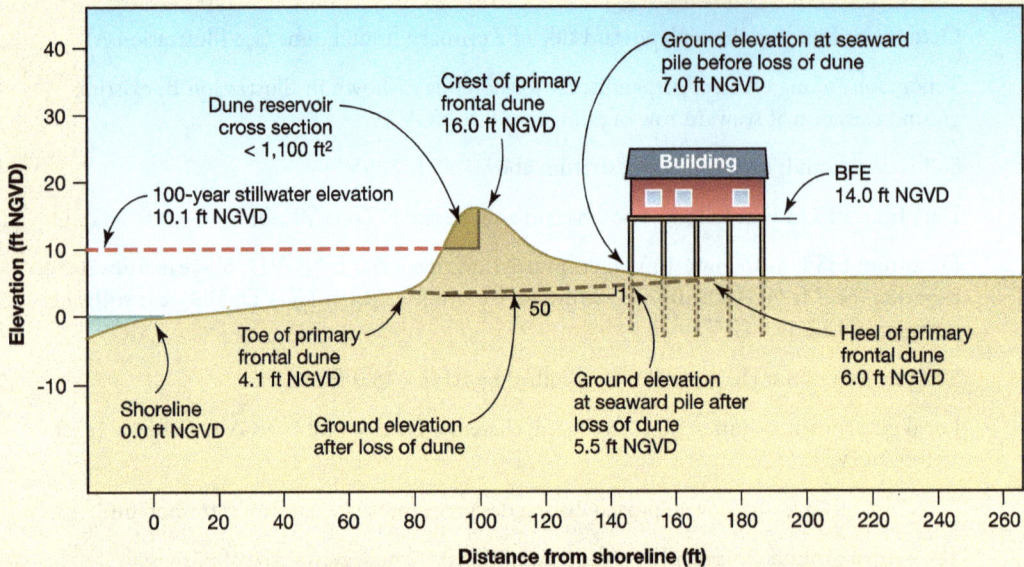

Illustration B. Primary frontal dune will be lost to erosion during a 100-year flood
because dune reservoir is less than 1,100 ft² (Section A of Illustration A)

Find:

The design stillwater flood depth (d_s) at the seaward row of piles for varying values of stillwater elevation, presence of freeboard, and consideration of the effects of future conditions (e.g., sea-level rise and long-term erosion).

The basis of the design flood for four scenarios are as follows:

1. 100-year stillwater elevation (NGVD). Future conditions not considered.

2. 100-year stillwater elevation (NGVD) plus freeboard. Future conditions not considered.

3. 100-year stillwater elevation (NGVD). Future conditions (sea-level rise and long-term erosion) in 50 years considered.

4. 500-year wave crest elevation (NGVD). Future conditions not considered.

Note: *Design stillwater flood depth (d_s) is determined using Equation A for scenarios in which a non-100-year frequency-based DFE is specified by the AHJ. Freeboard tied to the 100-year flood should not be used to increase d_s since load factors in ASCE 7 were developed for the 100-year nominal flood load.*

EXAMPLE 8.1. DESIGN STILLWATER FLOOD DEPTH CALCULATIONS (continued)

Σ EQUATION A

$$d_s = \left(\frac{DFE}{BFE} \right)(E_{SW}) - GS$$

where:

d_s = design stillwater flood depth

DFE = design flood elevation for a greater than 100-year flood event

BFE = base flood elevation

E_{SW} = design stillwater flood elevation in feet above datum (e.g. NGVD, NAVD)

GS = lowest eroded ground elevation, in feet above datum, adjacent to building, excluding effects of localized scour around foundations

Solution for Scenario #1: The design stillwater flood depth (d_s) at seaward row of pilings using the 100-year stillwater elevation can be calculated using Equation 8.1 as follows:

d_s = $E_{sw} - GS$

d_s = 10.1 ft NGVD – 5.5 ft NGVD

d_s = **4.6 ft**

Note: This is the same solution that is calculated in Example 8.4, #3

Solution for Scenario #2: The design stillwater flood depth (d_s) at seaward row of pilings using the 100-year stillwater elevation and freeboard will be calculated just as in Scenario #1–freeboard should not be included in the stillwater depth calculation but is used instead to raise the building to a higher-than-BFE level:

d_s = $E_{sw} - GS$

d_s = 10.1 ft NGVD – 5.5 ft NGVD

d_s = **4.6 ft**

Solution for Scenario #3: The design stillwater flood depth (d_s) at seaward row of pilings using the 100-year stillwater elevation and the future conditions of sea-level rise and long-term erosion can be calculated as follows:

Step 1: Increase 100-year stillwater elevation 50 years in the future to account for sea-level rise

E_{SW} = 10.1 ft NGVD + (0.01 ft/yr)(50 years) = 10.6 ft NGVD

Step 2: Calculate the lowest ground elevation in ft above the datum adjacent to the seaward row of pilings in 50 years

EXAMPLE 8.1. DESIGN STILLWATER FLOOD DEPTH CALCULATIONS (concluded)

- In 50 years, the front toe of the dune will translate horizontally toward the building by (50 yr) (2 ft/yr) = 100 ft landward

- Taking into account the 1:50 (*v:h*) slope of the eroded dune, the ground at the seaward row of piles will drop (100 ft)(1/50) = 2 ft over 50 years

$$GS = 5.5 \text{ ft} - 2 \text{ ft} = 3.5 \text{ ft NGVD in 50 years}$$

Step 3: Combine the effects of sea-level rise and erosion to calculate d_s

$$d_s = E_{sw} - GS$$
$$d_s = 10.6 \text{ ft NGVD} - 3.5 \text{ ft NGVD} = \textbf{7.1 ft}$$

Solution for Scenario #4: The design stillwater flood depth (d_s) at seaward row of pilings using the AHJ's 500-year wave crest elevation (DFE) can be calculated using Equation A of Example 8.1 as follows:

$$d_s = \left(\frac{DFE}{BFE}\right)(E_{SW}) - GS$$

$$d_s = \left(\frac{18 \text{ ft}}{14 \text{ ft}}\right)(10.1 \text{ ft}) - 5.5 \text{ ft} = 13.0 \text{ ft} - 5.5 \text{ ft} = \textbf{7.5 ft}$$

Note: Scenarios #1 through #4 show incremental increases in the design stillwater flood depth d_s, depending on how conservative the designer wishes to be in selecting the design scenario. As the design stillwater flood depth increases, the flood loads to which the building foundation must be designed also increase. The increase factor listed in Table A represents the square of the ratio of stillwater flood depth to the stillwater flood depth from Scenario #1(reference case).

Table A. Stillwater Flood Depths for Various Design Scenarios and Approximate Load Increase Factor from Increased Values of d_s

Scenario #	Design Condition	d_s (ft)	Approximate Load Increase Factor
#1 (reference case)	100-year	4.6	1.0
#2	100-year + freeboard	4.6	1.0
#3	100-year + future conditions	7.1	2.4
#4	500-year	7.5	2.7

Note: In subsequent examples, the building in Illustrations A and B and d_s in Scenario #1 are used.

8.5.4 Wave Setup Contribution to Flood Depth

Pre-1989 FIS reports and FIRMs do not usually include the effects of wave setup (d_{ws}), but some post-1989 FISs and FIRMs do. Because the calculation of design wave heights and flood loads depends on an accurate determination of the total stillwater flood depth, designers should review the effective FIS carefully, using the following procedure:

- Check the hydrologic analyses section of the FIS for mention of wave setup. Note the magnitude of the wave setup.

- Check the stillwater elevation table of the FIS for footnotes regarding wave setup. If wave setup is included in the listed BFEs but not in the 100-year stillwater elevation, add wave setup before calculating the design stillwater flood depth, the design wave height, the design flood velocity, flood loads, and localized scour. If wave setup is already included in the 100-year stillwater elevation, use the 100-year stillwater elevation to determine the design stillwater flood depth and other parameters. Wave setup should not be included in the 100-year stillwater elevation when calculating primary frontal dune erosion.

8.5.5 Design Breaking Wave Height

The design breaking wave height (H_b) at a coastal building site is one of the most important design parameters. Unless detailed analysis shows that natural or manmade obstructions will protect the site during a design event, wave heights at a site should be calculated as the heights of depth-limited breaking waves, which are equivalent to 0.78 times the design stillwater flood depth (see Figure 8-5). Note that 70 percent of the breaking wave height lies above the stillwater elevation. In some situations, such as steep ground slopes immediately seaward of a building, the breaking wave height can exceed 0.78 times the stillwater flood depth. In such instances, designers may wish to increase the breaking wave height used for design, with an upper limit for the breaking wave height equal to the stillwater flood depth.

8.5.6 Design Flood Velocity

Estimating design flood velocities (V) in coastal flood hazard areas is subject to considerable uncertainty. There is little reliable historical information concerning the velocity of floodwaters during coastal flood events. The direction and velocity of floodwaters can vary significantly throughout a coastal flood event. Floodwaters can approach a site from one direction during the beginning of a flood event and then shift

---

NOTE

Flood loads are applied to structures as follows:

- **Lateral hydrostatic loads** – at two-thirds depth point of stillwater elevation

- **Breaking wave loads** – at stillwater elevation

- **Hydrodynamic loads** – at mid-depth point of stillwater elevation

- **Debris impact loads** – at stillwater elevation

TERMINOLOGY: WAVE SETUP

Wave setup is an increase in the stillwater surface near the shoreline due to the presence of breaking waves. Wave setup typically adds 1.5 to 2.5 feet to the 100-year stillwater flood elevation and should be discussed in the FIS.

WARNING

This Manual does not provide guidance for estimating flood velocities during tsunamis. The issue is highly complex and site-specific. Designers should look for model results from tsunami inundation or evacuation studies.

to another direction (or several directions). Floodwaters can inundate low-lying coastal sites from both the front (e.g., ocean) and back (e.g., bay, sound, river). In a similar manner, flow velocities can vary from close to zero to high velocities during a single flood event. For these reasons, flood velocities should be estimated conservatively by assuming floodwaters can approach from the most critical direction relative to the site and by assuming flow velocities can be high (see Equation 8.2).

Σ EQUATION 8.2. DESIGN FLOOD VELOCITY

Lower bound $\qquad V = \dfrac{d_s}{t}$ $\qquad$ (Eq. 8.2a)

Upper bound $\qquad V = (g d_s)^{0.5}$ $\qquad$ (Eq. 8.2b)

where:

V = design flood velocity (ft/sec)

d_s = design stillwater flood depth (ft)

t = 1 sec

g = gravitational constant (32.2 ft/sec^2)

For design purposes, flood velocities in coastal areas should be assumed to lie between $V = (g d_s)^{0.5}$ (the expected upper bound) and $V = d_s/t$ (the expected lower bound). It is recommended that designers consider the following factors before deciding whether to use the upper- or lower-bound flood velocity for design:

■ Flood zone

■ Topography and slope

■ Distance from the source of flooding

■ Proximity to other buildings or obstructions

The upper bound should be taken as the design flood velocity if the building site is near the flood source, in Zone V, in Zone AO adjacent to Zone V, in Zone A subject to velocity flow and wave action, on steeply sloping terrain, or adjacent to other large buildings or obstructions that will confine or redirect floodwaters and increase local flood velocities. The lower bound is a more appropriate design flood velocity if the site is distant from the flood source, in Zone A, on flat or gently sloping terrain, or unaffected by other buildings or obstructions.

Figure 8-6 shows the velocity versus design stillwater flood depth relationship for non-tsunami, upper- and lower-bound velocities. Equation 8.2 shows the equations for the lower-bound and upper-bound velocity conditions.

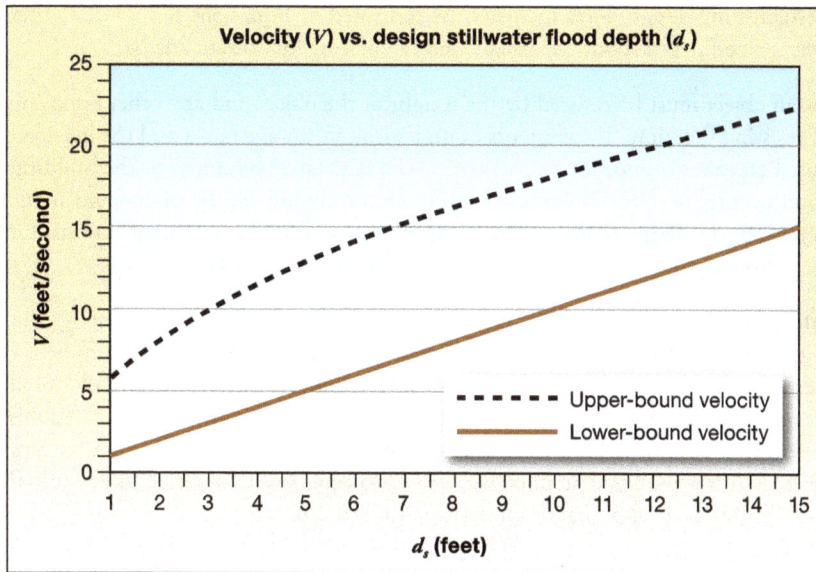

Figure 8-6.
Velocity versus design
stillwater flood depth

8.5.7   Hydrostatic Loads

Hydrostatic loads occur whenever floodwaters come into contact with a foundation, building, or building element. Hydrostatic loads can act laterally or vertically.

Lateral hydrostatic forces are generally not sufficient to cause deflection or displacement of a building or building element unless there is a substantial difference in water elevation on opposite sides of the building or component. This is why the NFIP requires that openings be provided in vertical walls that form enclosures below the BFE for buildings constructed in Zone A (see Section 5.2.3.2).

Likewise, vertical hydrostatic forces (buoyancy or flotation) are not generally a concern for properly constructed and elevated coastal buildings founded on adequate foundations. However, buoyant forces can have a significant effect on inadequately elevated buildings on shallow foundations. Such buildings are vulnerable to uplift from flood and wind forces because the weight of a foundation or building element is much less when submerged than when not submerged. For example, one cubic foot of a footing constructed of normal weight concrete weighs approximately 150 pounds. But when submerged, each cubic foot of concrete displaces a cubic foot of saltwater, which weighs about 64 pounds/cubic foot. Thus, the foundation's submerged weight is only 86 pounds if submerged in saltwater (150 pounds/cubic foot – 64 pounds/cubic foot = 86 pounds/cubic foot), or 88 pounds if submerged in fresh water (150 pounds/cubic foot – 62 pounds/cubic foot = 88 pounds/cubic foot). A submerged footing contributes approximately 40 percent less uplift resistance during flood conditions.

Section 3.2.2 of ASCE 7-10 states that the full hydrostatic pressure of water must be applied to floors and foundations when applicable. Sections 2.3.3 and 2.4.2 of ASCE 7-10 require factored flood loads to be considered in the load combinations that model uplift and overturning design limit states. For ASD, flood loads are increased by a factor of 1.5 in Zone V and Coastal A Zones (and 0.75 in coastal flood zones with base flood wave heights less than 1.5 feet, and in non-coastal flood zones). These load factors are applied to account for uncertainty in establishing design flood intensity. As indicated in Equations 8-3 and 8-4 (per Figure 8-7), the design stillwater flood depth should be used when calculating hydrostatic loads.

Any buoyant force (F_{buoy}) on an object must be resisted by the weight of the object and any other opposing force (e.g., anchorage forces) resisting flotation. The contents of underground storage tanks and the live load on floors should not be counted on to resist buoyant forces because the tanks may be empty or the building may be vacant when the flood occurs. Buoyant or flotation forces on a building can be of concern if the actual stillwater flood depth exceeds the design stillwater flood depth. Buoyant forces are also of concern for empty or partially empty aboveground tanks, underground tanks, and swimming pools.

Lateral hydrostatic loads are given by Equation 8.3 and illustrated in Figure 8-7. Note that f_{sta} (in Equation 8.3) is equivalent to the area of the pressure triangle and acts at a point equal to 2/3 d_s below the water surface (see Figure 8-7). Figure 8-7 is presented here solely to illustrate the application of lateral hydrostatic force. In communities participating in the NFIP, local floodplain ordinances or laws require that buildings in Zone V be elevated above the BFE on an open foundation and that the foundation walls of buildings in Zone A be equipped with openings that allow floodwater to enter so that internal and external hydrostatic pressures will equalize (see Section 5.2) and not damage the structure.

Vertical hydrostatic forces are given by Equation 8.4 and are illustrated by Figure 8-8.

EQUATION 8.3. LATERAL HYDROSTATIC LOAD

$$f_{sta} = \frac{1}{2}\gamma_w d_s^2 \tag{Eq. 8.3a}$$

where:

f_{sta} = hydrostatic force per unit width (lb/ft) resulting from flooding against vertical element

γ_w = specific weight of water (62.4 lb/ft³ for fresh water and 64.0 lb/ft³ for saltwater)

d_s = design stillwater flood depth (ft)

$$F_{sta} = f_{sta}(w) \tag{Eq. 8.3b}$$

where:

F_{sta} = total equivalent lateral hydrostatic force on a structure (lb)

f_{sta} = hydrostatic force per unit width (lb/ft) resulting from flooding against vertical element

w = width of vertical element (ft)

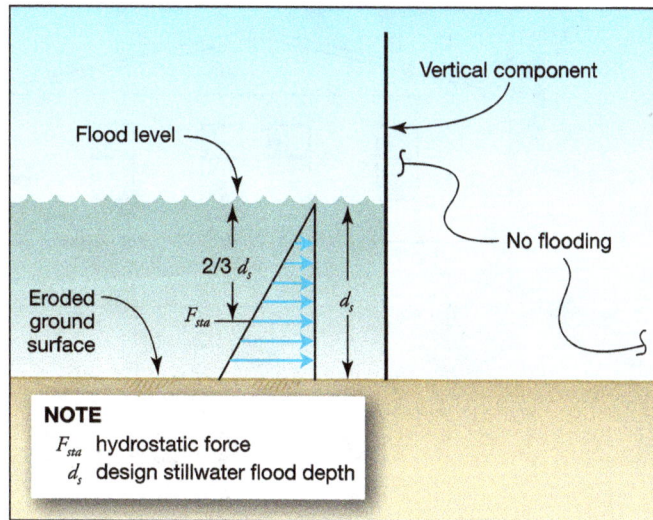

Figure 8-7.
Lateral flood force on a
vertical component

EQUATION 8.4. VERTICAL (BUOYANT) HYDROSTATIC
FORCE

$$F_{buoy} = \gamma_w (Vol)$$

(Eq. 8.4)

where:

F_{buoy} = vertical hydrostatic force (lb) resulting from the
displacement of a given volume of floodwater

γ_w = specific weight of water (62.4 lb/ft^3 for fresh water and
64.0 lb/ft^3 for saltwater)

Vol = volume of floodwater displaced by a submerged
object (ft^3)

Figure 8-8.
Vertical (buoyant) flood force; buoyancy forces are drastically reduced for open foundations (piles or piers)

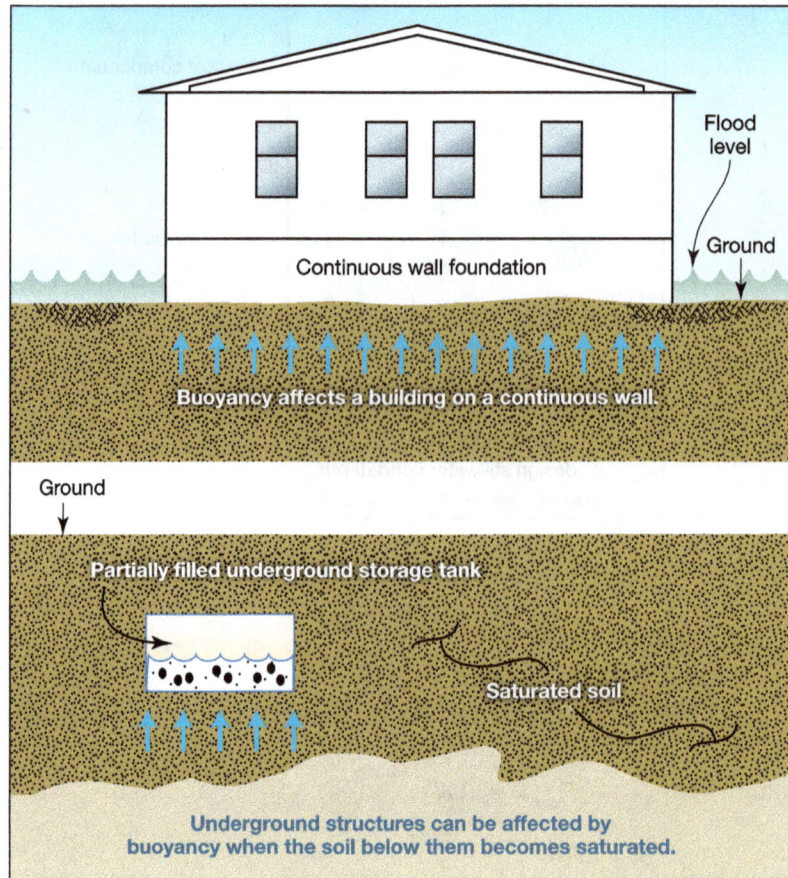

8.5.8   Wave Loads

Calculating wave loads requires the designer to estimate expected wave heights, which, for the purposes of this Manual, are limited by water depths at the site of interest. These data can be estimated using a variety of models. FEMA uses its Wave Height Analysis for Flood Insurance Studies (WHAFIS) model to estimate wave heights and wave crest elevations, and results from this model can be used directly by designers to calculate wave loads.

CROSS REFERENCE

For additional guidance in calculating wave loads, see ASCE 7-10.

Wave forces can be separated into four categories:

■ From nonbreaking waves – can usually be computed as hydrostatic forces against walls and hydrodynamic forces against piles

■ From breaking waves – short duration but large magnitude

■ From broken waves – similar to hydrodynamic forces caused by flowing or surging water

■ Uplift – often caused by wave run-up, deflection, or peaking against the underside of horizontal surfaces

The forces from breaking waves are the highest and produce the most severe loads. It is therefore strongly recommended that the breaking wave load be used as the design wave load.

The following three breaking wave loading conditions are of interest in residential design:

CROSS REFERENCE

For more information about FEMA's WHAFIS model, see http://www.fema.gov/plan/prevent/fhm/dl_wfis4.shtm.

■ Waves breaking on small-diameter vertical elements below the DFE (e.g., piles, columns in the foundation of a building in Zone V)

■ Waves breaking against walls below the DFE (e.g., solid foundation walls in Zone A, breakaway walls in Zone V)

■ Wave slam, where just the top of a wave strikes a vertical wall

8.5.8.1 Breaking Wave Loads on Vertical Piles

The breaking wave load on a pile can be assumed to act at the stillwater elevation and is calculated using Equation 8.5.

EQUATION 8.5. BREAKING WAVE LOAD ON VERTICAL PILES

$$F_{brkp} = \frac{1}{2} C_{db} \gamma_w D H_b^2$$

(Eq. 8.5)

where:

F_{brkp} = drag force (lb) acting at the stillwater elevation

C_{db} = breaking wave drag coefficient (recommended values are 2.25 for square and rectangular piles and 1.75 for round piles)

γ_w = specific weight of water (62.4 lb/ft³ for fresh water and 64.0 lb/ft³ for saltwater)

D = pile diameter (ft) for a round pile or 1.4 times the width of the pile or column for a square pile (ft)

H_b = breaking wave height (0.78 d_s), in ft, where d_s = design stillwater flood depth (ft)

Wave loads produced by breaking waves are greater than those produced by nonbreaking or broken waves. Example 8.3 shows the difference between the loads imposed on a vertical pile by nonbreaking waves and by breaking waves.

8.5.8.2 Breaking Wave Loads on Vertical Walls

Breaking wave loads on vertical walls are best calculated according to the procedure described in *Criteria for Evaluating Coastal Flood-Protection Structures* (Walton et al. 1989). The procedure is suitable for use in wave conditions typical during coastal flood and storm events. The relationship for breaking wave load per unit length of wall is shown in Equation 8.6.

NOTE

Equation 8.6 includes the hydrostatic component calculated using Equation 8.3. If Equation 8.6 is used, lateral hydrostatic force from Equation 8.3 should not be added to avoid double counting.

EQUATION 8.6. BREAKING WAVE LOAD ON VERTICAL WALLS

Case 1 (enclosed dry space behind wall):

$$f_{brkw} = 1.1 C_p \gamma_w d_s^2 + 2.4 \gamma_w d_s^2 \qquad \text{(Eq. 8.6a)}$$

Case 2 (equal stillwater elevation on both sides of wall):

$$f_{brkw} = 1.1 C_p \gamma_w d_s^2 + 1.9 \gamma_w d_s^2 \qquad \text{(Eq. 8.6b)}$$

where:

f_{brkw} = total breaking wave load per unit length of wall (lb/ft) acting at the stillwater elevation

C_p = dynamic pressure coefficient from Table 8-1

γ_w = specific weight of water (62.4 lb/ft³ for fresh water and 64.0 lb/ft³ for saltwater)

d_s = design stillwater flood depth (ft)

$$F_{brkw} = f_{brkw}(w) \qquad \text{(Eq. 8.6c)}$$

where:

F_{brkw} = total breaking wave load (lb) acting at the stillwater elevation

f_{brkw} = total breaking wave load per unit length of wall (lb/ft) acting at the stillwater elevation

w = width of wall (ft)

The procedure assumes that the vertical wall causes a reflected or standing wave to form against the seaward side of the wall and that the crest of the wave reaches a height of 1.2 d_s above the stillwater elevation. The resulting dynamic, static, and total pressure distributions against the wall and resulting loads are as shown in Figure 8-9.

Table 8-1. Value of Dynamic Pressure Coefficient (C_p) as a Function of Probability of Exceedance

C_p	Building Type	Probability of Exceedance
1.6	Buildings and other structures that represent a low hazard to human life or property in the event of failure	0.5
2.8	Coastal residential building	0.01
3.2	Buildings and other structures, the failure of which could pose a substantial risk to human life	0.002
3.5	High-occupancy building or critical facility or those designated as essential facilities	0.001

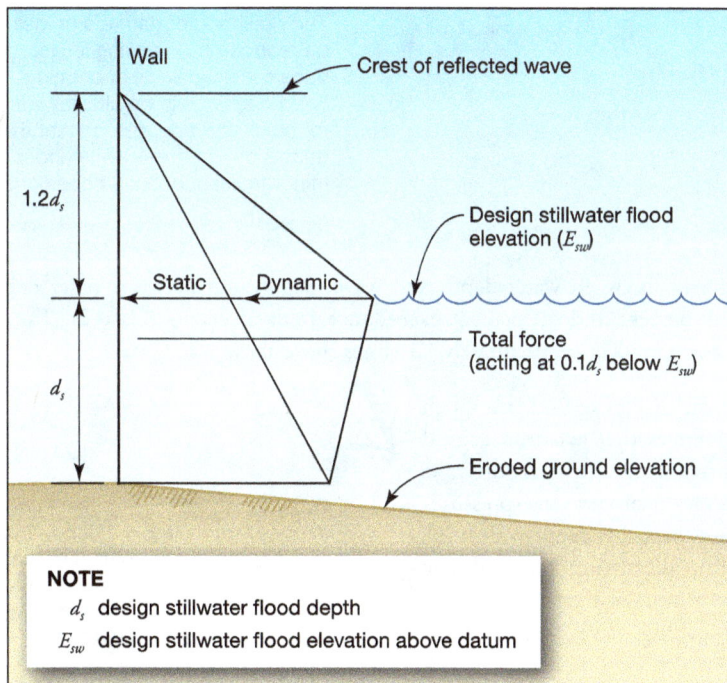

Figure 8-9.
Breaking wave pressure distribution against a vertical wall

Equation 8.6 includes two cases: (1) a wave breaks against a vertical wall of an enclosed dry space, shown in Equation 8.6a, and (2) the stillwater elevation on both sides of the wall is equal, shown in Equation 8.6b. Case 1 is equivalent to a situation in which a wave breaks against an enclosure in which there is no floodwater inside the enclosure. Case 2 is equivalent to a situation in which a wave breaks against a breakaway wall or a wall equipped with openings that allow floodwaters to equalize on both sides of the wall. In both cases, waves are normally incident (i.e., wave crests are parallel to the wall). If breaking waves are obliquely incident (i.e., wave crests are not parallel to the wall; see Figure 8-10), the calculated loads would be lower.

Figure 8-10.
Wave crests not parallel
to wall

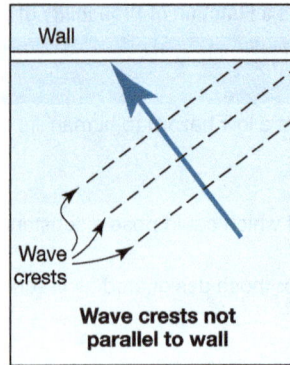

Wave crests not parallel to wall

Figure 8-11 shows the relationship between water depth and wave height, and between water depth and breaking wave force, for the 1 percent and 50 percent exceedance interval events (Case 2). The Case 1 breaking wave force for these two events is approximately 1.1 times those shown for Case 2.

The breaking wave forces shown in Figure 8-11 are much higher than the typical wind forces that act on a coastal building, even wind pressures that occur during a hurricane or typhoon. However, the duration of the wave pressures and loads is brief; peak pressures probably occur within 0.1 to 0.3 second after the wave breaks against the wall. See *Wave Forces on Inclined and Vertical Wall Surfaces* (ASCE 1995) for a discussion of breaking wave pressures and durations.

Post-storm damage inspections show that breaking wave loads have destroyed virtually all types of wood-frame walls and unreinforced masonry walls below the wave crest elevation. Only highly engineered, massive structural elements are capable of withstanding breaking wave loads. Damaging wave pressures and loads can be generated by waves much lower than the 3-foot wave currently used by FEMA to distinguish Zone A from Zone V. This fact was confirmed by the results of FEMA-sponsored laboratory tests of breakaway wall failures in which measured pressures

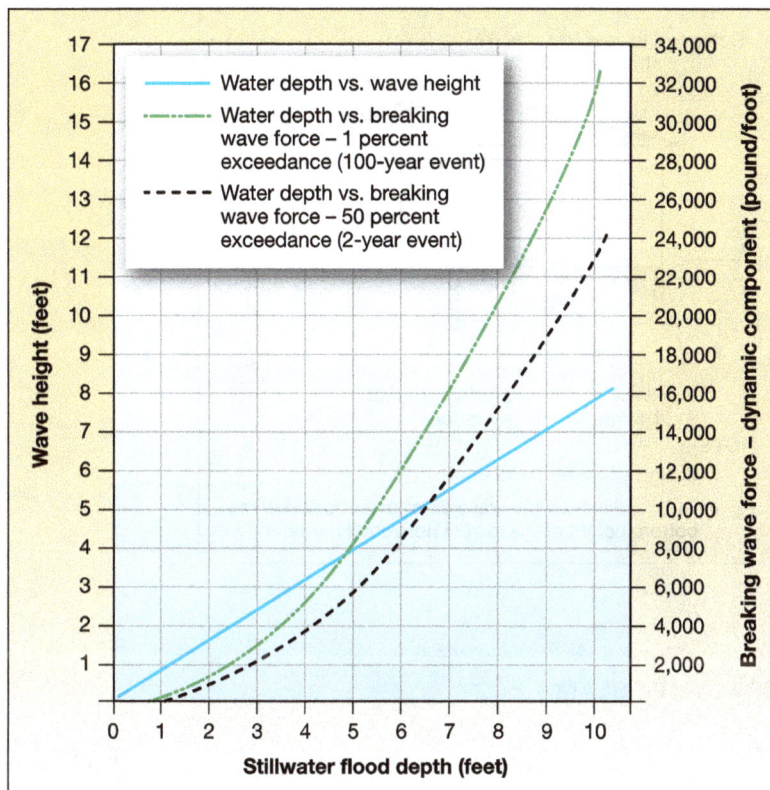

Figure 8-11.
Water depth versus
wave height, and water
depth versus breaking
wave force against, a
vertical wall

on the order of hundreds of pounds/ square foot were generated by waves that were only 12 to 18 inches high. See Appendix H for the results of the tests.

8.5.8.3 Wave Slam

The action of wave crests striking the elevated portion of a structure is known as "wave slam." Wave slam introduces lateral and vertical loads on the lower portions of the elevated structure (Figure 8-12). Wave slam force, which can be large, typically results in damaged floor systems (see Figure 3-26 in Chapter 3). This is one reason freeboard should be included in the design of coastal residential buildings. Lateral wave slam can be calculated using Equation 8.7, but vertical wave slam calculations are beyond the scope of this Manual.

Equation 8.7 is similar to Equation 8.8 (hydrodynamic load) with the wave crest velocity set at the wave celerity (upper-bound flow velocity, given by Equation 8.2b) and a wave slam coefficient instead of a drag coefficient. The wave slam coefficient used in Equation 8.7 is an effective slam coefficient, estimated using information contained in Bea et al. (1999) and McConnell et al. (2004).

Wave slam should not be computed for buildings that are elevated on solid foundation walls (the wave-load-on-wall calculation using Equation 8.6 includes wave slam) but should be computed for buildings that are elevated on piles or columns (wave loads on the piles or columns, and wave slam against the elevated building, can be computed separately and summed).

Figure 8-12.
Lateral wave slam against an elevated building

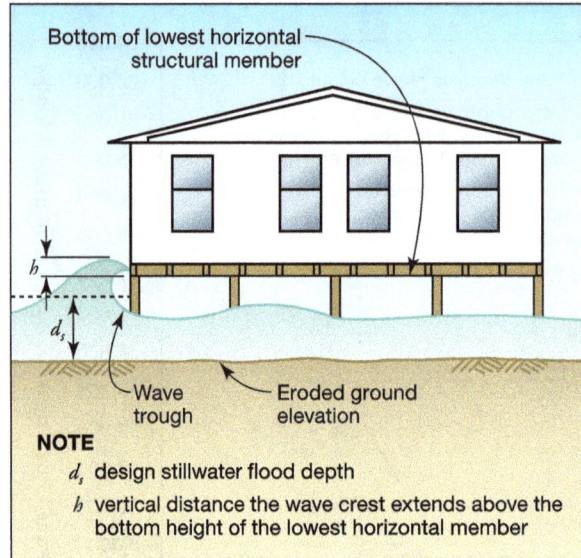

EQUATION 8.7. LATERAL WAVE SLAM

$$F_s = f_s w = \frac{1}{2}\gamma_w C_s d_s h w$$

(Eq. 8.7)

where:

F_s = lateral wave slam (lb)

f_s = lateral wave slam (lb/ft)

C_s = slam coefficient incorporating effects of slam duration and structure stiffness for typical residential structure (recommended value is 2.0)

γ_w = unit weight of water (62.4 lb/ft³ for fresh water and 64.0 lb/ft³ for saltwater)

d_s = stillwater flood depth (ft)

h = vertical distance (ft) the wave crest extends above the bottom of the floor joist or floor beam

w = length (ft) of the floor joist or floor beam struck by wave crest

EXAMPLE 8.2. WAVE SLAM CALCULATION

Given:

- Zone V building elevated on pile foundation near saltwater
- Bottom of floor beam elevation = 15.0 ft NGVD
- Length of beam (parallel to wave crest) = 50 ft
- Design stillwater elevation = 12.0 ft NGVD
- Eroded ground elevation = 5.0 ft NGVD
- C_s (wave slam coefficient; see Equation 8.7) = 2.0
- γ_w = specific weight of water (62.4 lb/ft³ for fresh water and 64.0 lb/ft³ for saltwater)
- A = (8 ft)(0.833 ft) = 6.664 ft²

Find:

1. Wave crest elevation
2. Vertical height of the beam subject to wave slam
3. Lateral wave slam acting on the elevated floor system

Solution for #1: The wave crest elevation can be calculated as 1.55 times the stillwater depth, above the eroded ground elevation

   Wave crest elevation = 5.0 ft NGVD + 1.55 (12.0 ft NGVD – 5.0 ft NGVD) = **15.9 ft NGVD**

Solution for #2: The vertical height of the beam subject to wave slam can be found as follows:

   Vertical height = wave crest elevation – bottom of beam elevation = 15.9 ft NGVD – 15.0 ft NGVD = **0.9 ft**

Solution for #3: Using Equation 8-7, the lateral wave slam acting on the elevated floor system can be found as follows:

$$F_s = f_s w = \frac{1}{2}\gamma C_s d_s h w = \left(\frac{1}{2}\right)(64 \text{ lb/ft}^3)\,(2.0)\,(7.0 \text{ ft})\,(0.9 \text{ ft})\,(50 \text{ ft}) = \mathbf{20{,}160\ lb}$$

8.5.9   Hydrodynamic Loads

As shown in Figure 8-13, water flowing around a building (or a structural element or other object) imposes loads on the building. In the figure, note that the lowest floor of the building is above the flood level and the loads imposed by flowing water affect only the foundation walls. However, open foundation systems, unlike that shown in Figure 8-13, can greatly reduce hydrodynamic loading. Hydrodynamic loads, which are a function of flow velocity and structural geometry, include frontal impact on the upstream face, drag along the sides, and suction on the downstream side. One of the most difficult steps in quantifying loads imposed by moving water is determining the expected flood velocity (see Section 8.5.6 for guidance on design flood velocities). In this Manual, the velocity of floodwater is assumed to be constant (i.e., steady-state flow). Hydrodynamic loads can be calculated using Equation 8.8.

Elevating above the DFE provides additional protection from hydrodynamic loads for elevated enclosed areas.

The drag coefficient used in Equation 8.8 is taken from the *Shore Protection Manual, Volume 2* (USACE 1984). Additional guidance is provided in Section 5.4.3 of ASCE 7-10 and in FEMA 259, *Engineering Principles and Practices for Retrofitting Floodprone Residential Buildings* (FEMA 2001). The drag coefficient is a function of the shape of the object around which flow is directed. When an object is something other than a round, square, or rectangular pile, the coefficient is determined by one of the following ratios (see Table 8-2):

1. The ratio of the width of the object (w) to the height of the object (h) if the object is completely immersed in water

2. The ratio of the width of the object (w) to the stillwater flood depth of the water (d_s) if the object is not fully immersed

Figure 8-13.
Hydrodynamic loads on a building

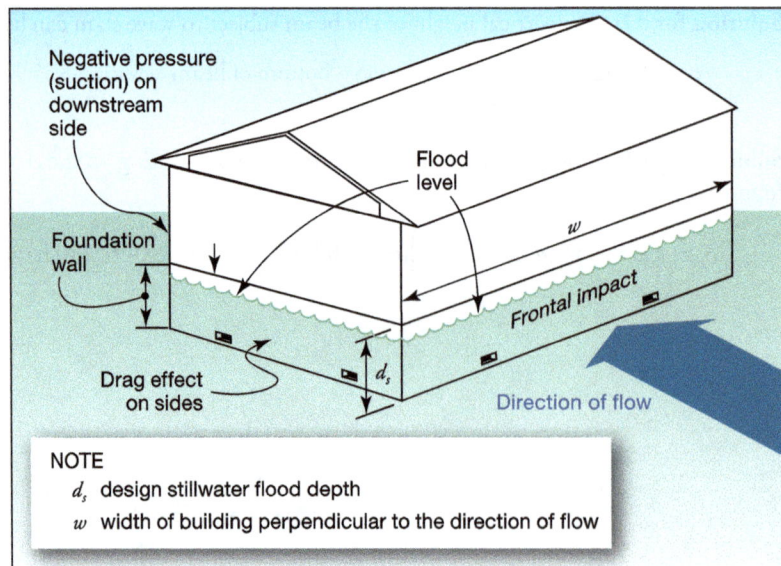

EQUATION 8.8. HYDRODYNAMIC LOAD (FOR ALL FLOW VELOCITIES)

$$F_{dyn} = \frac{1}{2}C_d \rho V^2 A$$

(Eq. 8.8)

where:

F_{dyn} = horizontal drag force (lb) acting at the stillwater mid-depth (half way between the stillwater elevation and the eroded ground surface)

C_d = drag coefficient (recommended values are 2.0 for square or rectangular piles and 1.2 for round piles; for other obstructions, see Table 8-2)

ρ = mass density of fluid (1.94 slugs/ft^3 for fresh water and 1.99 slugs/ft^3 for saltwater)

V = velocity of water (ft/sec); see Equation 8.2

A = surface area of obstruction normal to flow (ft^2) = $(w)(d_s)$(see Figure 8-13) or $(w)(h)$ if the object is completely immersed

Flow around a building or building element also creates flow-perpendicular forces (lift forces). When a building element is rigid, lift forces can be assumed to be small. When the element is not rigid, lift forces can be greater than drag forces. The equation for lift force is the same as that for hydrodynamic force except that the drag coefficient (C_d) is replaced with the lift coefficient (C_l). In this Manual, the foundations of coastal residential buildings are considered rigid, and hydrodynamic lift forces can therefore be ignored.

Equation 8.8 provides the total force against a building of a given surface area, A. Dividing the total force by either length or width yields a force per linear unit; dividing by surface area, A, yields a force per unit area. Example 8.3 shows the difference between the loads imposed on a vertical pile by nonbreaking and breaking waves. As noted in Section 8.5.8, nonbreaking wave forces on piles can be calculated as hydrodynamic forces.

Table 8-2. Drag Coefficients for Ratios of Width to Depth (w/d_s) and Width to Height (w/h)

Width-to-Depth Ratio (w/d_s or w/h)	Drag Coefficient (C_d)
1–12	1.25
13–20	1.3
21–32	1.4
33–40	1.5
41–80	1.75
81–120	1.8
>120	2.0

NOTE

Lift coefficients (C_l) are provided in *Introduction to Fluid Mechanics* (Fox and McDonald 1985) and in many other fluid mechanics textbooks.

EXAMPLE 8.3. HYDRODYNAMIC LOAD ON PILES VERSUS BREAKING WAVE LOAD ON PILES

Given:

- Building elevated on round-pile foundation near saltwater
- C_d (drag coefficient for nonbreaking wave on round pile; see Equation 8.8) = 1.2
- C_{db} (drag coefficient for breaking wave on round pile; see Equation 8.5) = 1.75
- D = 10 in. or 0.833 ft
- d_s = 8 ft
- Velocity ranges from 8 ft/sec to 16 ft/sec
- ρ = mass density of fluid (1.94 slugs/ft^3 for fresh water and 1.99 slugs/ft^3 for saltwater)
- γ_w = specific weight of water (62.4 lb/ft^3 for fresh water and 64.0 lb/ft^3 for saltwater)
- A = (8 ft)(0.833 ft) = 6.664 ft^2

Find:

1. The range of loads from hydrodynamic flow around a pile
2. Load from a breaking wave on a pile

Solution for #1: The hydrodynamic load from flow past a pile is calculated using Equation 8.8 as follows:

For a flood velocity of 8 ft/sec:

$$F_{nonbrkp} = \frac{1}{2} C_d \rho V^2 A$$

$$F_{nonbrkp} = \frac{1}{2}(1.2)\left(1.99 \text{ slugs/ft}^3\right)(8 \text{ ft/sec})^2 (6.664 \text{ ft}^2)$$

$$F_{nonbrkp} = 509 \text{ lb/pile}$$

For a flood velocity of 16 ft/sec:

$$F_{nonbrkp} = \frac{1}{2} C_d \rho V^2 A$$

$$F_{nonbrkp} = \left(\frac{1}{2}\right)(1.2)\left(1.99 \text{ slugs/ft}^3\right)(16 \text{ ft/sec})^2 (6.664 \text{ ft}^2)$$

$$F_{nonbrkp} = 2{,}037 \text{ lb/pile}$$

The range of loads from a nonbreaking wave: 509 lb/pile to 2,037 lb/pile

EXAMPLE 8.3. HYDRODYNAMIC LOAD ON PILES VERSUS BREAKING WAVE LOAD ON PILES (concluded)

Solution for #2: The load from a breaking wave on a pile is calculated with Equation 8.5 as follows:

$$F_{brkp} = \left(\frac{1}{2}\right)(1.75)\left(64.0 \text{ lb/ft}^3\right)(0.833 \text{ ft})(0.78)(8 \text{ ft}^2)$$

where:

H_b is the height of the breaking wave or $(0.78)d_s$

$F_{brkp} = $ **1,816 lb/pile**

Note: The load from the breaking wave is approximately 3.5 times the lower estimate of the hydrodynamic load. The upper estimate of the hydrodynamic load exceeds the breaking wave load only because of the very conservative nature of the upper flood velocity estimate.

8.5.10   Debris Impact Loads

Debris impact loads are imposed on a building by objects carried by moving water. The magnitude of these loads is very difficult to predict, but some reasonable allowance must be made for them. The loads are influenced by where the building is located in the potential debris stream, specifically if it is:

- Immediately adjacent to or downstream from another building

- Downstream from large floatable objects (e.g., exposed or minimally covered storage tanks)

- Among closely spaced buildings

A familiar equation for calculating debris loads is given in ASCE 7-10, *Commentary*. This equation has been simplified into Equation 8.9 using C_{Str}, the values of which are based on assumptions appropriate for the typical coastal buildings that are covered in this Manual. The parameters in Equation 8.9 are discussed below. See Chapter C5 of ASCE 7-10 for a more detailed discussion of the parameters.

Equation 8.9 contains the following uncertainties, each of which must be quantified before the effect of debris loading can be calculated:

- Size, shape, and weight (W) of the waterborne object

- Design flood velocity (V)

- Velocity of the waterborne object compared to the flood velocity

- Duration of the impact (Δt) (assumed to be equal to 0.03 seconds in the case of residential buildings is incorporated in C_{Str}, which is explained in more detail below)

- Portion of the building to be struck

Σ

EQUATION 8.9. DEBRIS IMPACT LOAD

$$F_i = WVC_DC_BC_{Str}$$ (Eq. 8.9)

where:

F_i = impact force acting at the stillwater elevation (lb)

W = weight of the object (lb)

V = velocity of water (ft/sec), approximated by $1/2(gd_s)^{1/2}$

C_D = depth coefficient (see Table 8-3)

C_B = blockage coefficient (taken as 1.0 for no upstream screening, flow path greater than 30 ft; see below for more information)

C_{Str} = Building structure coefficient (refer to the explanation of C_{Str} at the end of this section)

= 0.2 for timber pile and masonry column supported structures 3 stories or less in height above grade

= 0.4 for concrete pile or concrete or steel moment resisting frames 3 stories or less in height above grade

= 0.8 for reinforced concrete foundation walls (including insulated concrete forms)

Designers should consider locally adopted guidance because it may be based on more recent information than ASCE 7-10 or on information specific to the local hazards. Local guidance considerations may include the following:

Size, shape, and weight of waterborne debris. The size, shape, and weight of waterborne debris may vary according to region. For example, the coasts of Washington, Oregon, and other areas may be subject to very large debris in the form of whole trees and logs along the shoreline. The southeastern coast of the United States may be more subject to debris impact from dune crossovers and destroyed buildings than other areas. In the absence of information about the nature of potential debris, a weight of 1,000 pounds is recommended as the value of W. Objects with this weight could include portions of damaged buildings, utility poles, portions of previously embedded piles, and empty storage tanks.

Debris velocity. As noted in Section 8.5.6, flood velocity can be approximated within the range given by Equation 8.2. For calculating debris loads, the velocity of the waterborne object is assumed to be the same as the flood velocity. Although this assumption may be accurate for small objects, it may overstate debris velocities for large objects that drag on the bottom or that strike nearby structures.

Portion of the building to be struck. The object is assumed to be at or near the water surface level when it strikes the building and is therefore assumed to strike the building at the stillwater elevation.

Depth coefficient. The depth coefficient (C_D) accounts for reduced debris velocity as water depth decreases. For buildings in Zone A with stillwater flood depths greater than 5 feet or for buildings in Zone V, the depth coefficient = 1.0. For other conditions, the depth coefficient varies, as shown in Table 8-3.

Table 8-3. Depth Coefficient (C_D) by Flood Hazard Zone and Water Depth

Flood Hazard Zone and Water Depth	C_D
Floodway[a] or Zone V	1.0
Zone A, stillwater flood depth $\geq$ 5 ft	1.0
Zone A, stillwater flood depth = 4 ft	0.75
Zone A, stillwater flood depth = 2.5 ft	0.375
Zone A, stillwater flood depth $\leq$ 1 ft	0.00

(a) Per ASCE 24-05, a "floodway" is a "channel and that portion of the floodplain reserved to convey the base flood without cumulatively increasing the water surface elevation more than a designated height."

■ **Blockage coefficient.** The blockage coefficient (C_B) is used to account for the reduction in debris velocity expected to occur because of the screening provided by trees and other structures upstream from the structure or building on which the impact load is being calculated. The blockage coefficient varies, as shown in Table 8-4.

Table 8-4. Values of Blockage Coefficient C_B

Degree of Screening or Sheltering within 100 Ft Upstream	C_B
No upstream screening, flow path wider than 30 ft	1.0
Limited upstream screening, flow path 20-ft wide	0.6
Moderate upstream screening, flow path 10-ft wide	0.2
Dense upstream screening, flow path less than 5-ft wide	0.0

■ **Building structure coefficient.** The building structure coefficient, C_{str}, is derived from Equation C5-3, Chapter C5, ASCE 7-10. Coefficient values for C_{str}, (0.2, 0.4, and 0.8 as defined above for Equation 8.9) were generated by selecting input values recommended in ASCE 7-10, Chapter C5, with appropriate assumptions made to model typical coastal residential structures. The derived building structure coefficient formula with inputs is defined as follows:

$$C_{Str} = \frac{3.14 C_I C_O R_{max}}{2g\Delta t}$$

where:

C_I = importance coefficient = 1.0

C_O = orientation coefficient = 0.80

Δt = duration of impact = 0.03 sec

g = gravitational constant (32.2 ft/sec²)

R_{max} = maximum response ratio assuming approximate natural period, T, of building types as follows: for timber pile and masonry column, T = 0.75 sec; for concrete pile or concrete or steel moment resisting frames, T = 0.35 sec; and for reinforced concrete foundation walls, T = 0.2 sec. The ratio of impact duration (0.03 sec) to approximate natural period (T) is entered into Table C5-4 of ASCE 7-10 to yield the R_{max} value.

8.5.11 Localized Scour

Waves and currents during coastal flood conditions create turbulence around foundation elements, causing localized scour around those elements. Determining potential scour is critical in designing coastal foundations to ensure that failure does not occur as a result of the loss in either bearing capacity or anchoring resistance around the posts, piles, piers, columns, footings, or walls. Localized scour determinations will require knowledge of the flood conditions, soil characteristics, and foundation type.

At some locations, soil at or below the ground surface can be resistant to localized scour, and the scour depths calculated as described below would be excessive. When the designer believes the soil at a site may be scour-resistant, the assistance of a geotechnical engineer should be sought before calculated scour depths are reduced.

■ **Localized scour around vertical piles.** Generally, localized scour calculation methods in coastal areas are based largely on laboratory tests and empirical evidence gathered after storms.

The evidence suggests that the localized scour depth around a single pile or column or other thin vertical members is equal to approximately 1.0 to 1.5 times the pile diameter. In this Manual, a ratio of 2.0 is recommended (see Equation 8.10), consistent with the rule of thumb given in the *Coastal Engineering Manual* (USACE 2008). Figure 8-14 illustrates localized scour at a pile, with and without a scour-resistant terminating stratum.

Figure 8-14.
Scour at single vertical foundation member, with and without underlying scour-resistant stratum

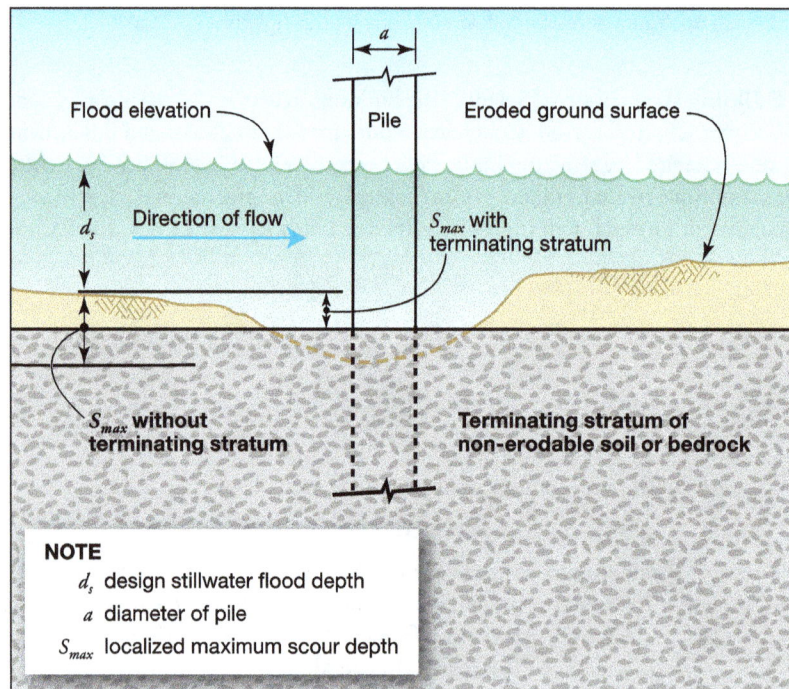

NOTE
d_s design stillwater flood depth
a diameter of pile
S_{max} localized maximum scour depth

Σ EQUATION 8.10. LOCALIZED SCOUR AROUND A SINGLE VERTICAL PILE

$$S_{max} = 2.0a$$

(Eq. 8.10)

where:

S_{max} = maximum localized scour depth (ft)

a = diameter of a round foundation element or the maximum diagonal cross-section dimension for a rectangular element

Observations after some hurricanes have shown cases in which localized scour around foundations far exceeded twice the diameter of any individual foundation pile. This was probably a result of flow and waves interacting with the group of foundation piles. In some cases, scour depressions were observed or reported to be 5 to 10 feet deep (see Figure 8-15). This phenomenon has been observed at foundations with or without slabs on grade but appears to be aggravated by the presence of the slabs.

Figure 8-15.
Deep scour around foundation piles, Hurricane Ike (Bolivar Peninsula, TX, 2008)

Some research on the interaction of waves and currents on pile groups suggests that the interaction is highly complex and depends on flow characteristics (depth, velocity, and direction), wave conditions (wave height, period, and direction), structural characteristics (pile diameter and spacing) and soil characteristics (Sumer et al. 2001). Conceptually, the resulting scour at a pile group can be represented as shown in Figure 8-16. In this Manual, the total scour depth under a pile group is estimated to be 3 times the single pile scour depth, plus an allowance for the presence of a slab or grade beam, as shown in Equation 8.11. The factor of 3 is consistent with data reported in the literature and post-hurricane observations.

Figure 8-16.
Scour around a group of foundation piles
SOURCE: ADAPTED FROM
SUMER ET AL. 2001

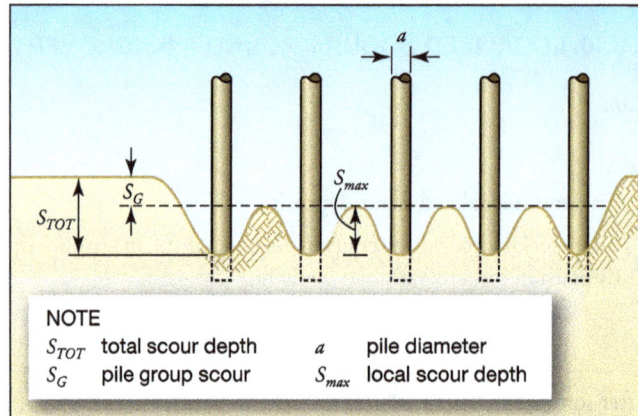

NOTE
S_{TOT} total scour depth a pile diameter
S_G pile group scour S_{max} local scour depth

EQUATION 8.11. TOTAL LOCALIZED SCOUR AROUND VERTICAL PILES

$S_{TOT} = 6a + 2$ ft (if grade beam and/or slab-on-grade present)   (Eq. 8.11a)

$S_{TOT} = 6a$ (if no grade beam or slab-on-grade present)   (Eq. 8.11b)

where:

S_{TOT} = total localized scour depth (ft)

a = diameter of a round foundation element or the maximum diagonal cross-section dimension for a rectangular element

$2\,ft$ = allowance for vertical scour due to presence of grade beam or slab-on-grade

One difficulty for designers is determining whether local soils and coastal flood conditions will result in pile group scour according to Equation 8.11. Observations after Hurricanes Rita and Ike suggest that such scour is widespread along the Gulf of Mexico shoreline in eastern Texas and southwestern Louisiana, and observations after Hurricanes Opal and Ivan suggest that it occurs occasionally along the Gulf of Mexico shoreline in Alabama and Florida. Deep foundation scour has also been observed occasionally on North Carolina barrier islands (Hurricane Fran) and American Samoa (September 2009 tsunami).

These observations suggest that some geographic areas are more susceptible than others, but deep foundation scour can occur at any location where there is a confluence of critical soil, flow, and wave conditions. Although these critical conditions cannot be identified precisely, designers should (1) be aware of the phenomenon, (2) investigate historical records for evidence of deep foundation scour around pile groups, and (3) design for such scour when the building site is low-lying, the soil type is predominantly silty, and the site is within several hundred feet of a shoreline.

Localized scour around vertical walls and enclosures. Localized scour around vertical walls and enclosed areas (e.g., typical Zone A construction) can be greater than that around single vertical piles, but it usually occurs at a corner or along one or two edges of the building (as opposed to under the entire building). See Figure 8-16.

Scour depths around vertical walls and enclosed areas should be calculated in accordance with Equation 8.12, which is derived from information in *Coastal Engineering Manual* (USACE 2006). The equation is based on physical model tests conducted on large-diameter vertical piles exposed to waves and currents ("large" means round and square objects with diameters/side lengths corresponding to several tens of feet in the real world, which is comparable to the coastal residential buildings considered in this Manual). Equation 8.12, like Equation 8.11, has no explicit consideration of soil type, so designers must consider whether soils are highly erodible and plan accordingly.

EQUATION 8.12. TOTAL SCOUR DEPTH AROUND VERTICAL WALLS AND ENCLOSURES

$$S_{TOT} = 0.15L \qquad\qquad \text{(Eq. 8.12)}$$

where:

S_{TOT} = total scour depth (ft), maximum value is 10 ft

L = horizontal length along the side of the building or obstruction exposed to flow and waves

8.5.12   Flood Load Combinations

Designers should be aware that not all of the flood loads described in Section 8.5 act at certain locations or against certain building types. Table 8-5 provides guidance for calculating appropriate flood loads in Zone V and Coastal A Zones (flood load combinations for the portion of Zone A landward of the Limit of Moderate Wave Action [LiMWA] are shown for comparison).

Table 8-5. Selection of Flood Loads for F_a in ASCE 7-10 Load Combinations for Global Forces

Description	Load Combination
Pile or open foundation in Zone V or Coastal A Zone	Greater of F_{brkp} or F_{dyn} (on front row of piles only) + F_{dyn} (on all other piles) + F_i (on one pile only)
Solid (perimeter wall) foundation	Greater of F_{brkw} or F_{dyn} + F_i (in one corner)

As discussed in Section 8.5.7, hydrostatic loads are included only when standing water will exert lateral or vertical loads on the building; these situations are usually limited to lateral forces being exerted on solid walls or buoyancy forces being exerted on floors and do not dominate in the Zone V or Coastal A Zone environment. Section 8.5.7 includes a discussion about how to include these hydrostatic flood loads in the ASCE 7-10 load combinations.

The guidance in ASCE 7-10, Sections 2.3 and 2.4 (Strength Design and Allowable Stress Design, respectively) also indicates which load combinations the flood load should be applied to. In the portion of Zone A landward of the LiMWA, the flood load F_a could either be hydrostatic or hydrodynamic loads. Both of these loads could be lateral loads; only hydrostatic will be a vertical load (buoyancy). When designing for global forces that will create overturning, sliding or uplift reactions, the designer should use F_a as the flood load that creates the most restrictive condition. In sliding and overturning, F_a should be determined by the type of expected flooding. Hydrostatic forces govern if the flooding is primarily standing water possibly saturating the ground surrounding a foundation; hydrodynamic forces govern if the flooding is primarily from moving water.

When designing a building element such as a foundation, the designer should use F_a as the greatest of the flood forces that affect that element (F_{sta} or F_{dyn}) + F_i (impact loads on that element acting at the stillwater level). The combination of these loads must be used to develop the required resistance that must be provided by the building element.

The designer should assume that breaking waves will affect foundation elements in both Zone V and Zone A. In determining total flood forces acting on the foundation at any given point during a flood event, it is generally unrealistic to assume that impact loads occur on all piles at the same time as breaking wave loads. Therefore, it is recommended that the load be calculated as a single wave impact load acting in combination with other sources of flood loads.

For the design of foundations in Zone V or Coastal A Zone, load combination cases considered should include breaking wave loads alone, hydrodynamic loads alone, and the greater of hydrodynamic loads and breaking wave loads acting in combination with debris impact loads. The value of flood load, F_a, used in ASCE 7-10 load combinations, should be based on the greater of F_{brk} or F_{dyn}, as applicable for global forces (see Table 8-5) or $F_i + (F_{brk}$ or $F_{dyn})$, as applicable for an individual building element such as a pile.

Example 8-4 is a summary of the information regarding flood loads and the effects of flooding on an example building.

EXAMPLE 8.4. FLOOD LOAD EXAMPLE PROBLEM

Given:

- Oceanfront building site on landward side of a primary frontal dune (see Example 8.1, Illustration A)

- Topography along transect perpendicular to shoreline is shown in Example 8.1, Illustration B; existing ground elevation at seaward row of pilings = 7.0 ft NGVD

- Soil is dense sand; no terminating stratum above –25 ft NGVD

- Data from FIRM are as follows: flood hazard zone at site is Zone VE, BFE = 14.0 ft NGVD

- Data from FIS are as follows: 100-year stillwater elevation = 10.1 ft NGVD, 10-year stillwater elevation = 5.0 ft NGVD

EXAMPLE 8.4. FLOOD LOAD EXAMPLE PROBLEM (continued)

- Local government requires 1.0 ft freeboard; therefore DFE = 14.0 ft NGVD (BFE) + 1.0 ft = 15.0 ft NGVD

- Building to be supported on 8-in. × 8-in. square piles, as shown in Illustration A

- Direction of wave and flow approach during design event is perpendicular to shoreline (see Illustration A)

- The assumption is no grade beam or slab-on-grade present

Find:

1. Primary frontal dune reservoir: determine whether dune will be lost or provide protection during design event

2. Eroded ground elevation beneath building resulting from storm erosion

3. Design flood depth (d_s) at seaward row of piles

4. Probable range of design event flow velocities

5. Local scour depth (S) around seaward row of piles

6. Total localized scour (S_{TOT}) around piles

7. Design event breaking wave height (H_b) at seaward row of piles

8. Hydrodynamic (velocity flow) loads (F_{dyn}) on a pile (not in seaward row)

9. Breaking wave loads (F_{brk}) on the seaward row of piles

10. Debris impact load (F_i) from a 1,000-lb object acting on one pile

Solution for #1: Whether the dune will be lost or provides protection can be determined as follows:

- The cross-sectional area of the frontal dune reservoir is above the 100-year stillwater elevation and seaward of the dune crest.

- The area (see Example 8.1, Illustration B) can be approximated as a triangle with the following area:

$$A = \frac{1}{2}bh$$

Where b is the base dimension and h is the height dimension of the approximate triangle:

$$= \frac{1}{2}(16 \text{ ft NGVD dune crest elevation} - 10.1 \text{ ft NGVD 100-year stillwater elevation})(15 \text{ ft})$$

A = 44 ft² but the area shown is slightly larger than that of the triangular area, so assume A = 50 ft²

EXAMPLE 8.4. FLOOD LOAD EXAMPLE PROBLEM (continued)

- According to this Manual, the cross-sectional area of the frontal dune reservoir must be at least 1,100 ft² to survive a 100-year flood event.

- 50 ft² <1,100 ft² and therefore, **the dune will be lost and provide no protection during the 100-year event.**

Solution for #2: The eroded ground elevation beneath building can be found as follows:

- Remove dune from transect by drawing an upward-sloping (1:50 *v:h*) line landward from the lower of the dune toe or the intersection of the 10-year stillwater elevation and the pre-storm profile.

- The dune toe is 4.1 ft NGVD. The intersection of the 10-year stillwater elevation and pre-storm profile is 5.0 ft NGVD.

- The dune toe is lower (4.1 ft NGVD < 5.0 ft NGVD).

- Draw a line from the dune toe (located 75 ft from the shoreline at an elevation of 4.1 ft NGVD) sloping upward at a 1:50 (*v:h*) slope and find where the seaward row of piles intersects this line.

$$\text{Elevation} = 4.1 \text{ ft NGVD} + (145 \text{ ft} - 75 \text{ ft})\left(\frac{1}{50}\right) = 5.5 \text{ ft NGVD}$$

Therefore, the eroded ground elevation at the seaward row of pilings = **5.5 ft NGVD**

Note: This value does not include local scour around the piles.

Solution for #3: The design stillwater flood depth (d_s) at seaward row of pilings can be calculated with Equation 8.1 as follows:

$$d_s = E_{sw} - GS$$

Using the 100-year stillwater elevation (NGVD):

d_s = 10.1 ft NGVD − 5.5 ft NGVD

d_s = **4.6 ft**

Note: This is the same solution as calculated in Example 8.1, Solution #1.

Solution for #4: Use Equations 8.2a and 8.2b to determine the range of design flow velocities (V) as follows:

- Lower-bound velocity:

$$V = \frac{d_s}{t}$$

$$V = \frac{4.6 \text{ ft}}{1 \text{ sec}}$$

EXAMPLE 8.4. FLOOD LOAD EXAMPLE PROBLEM (continued)

    Lower-bound $V = 4.6$ ft/sec

- Upper-bound velocity:

$$V = (gd_s)^{0.5}$$

Upper-bound $V = \left(32.2 \text{ ft/sec}^2\right)(4.6 \text{ ft})^{0.5} = 12.2$ ft/sec

The range of velocities: **4.6 ft/sec to 12.2 ft/sec**

Note: t *is assumed to be equal to 1 sec, as given in Equation 8.2.*

Solution for #5: Local scour depth (S) around seaward row of pilings can be found using Equation 8.10 as follows:

$$S = 2.0a$$

where:

$$a = \frac{\sqrt{7.5^2 \text{ in.} + 7.5^2 \text{ in.}}}{12 \text{ in./ft}} = \frac{10.6 \text{ in.}}{12 \text{ in./ft}} = 0.88 \text{ ft}$$

$$S = (2.0)(0.88 \text{ ft}) = \textbf{1.76 ft}$$

Solution for #6: To find the total localized scour (S_{TOT}) around piles, use Equation 8.11b as follows:

$$S_{TOT} = 6a = 6(0.88 \text{ ft}) = \textbf{5.28 ft}$$

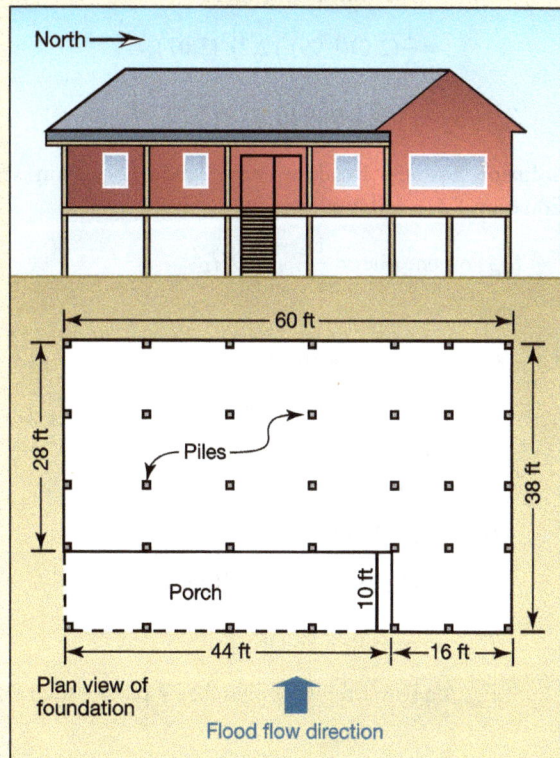

Illustration A.
Building elevation and plan view of pile foundation

EXAMPLE 8.4. FLOOD LOAD EXAMPLE PROBLEM (continued)

Solution for #7: Breaking wave height (H_b) at seaward row of pilings can be found as follows:

At seaward row of pilings, $H_b = (d_s)(0.78)$ where $d_s = 4.6$ ft from Solution #3

$$H_b = (4.6 \text{ ft})(0.78) = \textbf{3.6 ft}$$

Solution for #8: Hydrodynamic (velocity flow) loads (F_{dyn}) on a pile (not in seaward row) can be calculated using Equation 8.8 as follows:

On one pile: $F_{dyn} = \dfrac{1}{2} C_d \rho V^2 A$

where:

C_d = 2.0 for a square pile

ρ = 1.99 slugs/ft^3

A = $\dfrac{8 \text{ in.}}{12 \text{ in.}}(10.1 \text{ ft} - 5.5 \text{ ft}) = 3.07 \text{ ft}^2$

V = 12.2 ft/sec (because the building is on oceanfront, use the upper bound flow velocity for calculating loads)

$$F_{dyn} = \frac{1}{2}(2.0)(1.99)(12.2)^2(3.07)$$

F_{dyn} on one pile = **909 lb**

Solution for #9: Breaking wave loads (F_{brkp}) on seaward row of pilings can be found using Equation 8.5 as follows:

F_{brkp} on one pile $= \dfrac{1}{2} C_{db} \gamma_w D H_b^2$

where:

C_{db} = 2.25 for square piles

γ_w = 64.0 lb/ft^3 for saltwater

$D = \dfrac{8 \text{ in.}}{12 \text{ in.}}(1.4) = 0.93$ ft

H_b = 3.6 ft from Solution #7

$$F_{brkp} = \frac{1}{2}(2.25)\left(64.0 \text{ lb/ft}^3\right)(0.93 \text{ ft})(3.6 \text{ ft})^2$$

F_{brkp} on one pile = 868 lb

F_{brkp} on seaward row of piles (i.e., 7 piles) = (625 lb)(7) = **6,076 lb**

EXAMPLE 8.4. FLOOD LOAD EXAMPLE PROBLEM (concluded)

Solution for #10: Debris impact load (F_i) from a 1,000-lb object on one pile can be determined using Equation 8.9 as follows: $F_i = W V C_D C_B C_{Str}$

where:

W = 1,000 lb

C_D = 1.0

C_B = 1.0

C_{Str} = 0.2 (timber pile)

Debris impact load = $(1{,}000 \text{ lb})(12.2 \text{ ft/sec})(1.0)(1.0)(0.2)$

Debris impact load = **2,440 lb**

Note: C_D and C_B are each assumed to be 1.0.

The following worksheets will facilitate flood load computations.

Worksheet 1. Flood Load Computation Non-Tsunami Coastal A Zones (Solid Foundation)

Flood Load Computation Worksheet: Non-Tsunami Coastal A Zones (Solid Foundation)

OWNER'S NAME: _____   PREPARED BY: _____

ADDRESS: _____   DATE: _____

PROPERTY LOCATION: _____

Constants

γ_w = specific weight of water = 62.4 lb/ft³ for fresh water and 64.0 lb/ft³ for saltwater

ρ = mass density of fluid) = 1.94 slugs/ft³ for fresh water and 1.99 slugs/ft³ for saltwater

g = gravitational constant = 32.2 ft/sec²

Variables

d_s = design stillwater flood depth (ft) =

Vol = volume of floodwater displaced (ft³) =

V = velocity (fps) =

C_{db} = breaking wave drag coefficient =

H_b = breaking wave height (ft) =

C_p = dynamic pressure coefficient =

C_s = slam coefficient =

C_d = drag coefficient =

w = width of element hit by water (ft) =

h = vertical distance (ft) wave crest extends above bottom of member =

A = area of structure face (ft²) =

W = weight of object (lb) =

C_D = depth coefficient =

C_B = blockage coefficient =

C_{Str} = building structure coefficient =

a = diameter of round foundation element =

L = horizontal length alongside building exposed to waves (ft)

Summary of Loads

F_{sta} =

F_{buoy} =

F_{brkw} =

F_s =

F_{dyn} =

F_i =

S_{max} =

S_{TOT} =

Worksheet 1. Flood Load Computation Non-Tsunami Coastal A Zones (Solid Foundation) (concluded)

Equation 8.3 Lateral Hydrostatic Load (Flood load on one side only)
$F_{sta} = \dfrac{1}{2}\gamma_w d_s^{\,2} w =$
Equation 8.4 Vertical (Buoyancy) Hydrostatic Load
$F_{buoy} = \gamma_w (Vol) =$
Equation 8.6 Breaking Wave Load on Vertical Walls
$F_{brkw} = \left(1.1 C_p \gamma_w d_s^{\,2} + 2.4 \gamma_w d_s^{\,2}\right) w$ (if dry behind wall) =   or $F_{brkw} = \left(1.1 C_p \gamma_w d_s^{\,2} + 1.9 \gamma_w d_s^{\,2}\right) w$ (if stillwater elevation is the same on both sides of wall) =
Equation 8.7 Wave Slam
$F_S = \dfrac{1}{2}\gamma_w C_S d_s hw =$
Equation 8.8 Hydrodynamic Load
$F_{dyn} = \dfrac{1}{2} C_d \rho V^2 A =$
Equation 8.9 Debris Load
$F_i = W V C_D C_B C_{Str} =$
Equation 8.10 Localized Scour Around Single Vertical Pile
$S_{max} = 2a =$
Equation 8.11 Total Localized Scour Around Vertical Piles
$S_{TOT} = 6a + 2$ ft (if grade beam and/or slab-on-grade present) = $S_{TOT} = 6a$ (if no grade beam or slab-on-grade present) =
Equation 8.12 Total Scour Depth Around Vertical Walls and Enclosures
$S_{MAX} = 0.15L =$

Worksheet 2. Flood Load Computation Non-Tsunamic Zone V and Coastal A Zone (Open Foundation)

Flood Load Computation Worksheet: Non-Tsunami Zones V and Coastal A Zone (Open Foundation)

OWNER'S NAME: _____   PREPARED BY: _____

ADDRESS: _____   DATE: _____

PROPERTY LOCATION: _____

Constants

γ_w = specific weight of water = 62.4 lb/ft³ for fresh water and 64.0 lb/ft³ for saltwater

ρ = mass density of fluid = 1.94 slugs/ft³ for fresh water and 1.99 slugs/ft³ for saltwater

g = gravitational constant = 32.2 ft/sec²

Variables

d_s = design stillwater flood depth (ft) =

V = velocity (fps) =

C_{db} = breaking wave drag coefficient =

a, D = pile diameter (ft) =

H_b = breaking wave height (ft) =

C_p = dynamic pressure coefficient =

C_s = slam coefficient =

C_d = drag coefficient for piles =

w = width of element hit by water (ft) =

h = vertical distance (ft) wave crest extends above bottom of member =

W = debris object weight (lb) =

C_D = depth coefficient =

C_B = blockage coefficient =

C_{Str} = building structure coefficient =

L = horizontal length alongside building exposed to waves (ft) =

Summary of Loads

F_{brkp} =

F_s =

F_{dyn} =

F_i =

S_{max} =

S_{TOT} =

Equation 8.5 Breaking Wave Load on Vertical Piles

$$F_{brkp} = \frac{1}{2} C_{db} \gamma_w D H_b^2 =$$

Worksheet 2. Flood Load Computation Non-Tsunamic Zone V and Coastal A Zone (Open Foundation) (concluded)

Equation 8.7 Wave Slam
$F_S = \dfrac{1}{2}\gamma_w C_S d_s hw =$
Equation 8.8 Hydrodynamic Load
$F_{dyn} = \dfrac{1}{2} C_{dr} V^2 A =$
Equation 8.9 Debris Load
$F_i = W V C_D C_B C_{Str} =$
Equation 8.10 Localized Scour around Single Vertical Pile
$S_{max} = 2a =$
Equation 8.11 Total Localized Scour Around Vertical Piles
$S_{TOT} = 6a + 2$ ft (if grade beam and/or slab-on-grade present) $=$
$S_{TOT} = 6a$ (if no grade beam or slab-on-grade present) $=$

8.6   Tsunami Loads

In general, tsunami loads on residential buildings may be calculated in the same fashion as other flood loads; the physical processes are the same, but the scale of the flood loads is substantially different in that the wavelengths and runup elevations of tsunamis are much greater than those of waves caused by tropical and extratropical cyclones (see Section 3.2). If the tsunami acts as a rapidly rising tide, most of the damage is the result of buoyant and hydrostatic forces (see *Tsunami Engineering* [Camfield 1980]). When the tsunami forms a bore-like wave, the effect is a surge of water to the shore and the expected flood velocities are substantially higher than in non-tsunami conditions.

The tsunami velocities are very high and if realized at the greater water depths, would cause substantial damage to buildings in the path of the tsunami. Additional guidance on designing for tsunami forces including flow velocity, buoyant forces, hydrostatic forces, debris impact, and impulsive forces is provided in FEMA P646, *Guidelines for Design of Structures for Vertical Evacuation from Tsunami* (FEMA 2008b). For debris impact loads under tsunami conditions, see Section 6.5.6 of FEMA P646, which recommends an alternative to Equation 8.6 in this Manual for calculating tsunami debris impact loads.

8.7   Wind Loads

ASCE 7-10 is the state-of-the-art wind load design standard. It contains a discussion of the effects of wind pressure on a variety of building types and building elements. Design for wind loads is essentially the same whether the winds are due to hurricanes, thunderstorms, or tornadoes.

Important factors that affect wind load design pressures include:

■ Location of the building site on wind speed maps

- Topographic effects (hills and escarpments), which create a wind speedup effect

- Building risk category (one- and two-family dwellings are assigned to Risk Category II; accessory structures may be assigned to Risk Category I) (see Section 6.2.1.1)

- Building height and shape

- Building enclosure category: enclosed, partially enclosed or open

- Terrain conditions, which determine building exposure category

The effects of wind on buildings can be summarized as follows:

- Windward walls and windward surfaces of steep-sloped roofs are acted on by inward-acting, or positive pressures. See Figure 8-17.

- Leeward walls and leeward surfaces of steep-sloped roofs and both windward and leeward surfaces of low-sloped roofs are acted on by outward-acting, or negative pressures. See Figure 8-17.

- Air flow separates at sharp edges and at locations where the building geometry changes.

- Localized suction, or negative, pressures at eaves, ridges, and the corners of roofs and walls are caused by turbulence and flow separation. These pressures affect loads on components and cladding (C&C) and elements of the main wind force resisting system (MWFRS).

> **NOTE**
>
> Basic mapped wind speeds in ASCE 7-10 for Category II structures (residential buildings) are higher than those in ASCE 7-05 because they represent ultimate wind speeds or strength-based design wind speeds. Load factors for wind in ASCE 7-10 are also different from those in ASCE 7-05. In ASCE 7-10, the wind load factor in the load combinations for LRFD strength design (LRFD) is 1.0 (but ASCE 7-05 provides a load factor of 1.6), and the ASD wind load factor in the load combinations for allowable stress design (ASD) for wind is 0.6 (but ASCE 7-05 provides a load factor of 1.0).

The phenomena of localized high pressures occurring at locations where the building geometry changes is accounted for by the various pressure coefficients in the equations for both MWFRS and C&C. Internal pressures must be included in the determination of net wind pressures and are additive to (or subtractive from) the external pressures. Openings and the natural porosity of building elements contribute to internal

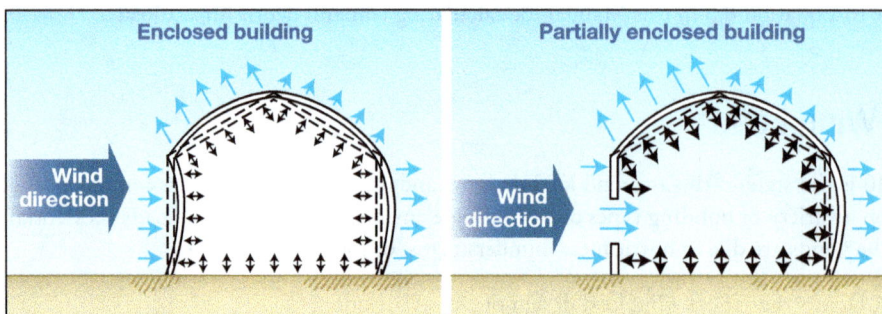

Figure 8-17.
Effect of wind on an enclosed building and a building with an opening

pressure. The magnitude of internal pressures depends on whether the building is enclosed, partially enclosed, or open, as defined in ASCE 7-10. Figure 8-17 shows the effect of wind on an enclosed and partially enclosed building.

In wind-borne debris regions (as defined in ASCE 7-10), in order for a building to be considered enclosed for design purposes, glazing must either be impact-resistant or protected with shutters or other devices that are impact-resistant. This requirement also applies to glazing in doors.

Methods of protecting glazed openings are described in ASCE 7-10 and in Chapter 11 of this Manual.

8.7.1 Determining Wind Loads

In this Manual, design wind pressures for MWFRS are based on the results of the envelope procedure for low-rise buildings. A low-rise building is defined in ASCE 7-10. The envelope procedure in ASCE 7-10 is only one of several for determining MWFRS pressures in ASCE 7-10, but it is the procedure most commonly used for designing low-rise residential buildings. The envelope procedure for low-rise buildings is applicable for enclosed and partially enclosed buildings with a mean roof height (h) of less than or equal to 60 feet and where mean roof height (h) does not exceed the smallest horizontal building dimension.

Figure 8-18 depicts the distribution of external wall and roof pressures and internal pressures from wind. The figure also shows the mean roof height, which is defined in ASCE 7-10

TERMINOLOGY: HURRICANE-PRONE REGIONS

In the United States and its territories, hurricane-prone areas are defined by ASCE 7-10 as (1) the U.S. Atlantic Ocean and Gulf of Mexico Coasts where the basic wind speed for Risk Category II buildings is greater than 115 mph and (2) Hawaii, Puerto Rico, Guam, the Virgin Islands, and American Samoa.

FORMULA

The following formula converts ASCE 7-05 wind speeds to ASCE 7-10 Risk Category II wind speeds.

$$\text{ASCE 7-10} = \left(\text{ASCE 7-05}\right)\left(\sqrt{1.6}\right)$$

For conversion from ASCE 7-10 to ASCE 7-05, use:

$$\text{ASCE 7-05} = \frac{\text{ASCE 7-10}}{\sqrt{1.6}}$$

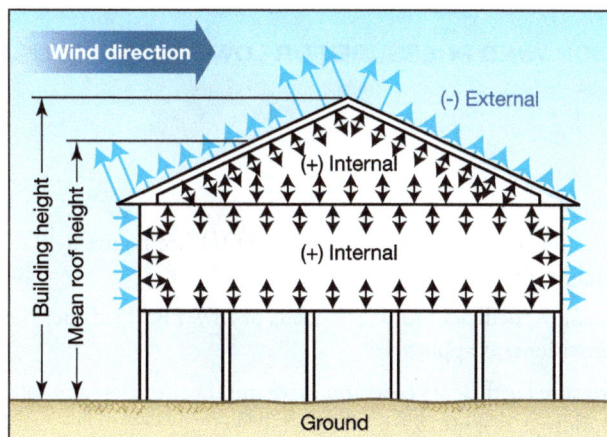

Figure 8-18.
Distribution of roof, wall, and internal pressures on one-story, pile-supported building

as "the average of the roof eave height and the height to the highest point on the roof surface ..." Mean roof height is not the same as building height, which is the distance from the ground to the highest point.

For calculating both MWFRS and C&C pressures, velocity pressures (q) should be calculated in accordance with Equation 8.13. Velocity pressure varies depending on many factors including mapped wind speed at the site, height of the structure, local topographic effects, and surrounding terrain that affects the exposure coefficient.

> **NOTE**
>
> ASCE 7-10 *Commentary* states that where a single component, such as a roof truss, comprises an assemblage of structural elements, the elements of that component should be analyzed for loads based on C&C coefficients, and the single component should be analyzed for loads as part of the MWFRS.

EQUATION 8.13. VELOCITY PRESSURE

$$q_z = 0.00256 K_z K_{zt} K_d V^2 \qquad \text{(Eq. 8.13)}$$

where:

q_z = velocity pressure evaluated at height z (psf)

K_z = velocity pressure exposure coefficient evaluated at height z

K_{zt} = topographic factor

K_d = wind directionality factor

V = basic wind speed (mph) (3-sec gust speed at 33 ft above ground in Exposure Category C)

The design wind pressure is calculated from the combination of external and internal pressures acting on a building element. This combination of pressures for both MWFRS and C&C loads in accordance with provisions of ASCE 7-10 is represented by Equation 8.14.

EQUATION 8.14. DESIGN WIND PRESSURE FOR LOW-RISE BUILDINGS

$$p = q_h \left[GC_{pf} - GC_{pi} \right] \qquad \text{(Eq. 8.14)}$$

where:

p = design wind pressure

q_h = velocity pressure evaluated at mean roof height (h) (see Figure 8-18 for an illustration of mean roof height)

GC_{pf} = external pressure coefficient for C&C loads or MWFRS loads per the low-rise building provisions, as applicable

GC_{pi} = internal pressure coefficients based on exposure classification as applicable; GC_{pi} for enclosed buildings is +/- 0.18

Figure 8-19 depicts how net suction pressures can vary across different portions of the building. Central portions of the walls represent the location of the least suction, while wall corners, the roof ridge, and the roof perimeter areas have potential for suction pressures that are 1.3, 1.4, and 2 times the central wall areas, respectively. Wall areas and roof areas that experience the largest suction pressures are shown as edge zones in Figure 8-19. The variation of pressures for different portions of the building is based on an enclosed structure (e.g., GC_{pi} = +/- 0.18) and use of external pressure coefficients of the low-rise building provisions.

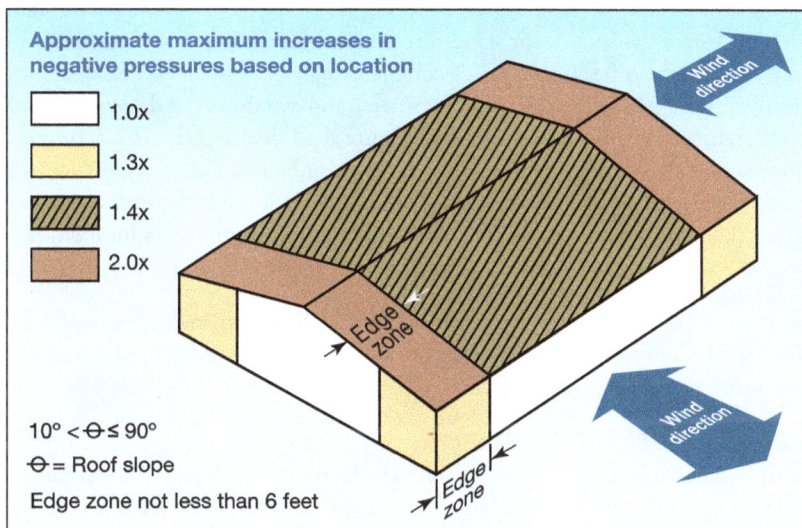

Figure 8-19. Variation of maximum negative MWFRS pressures based on envelope procedures for low-rise buildings

To simplify design for wind, as well as establish consistency in the application of the wind design provisions of ASCE 7-10, several consensus standards with prescriptive designs tabulate maximum wind loads for the design of specific building elements based on wind pressures (both MWFRS and C&C are often referred to as "prescriptive" standards because they prescribe or tabulate load requirements for pressures) in accordance with ASCE 7-10. These standards, which are referenced in the 2012 IRC, are specific building applications based on factors such as wind speed, exposure, and height above grade. Examples of prescriptive standards for wind design that are referenced in the 2012 IRC are:

- ICC 600-2008, *Standard for Residential Construction in High-Wind Regions* (ICC 2008)

- ANSI/AF&PA, *Wood Frame Construction Manual* (WFCM) (AF&PA 2012)

- ANSI/AISI-S230, *Standard for Cold-Formed Steel Framing-Prescriptive Method for One and Two Family Dwellings* (AISI 2007)

Tabulated wind load requirements in these standards often use conservative assumptions for sizing members and connections. Therefore, load requirements are often more conservative than those developed by direct application of ASCE 7-10 pressures when design loads can be calculated for each element's unique characteristics.

8.7.2 Main Wind Force Resisting System

The MWFRS consists of the foundation; floor supports (e.g., joists, beams); columns; roof rafters or trusses; and bracing, walls, and diaphragms that assist in transferring loads. ASCE 7-10 defines the MWFRS as "… an assemblage of structural elements assigned to provide support and stability for the overall structure. The system generally receives wind load from more than one surface." Individual MWFRS elements of shear walls and roof diaphragms (studs and cords) may also act as components and should also be analyzed under the loading requirements of C&C.

For a typical building configuration with a gable roof, the wind direction is perpendicular to the roof ridge for two cases and parallel to the ridge in the other two cases. A complete analysis of the MWFRS includes determining windward and leeward wall pressures, side wall pressures, and windward and leeward roof pressures for wind coming from each of four principal directions. Figure 8-18 depicts pressures acting on the building structure for wind in one direction only. The effect of the combination of pressures on the resulting member and connection forces is of primary interest to the designer. As a result, for each direction of wind loading, structural calculations are required to determine the maximum design forces for members and connections of the building structure.

Prescriptive standards can be used to simplify the calculation of MWFRS design loads. Examples of prescriptive MWFRS design load tables derived from the application of ASCE 7-10 wind load provisions are included in this Manual for the purpose of illustration, as follows:

- **Roof uplift connector loads** (see Table 8-6). The application of ASCE 7-10 provisions and typical assumptions used to derive the tabulated load values are addressed in Example 8.5. Equation 8.13 for velocity pressure and Equation 8.14 for determining design wind pressure are used to arrive at the design uplift connector load. The roof uplift connection size is based on moment balance of forces acting on both the windward and leeward side of the roof. The uplift load is used to size individual connectors and also provides the distributed wind uplift load acting at the buildings perimeter walls. Note that while wind speeds are based on 700-year Mean Recurrence Interval, the resulting uplift loads are based on ASD design.

- **Diaphragm loads due to wind acting perpendicular to the ridge** (see Table 8-7). Application of the ASCE 7-10 provisions and typical assumptions used to derive the tabulated load values are addressed in Example 8.6. The diaphragm load is based on wind pressures simultaneously acting on both the windward and leeward side of the building. The diaphragm load is used to size the diaphragm for resistance to wind and is also used for estimating total lateral forces for a given wind direction based on combining diaphragm loads for the roof and wall(s) as applicable. Total lateral forces from wind for a given direction can be used for preliminary sizing of the foundation and for determining shear wall capacity requirements.

The example loads in Table 8-6 and Table 8-7, which are based on ASCE 7-10 envelope procedures for low-rise buildings, are used in Examples 8.7 and 8.8 to illustrate their application in the wind design of select load path elements. Tables 8-6 and 8-7 and Examples 8.5 and 8.6 are derived from wind load procedures in the WFCM (AF&PA 2012). Tables 8-6 and 8-7 are not intended to replace requirements of the building code or applicable reference standards for the actual design of a building to resist wind.

Table 8-6. Roof Uplift Connector Loads (Based on ASD Design) at Building Edge Zones, plf (33-ft mean roof height, Exposure C)

Roof Span (ft)	Wind Speed[a] (mph)								
	110	115	120	130	140	150	160	170	180
	Roof uplift connector load[b][c][d] (plf)								
24	189	215	241	298	358	424	494	568	647
32	237	269	303	374	451	534	622	716	816
40	285	324	364	450	544	643	750	864	985
48	333	379	426	527	636	753	879	1,012	1,154

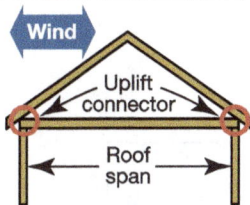

(a) 700-year wind speed, 3-sec gust.

(b) Uplift connector loads are based on 33-ft mean roof height, Exposure C, roof dead load of 10 psf, and roof overhang length of 2 ft (see Example 8.5).

(c) Uplift connector loads are tabulated in pounds per linear ft of wall. Individual connector loads can be calculated for various spacing of connectors (e.g., for spacing of connectors at 2 ft o.c., the individual connector load would be 2 ft times the tabulated value).

(d) Tabulated uplift connector loads are conservatively based on a 20-degree roof slope. Reduced uplift forces may be calculated for greater roof slopes.

Table 8-7. Lateral Diaphragm Load from Wind Perpendicular to Ridge, plf (33-ft mean roof height, Exposure C)

Roof Span (ft)	Wind Speed[a] (mph)								
	110	115	120	130	140	150	160	170	180
	Roof diaphragm load[b][c] for 7:12 roof slope (plf)								
24	138	151	164	192	223	256	291	329	369
32	161	176	191	224	260	299	340	384	430
40	186	203	221	259	301	345	393	443	497
48	210	230	250	294	341	391	445	503	563
	Floor diaphragm load (plf)								
Any	154	168	183	214	249	286	325	367	411

Legend

■ Tributary area for roof diaphram

□ Tributary area for floor diaphram

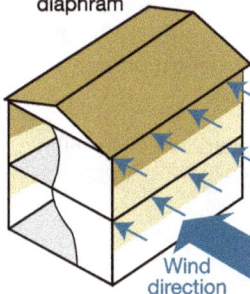

Same figure as Example 8.6, Illustration A

(a) 700-year wind speed, 3-sec gust.

(b) Lateral diaphragm loads are based on 33-ft mean roof height, Exposure C, and wall height of 8 ft (see Example 8.6). Tabulated roof diaphragm loads are for a 7:12 roof slope. Larger loads can be calculated for steeper roof slopes and smaller loads can be calculated for shallower roof slopes.

(c) Total shear load equals the tabulated unit lateral load by the building length perpendicular to the wind direction.

EXAMPLE 8.5. ROOF UPLIFT CONNECTOR LOADS

Given:

- Roof span of 24 ft with 2-ft overhangs

- Roof/ceiling dead load of 10 psf

- Wind load based on 150 mph, Exposure C at 33-ft mean roof height

- Building is enclosed

- K_z = 1.0 (velocity pressure exposure coefficient evaluated at height of 33 ft)

- K_{zt} = 1.0 (topographic factor)

- K_d = 0.85 (wind directionality)

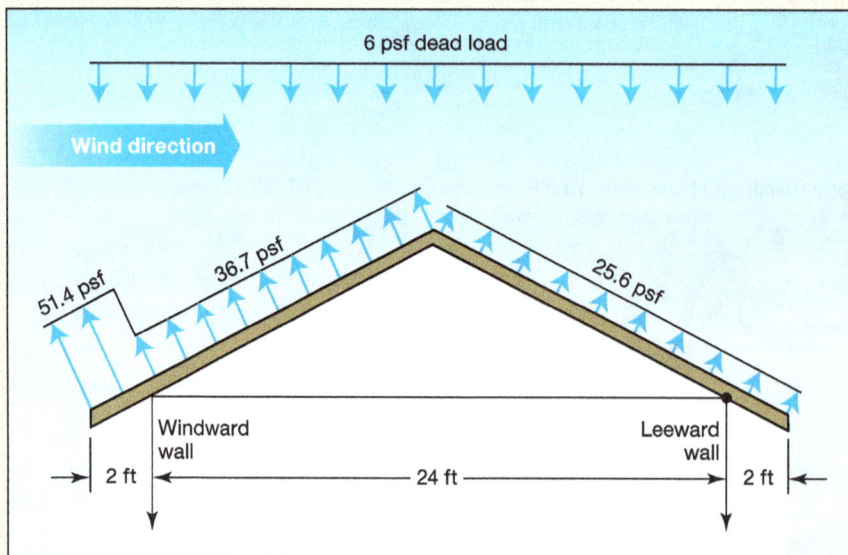

Illustration A. Roof-to-wall uplift connection loads from wind forces

Find: The roof-to-wall uplift connection load using the envelope procedure for low-rise buildings (see Figure 28.4-1 in ASCE 7-10).

Solution: The roof-to-wall uplift connection load can be found using the envelope procedure for low-rise buildings as follows:

- The velocity pressure (q) for the site conditions is determined from Equation 8.13 as follows:

$$q_h = 0.00256 K_z K_{zt} K_d V^2$$

$$q_h = 0.00256(1.0)(1.0)(0.85)(150 \text{ mph})^2$$

$$q_h = 48.96 \text{ psf}$$

EXAMPLE 8.5. ROOF UPLIFT CONNECTOR LOADS (continued)

- For ASD, multiply by the ASD wind load factor of 0.6, which comes from Load Combination 7 (See Section 8.10) 0.6D + 0.6W:

$$q_h = 48.96 \text{ psf}(0.6) = 29.38 \text{ psf}$$

The largest uplift forces occur for a roof slope of 20 degrees where wind is perpendicular to the ridge. The addition of an overhang also increases the roof-to-wall uplift connection load. For the windward overhang, a pressure coefficient of 0.68 is used based on the gust factor of 0.85 and pressure coefficient of 0.80 from ASCE 7-10. Otherwise, pressure coefficients for other elements of the roof are based on $GC_{pi} = 0.18$ and GC_{pf} from the edge zone coefficients shown in Figure 28.4-1 of ASCE 7-10.

Pressures and moments given below contain subscripts for their location:

- W = windward

- L = leeward

- O = overhang

- R = roof

The design wind pressure is determined from Equation 8.14 as follows:

$$p = q_h(GC_{pf} - GC_{pi})$$

$$p_{WO} = 29.38 \text{ psf}(-1.07 - 0.68) = -51.4 \text{ psf}$$

$$p_{WR} = 29.38 \text{ psf}(-1.07 - 0.18) = -36.7 \text{ psf}$$

$$p_{LR} = 29.38 \text{ psf}(-0.69 - 0.18) = -25.6 \text{ psf}$$

$$p_{LO} = 29.38 \text{ psf}(-0.69 - 0.18) = -25.6 \text{ psf}$$

- The roof/ceiling dead load is adjusted for the load case where dead load is used to resist uplift forces as follows:

Dead load = 10 psf(0.6) = 6 psf where 0.6 is the ASD load factor for dead load in the applicable load combination

- Wind loads on the roof have both a horizontal and vertical component. The uplift connector force, located at the windward wall, can be determined by summing moments about the leeward roof-to-wall connection and solving for the connector force that will maintain moment equilibrium. Clockwise moments are considered positive.

Moment (M) created by windward overhang pressures is solved as follows:

Vertical component, windward overhang (VWO):

EXAMPLE 8.5. ROOF UPLIFT CONNECTOR LOADS (continued)

$$M_{VWO} = [(51.4 \text{ psf } \cos(20)]\left(\frac{2 \text{ ft}}{\cos(20)}\right) + (-6 \text{ psf})(2 \text{ ft})](1 \text{ ft} + 24 \text{ ft}) = 2{,}270 \text{ ft-lb}$$

Horizontal component, windward overhang (*HWO*):

$$M_{HWO} = [-51.4 \text{ psf } \sin(20)]\left(\frac{2 \text{ ft}}{\cos(20)}\right)\left(-\frac{2\tan(20)}{2}\right) = 13.6 \text{ ft-lb}$$

Moment (*M*) created by windward roof pressures is solved as follows:

Vertical component, windward roof (*VWR*):

$$M_{VWR} = [(36.7 \text{ psf } \cos(20)]\left(\frac{12 \text{ ft}}{\cos(20)}\right) + (-6 \text{ psf})(12 \text{ ft})](18 \text{ ft}) = 6{,}631.2 \text{ ft-lb}$$

Horizontal component, windward roof (*HWR*):

$$M_{HWR} = [-36.7 \text{ psf } \sin(20)]\left(\frac{12}{\cos(20)}\right)\left(\frac{12\tan(20)}{2}\right) = -349.7 \text{ ft-lb}$$

Moment (*M*) created by leeward roof pressures is solved as follows:

Vertical component, leeward roof (*VLR*):

$$M_{VLR} = [(25.6 \text{ psf } \cos(20)]\left(\frac{12}{\cos(20)}\right) + (-6 \text{ psf})(12 \text{ ft})](6 \text{ ft}) = 1{,}411.2 \text{ ft-lb}$$

Horizontal component, leeward roof (*HLR*):

$$M_{HLR} = [25.6 \text{ psf } \sin(20)]\left(\frac{12}{\cos(20)}\right)\left(\frac{12\tan(20)}{2}\right) = 243.9 \text{ ft-lb}$$

Moment(*M*) created by leeward overhang pressures is solved as follows:

Vertical component, leeward overhang (*VLO*):

$$M_{VLO} = [(25.6 \text{ psf } \cos(20)]\left(\frac{2 \text{ ft}}{\cos(20)}\right) + (-6 \text{ psf})(2 \text{ ft})](-1 \text{ ft}) = -39.2 \text{ ft-lb}$$

Horizontal component, leeward overhang (*HLO*):

$$M_{HLO} = [25.6 \text{ psf } \sin(20)]\left(\frac{2 \text{ ft}}{\cos(20)}\right)\left(\frac{-2\tan(20)}{2}\right) = -6.8 \text{ ft-lb}$$

The total overturning moment per ft of roof width = 10,174.3 ft-lb

EXAMPLE 8.5. ROOF UPLIFT CONNECTOR LOADS (concluded)

Solving for uplift load: F_w = 10174.3 ft-lb/roof span ft = 10,174.3 ft-lb/24 ft = 424 lb

Assuming the uplift forces are calculated for a 1-ft-wide section of the roof, the unit uplift connector force can be expressed as f_w = **424 plf.**

Note: *This solution matches the information in Table 8-6.*

EXAMPLE 8.6. LATERAL DIAPHRAGM LOADS FROM WIND PERPENDICULAR TO RIDGE

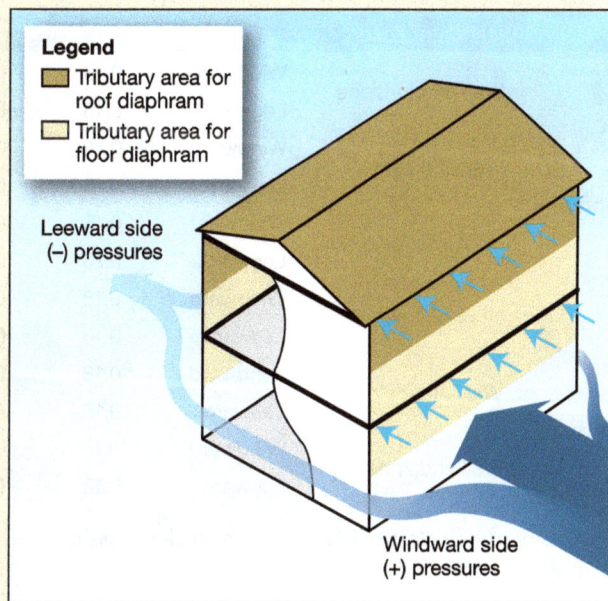

Legend
- Tributary area for roof diaphram
- Tributary area for floor diaphram

Leeward side (–) pressures

Windward side (+) pressures

Illustration A. Lateral diaphragm loads from wind perpendicular to building ridge

Given:

- Roof span of 24 ft

- 7:12 roof pitch

- The wind load is based on 150 mph, Exposure C at 33-ft mean roof height

- The building is enclosed

- From Example 8.5, for the same site condition, the ASD velocity pressure q = 29.38 psf

EXAMPLE 8.6. LATERAL DIAPHRAGM LOADS FROM WIND PERPENDICULAR TO RIDGE (continued)

Find: The roof diaphragm load using the envelope procedure for low-rise buildings (see Figure 28.4-1 in ASCE 7-10).

Solution: The roof diaphragm load using the envelope procedure for low-rise buildings can be found as follows:

- Lateral loads (see Illustration A) into the roof diaphragm are a function of roof slope and wall loads tributary to the roof diaphragm.

- Pressure coefficients for elements of the roof GC_{pi} and GC_{pf} are given in Table A.

Table A. Pressure Coefficients for Roof and Wall Zones

Diaphragm Zone			GC_{pi}	GC_{pf}
Roof diaphragm	Wall interior zone	Windward	0.18	0.56
		Leeward	0.18	−0.37
	Wall end zone	Windward	0.18	0.69
		Leeward	0.18	−0.48
	Roof interior zone	Windward	0.18	0.21
		Leeward	0.18	−0.43
	Roof end zone	Windward	0.18	−0.53
		Leeward	0.18	0.27
Floor diaphragm	Wall interior zone	Windward	0.18	0.53
		Leeward	0.18	−0.43
	Wall end zone	Windward	0.18	0.80
		Leeward	0.18	−0.64

- GC_{pi} is determined using the Enclosure Classification (enclosed building in this example) and Table 26.11-1 from ASCE 7-10

- GC_{pf} is determined using Figure 28.4-1 in ASCE 7-10

- Both interior zone and end zone coefficients are used to establish an average pressure on the wall and roof.

The design wind pressure is determined from Equation 8.14 ($q = q_h$ in this case) as follows:

$$p = q \, | GC_{pf} - GC_{pi} |$$

Step 1: Roof Diaphragm

- L = leeward

- W = windward

EXAMPLE 8.6. LATERAL DIAPHRAGM LOADS FROM WIND PERPENDICULAR TO RIDGE (continued)

- w = wall

Wall interior zone

$$p_{Ww} = 29.38 \text{ psf}(0.56 - 0.18) = 11.16 \text{ psf}$$

$$p_{Lw} = 29.38 \text{ psf}(-0.37 - 0.18) = -16.16 \text{ psf}$$

$Sum = 11.16 \text{ psf} + \left| -16.16 \text{psf} \right| = 27.3 \text{ psf}$ (note that leeward and windward forces are acting in the same direction)

Wall end zone

$$p_{Ww} = 29.38 \text{ psf}(0.69 - 0.18) = 14.98 \text{ psf}$$

$$p_{Lw} = 29.38 \text{ psf}(-0.48 - 0.18) = -19.39 \text{ psf}$$

$Sum = 14.98 \text{ psf} + \left| -19.39 \text{ psf} \right| = 34.4 \text{ psf}$ (note that leeward and windward forces are acting in the same direction)

Under the procedures and notes shown in Figure 28.4-1 of ASCE 7-10, end zones extend a minimum of 3 ft at each end of the wall. For long or tall walls, end zone lengths are based on 10 percent of the least horizontal dimension or 40 percent of the mean roof height, whichever is smaller, but not less than either 4 percent of the least horizontal dimension or 3 ft at each end of the wall. The end zone width where the pressures are applied is 3 ft.

The average pressure on the wall is:

$$P = \frac{[34.4 \text{ psf}(6 \text{ ft}) + 27.3 \text{ psf}(24 \text{ ft} - 6 \text{ ft})]}{24 \text{ ft}} = 29.1 \text{ psf}$$

where:

24 ft = building length assumed to be equal to the roof span for purposes of accounting for average effects of pressure differences at end zones and interior zones

Roof interior zone

$$p_{Ww} = 29.38 \text{ psf}(0.21 - 0.18) = 0.88 \text{ psf}$$

$$p_{Lw} = 29.38 \text{ psf}(-0.43 - 0.18) = -17.92 \text{ psf}$$

$Sum = 0.88 \text{ psf} + \left| -17.92 \text{ psf} \right| = 18.8 \text{ psf}$ (note that leeward and windward forces are acting in the same direction)

Roof end zone

$$p_{Ww} = 29.38 \text{ psf}(0.27 - 0.18) = 2.64 \text{ psf}$$

EXAMPLE 8.6. LATERAL DIAPHRAGM LOADS FROM WIND PERPENDICULAR TO RIDGE (concluded)

$$p_{Lw} = 29.38 \text{ psf}(-0.53 - 0.18) = -20.86 \text{ psf}$$

$Sum = 2.64 \text{ psf} + |-20.86 \text{ psf}| = 23.5 \text{ psf}$ (note that leeward and windward forces are acting in the same direction)

The average pressure on the roof is:

$$P = \frac{23.5 \text{ psf}(6 \text{ ft}) + 18.8 \text{ psf}(24 \text{ ft} - 6 \text{ ft})}{24 \text{ ft}} = 19.98 \text{ psf}$$

The roof diaphragm will take its load plus half the load of the 8-ft-tall wall below.

$$w_{roof} = \frac{1}{2}(29.1 \text{ psf})(8 \text{ ft}) + 19.98 \text{ psf}(7 \text{ ft}) = \mathbf{256.3 \text{ plf}}$$

Step 2: Floor Diaphragm

- The floor diaphragm loads are based on the maximum MWRFS coefficients associated with a 20-degree roof slope. It is assumed that the floor diaphragm tributary area is the height of one 8-ft wall plus 1 ft to account for floor framing depth.

Wall interior zone

$$p_{Ww} = 29.38 \text{ psf}(0.53 - 0.18) = 10.28 \text{ psf}$$

$$p_{Lw} = 29.38 \text{ psf}(-0.43 - 0.18) = -17.92 \text{ psf}$$

$Sum = 10.28 \text{ psf} + |-17.92 \text{ psf}| = 28.2 \text{ psf}$ (note that leeward and windward forces are acting in the same direction)

Wall end zone

$$p_{Ww} = 29.38 \text{ psf}(0.80 - 0.18) = 18.22 \text{ psf}$$

$$p_{Lw} = 29.38 \text{ psf}(-0.64 - 0.18) = -24.09 \text{ psf}$$

$Sum = 18.22 \, psf + |-24.09 \, psf| = 42.3 \, psf$ (note that leeward and windward forces are acting in the same direction)

The average pressure on the wall is:

$$P = \frac{42.3 \text{ psf}(6 \text{ ft}) + 28.2 \text{ psf}(24 \text{ ft} - 6 \text{ ft})}{24 \text{ ft}} = 31.73 \text{ psf}$$

The floor diaphragm load is based on a 9-ft tributary height obtained from adding the height of one 8-ft wall plus 1 ft to account for the floor framing depth.

$$w_{floor} = 31.73 \text{ psf}(9 \text{ ft}) = \mathbf{286 \text{ plf}}$$

Note: This solution matches the information in Table 8-7.

8.7.3   Components and Cladding

ASCE 7-10 defines components and cladding (C&C) as "... elements of the building envelope that do not qualify as part of the MWFRS." These elements include roof sheathing, roof coverings, exterior siding, windows, doors, soffits, fascia, and chimneys. The design and installation of the roof sheathing attachment may be the most critical consideration because the attachment location for the sheathing is where the uplift load path begins (load path is described more fully in Chapter 9 of this Manual).

C&C pressures are determined for various "zones" of the building. ASCE 7-10 includes illustrations of those zones for both roofs and walls. Illustrations for gable, monoslope, and hip roof shapes are presented. The pressure coefficients vary according to roof pitch (from 0 degrees to 45 degrees) and effective wind area (defined in ASCE 7-10).

C&C loads act on all elements exposed to wind. These elements and their attachments must be designed to resist these forces to prevent failure that could lead to breach of the building envelope and create sources of flying debris. Examples of building elements and their connections subject to C&C loads include the following:

- Exterior siding

- Roof sheathing

- Roof framing

- Wall sheathing

- Wall framing (e.g., studs, headers)

- Wall framing connections (e.g., stud-to-plate, header-to-stud)

- Roof coverings

- Soffits and overhangs

- Windows and window frames

- Skylights

- Doors and door frames, including garage doors

- Wind-borne debris protection systems

- Any attachments to the building (e.g., antennas, chimneys, roof and ridge vents, roof turbines)

Furthermore, individual MWFRS elements of shear walls and roof diaphragms (studs and chords) may also act as components and should also be analyzed under the loading requirements of C&C.

Figure 8-20 shows the locations of varying localized pressures on wall and roof surfaces. The magnitude of roof uplift and wall suction pressures is based on the most conservative wind pressures in each location for given roof types and slopes in accordance with Figure 30.4 of ASCE 7-10. As noted previously, prescriptive

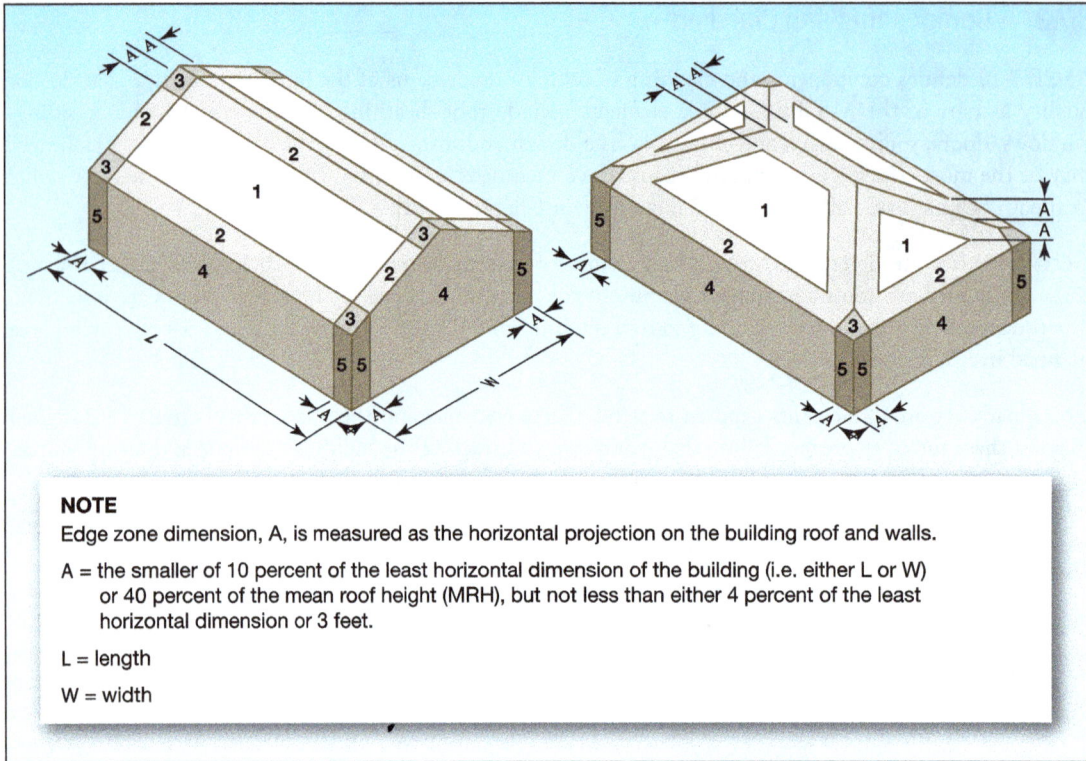

NOTE

Edge zone dimension, A, is measured as the horizontal projection on the building roof and walls.

A = the smaller of 10 percent of the least horizontal dimension of the building (i.e. either L or W) or 40 percent of the mean roof height (MRH), but not less than either 4 percent of the least horizontal dimension or 3 feet.

L = length

W = width

Figure 8-20.
Components and cladding wind pressures

standards can be used to simplify the calculation of C&C design loads. Examples of prescriptive C&C design load tables for purposes of illustration are included in this Manual as follows:

- **Roof and wall suction pressures** (see Table 8-8). Application of ASCE 7-10 provisions and typical assumptions used to derive the tabulated load values are addressed in Example 8.7. Suction pressures are used to size connections between sheathing and framing and to size the sheathing material itself for wind induced bending. In ASCE 7-10, there is no adjustment for effective wind areas less than 10 square feet; therefore, sheathing suction loads are based on an effective wind area of 10 square feet.

- **Lateral connector loads from wind and building end zones** (see Table 8-9). Application of ASCE 7-10 provisions and typical assumptions used to derive the tabulated load values are addressed in Example 8.8. Lateral connector loads from wind are used to size the connection from wall stud-to-plate, wall plate-to-floor, and wall plate-to-roof connections to ensure that higher C&C loads acting over smaller wall areas can be adequately resisted and transferred into the roof or floor diaphragm. In ASCE 7-10, the effective wind area for a member is calculated as the span length times an effective width of not less than one-third the span length. For example, the effective area for analysis is calculated as $h^2/3$ where h represents the span (or height) of the wall stud.

Example load tables and example problems are derived from more wind load procedures provided in the WFCM-2012 load Tables 8-8 and 8-9 are not intended to replace requirements of the building code or reference standard for the actual design of C&C attachments for a building.

Table 8-8. Roof and Wall Sheathing Suction Loads (based on ASD design), psf (33-ft mean roof height, Exposure C)

Sheathing Location	Wind Speed[a] (mph)								
	110	115	120	130	140	150	160	170	180
	Roof, suction pressure[b][c] (psf)								
Zone 1	18.6	20.4	22.2	26.0	30.2	34.7	39.4	44.5	44.9
Zone 2	31.3	34.2	37.2	43.7	50.7	58.2	66.2	74.7	83.8
Zone 2 Overhang	34.8	38.0	41.4	48.5	56.3	64.6	73.5	83.0	93.1
Zone 3	47.1	51.5	56.0	65.8	76.3	87.5	99.6	112.4	126.1
Zone 3 Overhang	58.5	63.9	69.6	81.6	94.7	108.7	123.7	139.6	156.5
	Wall, suction pressure[b][c] (psf)								
Zone 4	20.2	22.1	24.1	28.2	32.8	37.6	42.8	48.3	54.1
Zone 5	25.0	27.3	29.7	34.9	40.4	46.4	52.8	59.6	66.8

(a) 700-year wind speed, 3-sec gust.

(b) Roof and wall sheathing suction loads are based on 33-ft mean roof height and Exposure C (see Example 8.7).

(c) Loads based on minimum effective area of 10 ft².

Table 8-9. Lateral Connector Loads from Wind at Building End Zones (Based on ASD Design), plf (33-ft mean roof height, Exposure C)

Wall Height (ft)	Wind Speed[a] (mph)								
	110	115	120	130	140	150	160	170	180
	Lateral connector loads[b][c][d] for wall zone 5 (plf)								
8	92	101	110	129	150	172	196	221	248
10	110	120	131	154	179	205	233	263	295
12	127	139	151	177	206	236	269	303	340
14	143	156	170	200	231	266	302	341	383
16	158	173	188	221	256	294	335	378	423

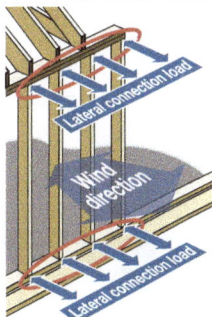

(a) 700-year wind speed, 3-sec gust.

(b) Lateral connector loads are based on 33-ft mean roof height and Exposure C (see Example 8.8).

(c) Lateral connector loads are tabulated in pounds per linear ft of wall. Individual connector loads can be calculated for various spacing of connectors (e.g., for spacing of connectors at 2 ft o.c., the individual connector load would be 2 ft times the tabulated value).

(d) Loads based on minimum area of (wall height)^{2/3}

EXAMPLE 8.7. ROOF SHEATHING SUCTION LOADS

Given:

- The wind load is based on 150 mph, Exposure C at 33-ft mean roof height
- The building is enclosed
- From Example 8.4, for the same site condition, the ASD velocity pressure q = 29.38 psf
- The internal pressure coefficient for roof and wall sheathing is GC_{pi}= 0.18

Find: Roof sheathing and wall sheathing suction loads using the C&C coefficients specified in Figure 30.4 of ASCE 7-10.

For cladding and fasteners, the effective wind area should not be greater than the area that is tributary to an individual fastener. In ASCE 7-10, there is no adjustment for wind areas less than 10 ft^2; therefore, sheathing suction loads are based on an effective wind area of 10 ft^2 for different zones on the roof.

Solution: The roof sheathing and wall sheathing suction loads can be determined using the C&C coefficients specified in Figure 30.4 of ASCE 7-10, as follows:

- The design wind pressure is determined from Equation 8.14 (where q = q_h in this case) as follows:

$$p = q \,|\, GC_{pf} - GC_{pi} \,|$$

- Determine the roof sheathing suction load pressure coefficients using Figure 30.4 of ASCE 7-10 as follows:

Step 1: Roof sheathing suction loads pressure coefficients

Pressure coefficient equations developed from C&C, Figure 30.4, graphs of ASCE 7-10 coefficients:

$$\text{Zone 1: } GC_{pf} = 0.9 - 0.1 \left[\frac{\log\left(\dfrac{A}{100}\right)}{\log\left(\dfrac{10}{100}\right)} \right] = -1.0$$

$$\text{Zone 2: } GC_{pf} = -1.1 - 0.7 \left[\frac{\log\left(\dfrac{A}{100}\right)}{\log\left(\dfrac{10}{100}\right)} \right] = -1.8$$

Zone 2 Overhang: $GC_{pf} = -2.2$

EXAMPLE 8.7. ROOF SHEATHING SUCTION LOADS (concluded)

$$\text{Zone 3: } GC_{pf} = -1.7 - 1.1 \left[\frac{\log\left(\frac{A}{100}\right)}{\log\left(\frac{10}{100}\right)} \right] = -2.8$$

$$\text{Zone 3 Overhang: } GC_{pf} = 2.5 - 1.2 \left[\frac{\log\left(\frac{A}{100}\right)}{\log\left(\frac{10}{100}\right)} \right] = -3.7$$

Step 2: Wall sheathing suction loads pressure coefficient

Pressure coefficient equations developed from C&C, Figure 30.4 graphs of ASCE 7-10 coefficients:

$$\text{Zone 4: } GC_{pf} = -0.8 - 0.3 \left[\frac{\log\left(\frac{A}{500}\right)}{\log\left(\frac{10}{500}\right)} \right] = -1.1$$

$$\text{Zone 5: } GC_{pf} = -0.8 - 0.6 \left[\frac{\log\left(\frac{A}{500}\right)}{\log\left(\frac{10}{500}\right)} \right] = -1.4$$

Step 3: For all zones – internal pressure coefficient

$GC_{pi} = +/-0.18$

Step 4: Calculate roof sheathing and wall sheathing suction pressures for all zones using Equation 8.14

Zone 1: $p = 29.38 \text{ psf}(-1 - 0.18) = $ **−34.7 psf**

Zone 2: $p = 29.38 \text{ psf}(-1.8 - 0.18) = $ **−58.2 psf**

Zone 2 Overhang: $p = 29.38 \text{ psf}(-2.2) = $ **−64.6 psf**

Zone 3: $p = 29.38 \text{ psf}(-2.8 - 0.18) = $ **−87.5 psf**

Zone 3 Overhang: $p = 29.38 \text{ psf}(-3.7) = $ **−108.7 psf**

Zone 4: $p = 29.38 \text{ psf}(-1.1 - 0.18) = $ **−37.6 psf**

Zone 5: $p = 29.38 \text{ psf}(-1.4 - 0.18) = $ **−46.4 psf**

Note: *This solution matches the information in Table 8-8.*

EXAMPLE 8.8. LATERAL CONNECTION FRAMING LOADS FROM WIND

Given:

- Wind load is based on 150 mph, Exposure C at 33-ft mean roof height, and wall and diaphragm framing as shown in Illustration A

- Building is enclosed

- Wall height is 10 ft

- Stud spacing is 16 in. o.c.

- Sheathing effective area is 10 ft^2

- ASD velocity pressure q = 29.38 psf (from Example 8.5)

- Wall suction equations for Zone 4 and Zone 5 are provided in Example 8.7

- Internal pressure coefficient for wall sheathing is GC_{pi} = +/- 0.18

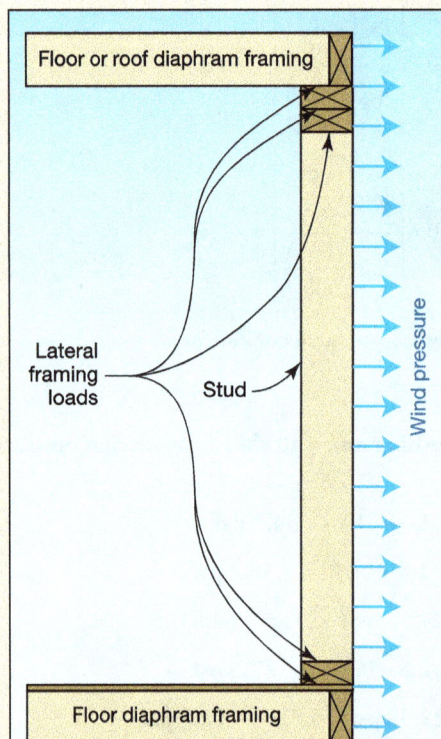

Illustration A. Lateral connector loads for wall-to-roof and wall-to-floor connections

Find: Framing connection requirements at the top and base of the wall.

Solution: The connector load can be determined as follows:

EXAMPLE 8.8. LATERAL CONNECTION FRAMING LOADS FROM WIND (concluded)

- C&C coefficients are used

- When determining C&C pressure coefficients, the effective wind area equals the tributary area of the framing members

- For long and narrow tributary areas, the area width may be increased to one-third the framing member span to account for actual load distributions

- The larger area results in lower average wind pressures

- The increase in width for long and narrow tributary areas applies only to calculation of wind pressure coefficients

- Determine the tributary area and pressure coefficient GC_{pf} for the wall sheathing: Stud effective wind area equals 13.33 ft². The minimum required area for analysis is $h^2/3 = 33.3$ ft², where h is 10 ft

- In accordance with ASCE 7-10, the pressure coefficient, GC_{pf}, for wall sheathing can be determined based on a minimum effective wind area of 33.3 ft² as follows:

$$\text{Zone 5: } GC_{pf} = -0.8 - 0.6 \left[\frac{\log\left(\dfrac{33.3}{500}\right)}{\log\left(\dfrac{10}{500}\right)} \right] = -1.22$$

The design wind pressure is determined as follows from Equation 8.14: $p = q(GC_{pf} - GC_{pi})$

$$\text{Zone 5: } p = 29.38 \text{ psf}(-1.22 - 0.18) = -41.13 \text{ psf}$$

The required capacity of connectors assuming load is based on half the wall height:

$$\text{Zone 5: } w = 41.13 \text{ psf}\left(\frac{10 \text{ ft}}{2}\right) = \textbf{205 plf}$$

Note: This solution matches the information in Table 8-9.

8.8 Tornado Loads

Tornadoes have wind speeds that vary based on the magnitude of the event; more severe tornadoes have wind speeds that are significantly greater than the minimum design wind speeds required by the building code. Designing an entire building to resist tornado-force winds of EF3 or greater based on the Enhanced Fujita tornado damage scale (in EF2 tornadoes, large trees are snapped or uprooted) may be beyond the realm

WARNING

Safe rooms should be located outside known flood-prone areas, including the 500-year floodplain, and away from any potential large debris sources. See Figure 5-2 of FEMA 320 for more direction regarding recommended siting for a safe room.

of practicality and cost-effectiveness, but this does not mean that solutions that provide life-safety protection cannot be achieved while maintaining cost-effectiveness.

A more practical approach is to construct an interior room or space that is "hardened" to resist not only tornado-force winds but also the impact of wind-borne missiles. FEMA guidance on safe rooms can be found in FEMA 320, *Taking Shelter from the Storm: Building a Safe Room for Your Home or Small Business* (FEMA 2008c), which provides prescriptive design solutions for safe rooms of up to 14 feet x 14 feet. These solutions can be incorporated into a structure or constructed as a nearby stand-alone safe room to provide occupants with a place of near-absolute protection. The designs in FEMA 320 are based on wind pressure calculations that are described in FEMA 361, *Design and Construction Guidance for Community Safe Rooms* (FEMA 2008a). FEMA 361 focuses on larger community safe rooms, but the process of design and many of the variables are the same for smaller residential safe rooms.

An additional reference, ANSI/ICC 500-2008 complements the information in FEMA 320 and FEMA 361 and is referenced in the 2012 IBC and 2012 IRC.

Safe rooms can be designed to resist both tornado and hurricane hazards, and though many residents of coastal areas are more concerned with hurricanes, tornadoes can be as prevalent in coastal areas as they are in inland areas such as Oklahoma, Kansas, and Missouri. Constructing to minimum requirements of the building code does not include the protection of life-safety or property of occupants from a direct hit of large tornado events. Safe rooms are not recommended in flood hazard areas.

8.9   Seismic Loads

This Manual uses the seismic provisions of ASCE 7-10 to illustrate a method for calculating the seismic base shear. To simplify design, the effect of dynamic seismic ground motion accelerations can be considered an equivalent static lateral force applied to the building. The magnitude of dynamic motion, and therefore the magnitude of the equivalent static design force, depends on the building characteristics, and the spectral response acceleration parameter at the specific site location.

The structural configuration in Figure 8-21 is called an "inverted pendulum" or "cantilevered column" system. This configuration occurs in elevated pile-supported buildings where almost all of the weight is at the top of the piles. For a timber frame cantilever column system, ASCE 7-10 assigns a response modification factor (R) equal to 1.5 (e.g., R = 1.5). For wood frame, wood structural panel shear walls, ASCE 7-10 assigns an R factor equal to 6.5. The R factor of 1.5 can be conservatively used to determine shear for the design of all elements and connections of the structure. An R factor of 1.5 is not permitted for use in Seismic Design Categories E and F per ASCE 7-10.

ASCE 7-10 contains procedures for the seismic design of structures with different structural systems stacked vertically within a single structure. Rules for vertical combinations can be applied to enable the base of the structure to be designed for shear forces associated with R=1.5 and the upper wood frame, wood structural panel shear wall structure to be designed for reduced shear forces associated with R=6.5.

ASCE 7-10 also provides R factors for cantilever column systems using steel and concrete columns. A small reduction in shear forces for steel piles or concrete columns could be obtained by using what ASCE 7-10 calls a "steel special cantilever column system" or a "special reinforced concrete moment frame," both of which have an R = 2.5. However, these systems call for additional calculations, connection design, and

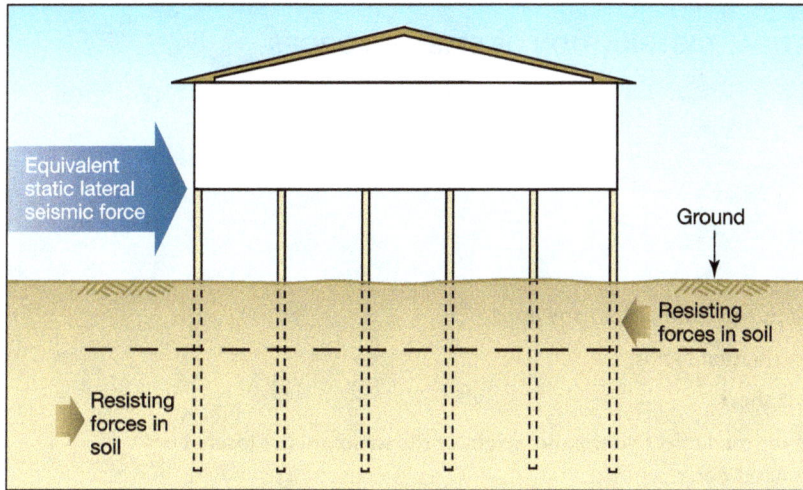

detailing, which are not commonly done for low-rise residential buildings. An engineer experienced in seismic design should be retained for this work, and builders should expect larger pile and column sizes and more reinforcing than is normally be required in a low-seismic area.

Total seismic base shear can be calculated using the Equivalent Lateral Force (ELF) procedure of ASCE 7-10 in accordance with Equation 8.15.

Σ

EQUATION 8.15. SEISMIC BASE SHEAR BY EQUIVALENT LATERAL FORCE PROCEDURE

$$V = C_s W$$ (Eq. 8.15a)

$$C_s = \frac{S_{DS}}{(R/I)}$$ (Eq. 8.15b)

where:

V = seismic base shear

C_s = seismic response coefficient

S_{DS} = design spectral response acceleration parameter in the short period range

R = response modification factor

I = occupancy importance factor

W = effective seismic weight

Lateral seismic forces are distributed vertically through the structure in accordance with Equation 8.16, taken from ASCE 7-10.

Σ

EQUATION 8.16. VERTICAL DISTRIBUTION OF SEISMIC FORCES

$$F_x = C_{vx}V \qquad\qquad\qquad\qquad\text{(Eq. 8.16a)}$$

$$C_{vx} = \frac{w_x h_x^k}{\displaystyle\sum_{i=1}^{n} w_i h_i^k} \qquad\qquad\qquad\qquad\text{(Eq. 8.16b)}$$

where:

F_x = lateral seismic force induced at any level

C_{vx} = vertical distribution factor

V = seismic base shear

w_i and w_x = portion of the total effective seismic weight of the structure (w) located or assigned to level i or x

h_i and h_x = height from the base to Level i or x

k = exponent related to the structure period; for structures having a period of 0.5 sec or less, $k=1$

The calculated seismic force at each story must be distributed into the building frame. The horizontal shear forces and related overturning moments are taken into the foundation through a load path of horizontal floor and roof diaphragms, shear walls, and their connections to supporting structural elements. A complete seismic analysis includes evaluating the structure for vertical and plan irregularities, designing elements and their connections in accordance with special seismic detailing, and considering structural system drift criteria. Example 8.9 illustrates the use of basic seismic calculations.

EXAMPLE 8.9. SEISMIC LOAD

Given:

- S_{DS} for the site = $2/3F_aS_s$, which is determined to be 2/3(1.2)(0.50) = 0.4

- The building structure as shown in Illustration A. Dead load for the building is as follows:

    Roof and ceiling = 10 lb/ft²

    Exterior walls = 10 lb/ft²

    Interior Walls = 8 lb/ft²

    Floor = 10 lb/ft²

    Piles = 409 lb each

EXAMPLE 8.9. SEISMIC LOAD (continued)

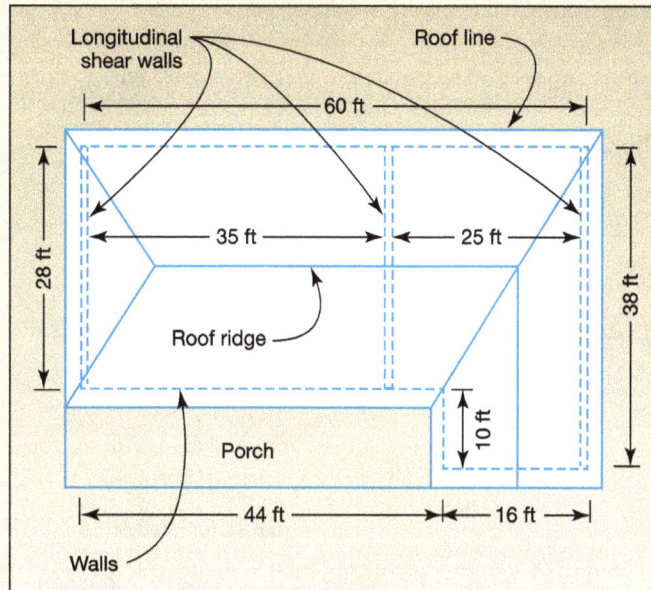

Illustration A. Building elevation and plan view of roof showing
longitudinal shearwalls; dimensions are wall-to-wall and do not
include the 2-ft roof overhang

Find (using ASCE 7-10 ELF procedure):

1. The total shear wall force

2. The shear force at the top of the pile foundation

Solution for #1: The total shear wall force using the ASCE 7-10 ELF procedure can be determined
as follows:

- Calculate effective seismic weight:

  Roof/ceiling: $(10 \text{ lb/ft}^2)(2,390 \text{ ft}^2) = 23,900 \text{ lb}$

  Exterior walls: $(10 \text{ lb/ft}^2)(1,960 \text{ ft}^2) = 19,600 \text{ lb}$

  Interior partitions: $(8 \text{ lb/ft}^2)(2,000 \text{ ft}^2) = 16,000 \text{ lb}$

  Floor = $(10 \text{ lb/ft}^2)(2,160 \text{ ft}^2) = 21,600 \text{ lb}$

  Piles: $(409 \text{ lb/pile})(31 \text{ piles}) = 12,679 \text{ lb}$

  Total effective seismic weight:
  $W = 23,900 + 19,600 \text{ lb} + 16,000 \text{ lb} + 21,600 \text{ lb} + 12,679 \text{ lb} = 93,454 \text{ lb}$

EXAMPLE 8.9. SEISMIC LOAD (continued)

- Seismic forces are distributed vertically as follows:

  Roof level:

  Effective seismic weight, $w_{x\,roof}$ = 23,900 lb + (0.5)(19,600) lb + (0.5)(16,000/2) lb = 41,700 lb

  Height from base, $h_{x\,roof}$ = 18 ft

  $w_{x\,roof}(h_{x\,roof})$ = 750,600 ft-lb

  Floor level:

  Effective seismic weight, $w_{x\,floor}$ = 19,600/2 lb + 16,000/2 lb + 21,600 lb + 12,679 lb = 52,079 lb

  Height from base: $h_{x\,floor}$ = 8 ft

  $w_{x\,floor}(h_{x\,floor})$ = (52,079 lb)(8 ft) = 416,632 ft-lb

  $$C_{vx,roof} = \frac{750,600 \text{ ft-lb}}{750,600 \text{ ft-lb} + 416,632 \text{ ft-lb}} = 0.64 \text{ from Equation 8.16}$$

  $$C_{vx,floor} = \frac{416,632 \text{ ft-lb}}{750,000 \text{ ft-lb} + 416,632 \text{ ft-lb}} = 0.36 \text{ from Equation 8.16}$$

  The force in the shear walls and at the top of the piles will vary by the R factor for the shear wall system and the pile system (e.g., cantilevered column system).

- Lateral seismic force at the roof level for design of wood-frame shear walls (R = 6.5):

  $$F_{x\,roof} = C_{vx\,roof}V = C_{vx\,roof}\frac{S_{DS}}{R/I}W \text{ using Equation 8.15 for } V \text{ substituted into Equation 8.16}$$

  $$F_{x,roof} = \left[\frac{\dfrac{(0.64)(0.4)}{6.5}}{1.0}\right](93,454 \text{ lb}) = 3,681 \text{ lb}$$

  where:

       R = 6.5 for light-frame walls with plywood

       I = 1.0 for residential structure

  The design shear force for the shear walls is based on the lateral seismic force at the roof level. Total seismic force for shear wall design is **3,681 lb**

Solution for #2: The shear force to the top of the pile foundation (i.e., cantilevered column system, R = 1.5) can be determined as follows:

- Roof level

  $$F_{x\,roof} = C_{vx\,roof}V = C_{vx\,roof}\frac{S_{DS}}{R/I}W \text{ using Equation 8.15 for } V \text{ substituted into Equation 8.16}$$

EXAMPLE 8.9. SEISMIC LOAD (concluded)

$$F_{x,roof} = \left[\frac{\frac{(0.64)(0.4)}{1.5}}{1.0} \right] (93,454 \text{ lb}) = 15,949 \text{ lb}$$

- Floor level

$$F_{x\ floor} = C_{vx\ floor}V = C_{vx\ floor} \frac{S_{DS}}{R/I} W \quad \text{using Equation 8.15 for } V \text{ substituted into Equation 8.16}$$

$$F_{x,floor} = \left[\frac{\frac{(0.36)(0.4)}{1.5}}{1.0} \right] (93,454 \text{ lb}) = 8,972 \text{ lb}$$

where:

   R = 1.5 for cantilevered column system. For vertically mixed seismic-force-resisting systems, ASCE 7-10 allows a lower R to be used below a higher R value.

   I = 1.0 for a residential structure

Total shear at the floor is based on the sum of the force at the roof level and the floor level:

$$F_{floor} = 15,949 \text{ lb} + 8,972 \text{ lb} = \textbf{24,921 lb}$$

8.10   Load Combinations

It is possible for more than one type of natural hazard to occur at the same time. Floods can occur at the same time as a high-wind event, which happens during most hurricanes. Heavy rain, high winds, and flooding conditions can occur simultaneously. ASCE 7-10 addresses the various load combination possibilities.

The following symbols are used in the definitions of the load combinations:

   D = dead load

   L = live load

   E = earthquake load

   F = load due to fluids with well-defined pressures and maximum heights (e.g., fluid load in tank)

   F_a = flood load

   H = loads due to weight and lateral pressures of soil and water in soil

   L_r = roof live load

S = snow load

R = rain load

T = self-straining force

W = wind load

Loads combined using the ASD method are considered to act in the following combinations for buildings in Zone V and Coastal A Zone (Section 2.4.1 of ASCE 7-10), whichever produces the most unfavorable effect on the building or building element:

Combination No. 1: D

Combination No. 2: $D + L$

Combination No. 3: $D + (Lr \text{ or } S \text{ or } R)$

Combination No. 4: $D + 0.75L + 0.75(Lr \text{ or } S \text{ or } R)$

Combination No. 5: $D + (0.6W \text{ or } 0.7E)$

Combination No. 6a: $D + 0.75L + 0.75(0.6W) + 0.75(Lr \text{ or } S \text{ or } R)$

Combination No. 6b: $D + 0.75L + 0.75(0.7E) + 0.75S$

Combination No. 7: $0.6D + 0.6W$

Combination No. 8: $0.6D + 0.7E$

When a structure is located in a flood zone, the following load combinations should be considered in addition to the basic combinations in Section 2.4.1 of ASCE 7-10:

In Zone V or Coastal A Zone, $1.5F_a$ should be added to load combinations Nos. 5, 6, and 7, and E should be set equal to zero in Nos. 5 and 6

In the portion of Zone A landward of the LiMWA, $0.75F_a$ should be added to combinations Nos. 5, 6, and 7, and E should be set equal to zero in Nos. 5 and 6.

The ASCE 7-10 *Commentary* states "Wind and earthquake loads need not be assumed to act simultaneously. However, the most unfavorable effects of each should be considered in design, where appropriate."

The designer is cautioned that F is intended for fluid loads in tanks, not hydrostatic loads. F_a should be used for all flood loads, including hydrostatic loads, and should include the various components of flood loads as recommended in Section 8.5.11 in this chapter. It is important to note that wind and seismic loads acting on a building produce effects in both the vertical and horizontal directions. The load combinations discussed in this section must be evaluated carefully, with consideration given to whether a component of the wind or seismic load acts in the same vertical or horizontal direction as other loads in the combination. In some cases, gravity loads (dead and live loads) may counteract the effect of the wind or seismic load, either vertically or horizontally. Building elements submerged in water have a reduced effective weight due to buoyancy. Example 8.10 illustrates the use of load combinations for determining design loads.

EXAMPLE 8.10. LOAD COMBINATION EXAMPLE PROBLEM

Given:

Use the flood loads from Example 8.3:

- $F_{sta} = 0$
- $F_{dyn} = 909$ lb
- $F_{brkp} = 625$ lb
- $F_i = 2,440$ lb
- $d_s = 4.6$ ft

Use for wind loads:

- Roof span = 28 ft
- Roof pitch = 7:12
- Wall height = 10 ft
- Wind uplift load = 33,913 lb (pre-factored with 0.6)
- Exposure Category D (multiply Exposure C wind loads by 1.18 at 33 ft mean roof height)
  - 1.18 is a conservative value because while the higher Exposure Category D has been factored, the lower roof height (24 ft versus 33 ft) has not. Refer to ASCE 7-10, Figure 28.6-1 for guidance.

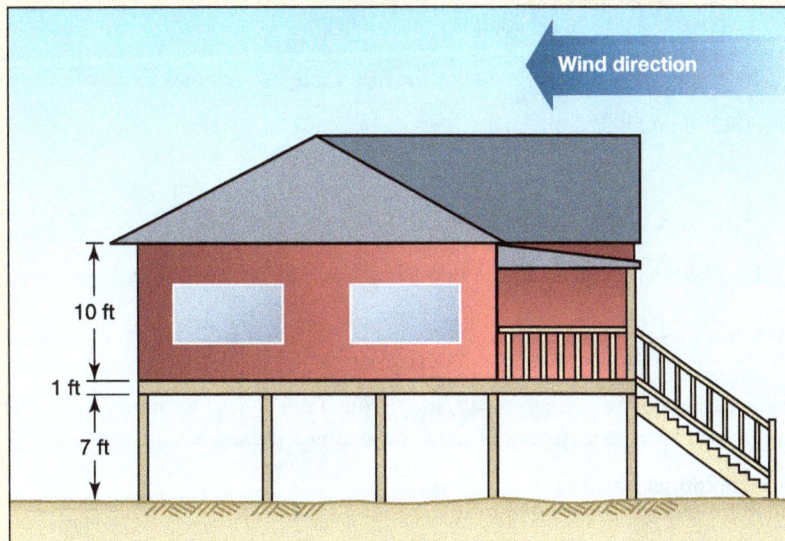

Illustration A. Side view of building

EXAMPLE 8.10. LOAD COMBINATION EXAMPLE PROBLEM (continued)

Use for dead load:

- 95,090 lb for house and piles

Use for buoyancy load:

- 9,663 lb

The locations given in Illustration B for the forces.

Find:

1. Calculate maximum horizontal wind load that occurs perpendicular to the ridge and the floor for the example building

2. Find the horizontal load required for foundation design

3. Calculate global overturning moment due to horizontal loads and wind uplift (see Illustration B)

Solution for #1: To determine the horizontal wind load perpendicular to the ridge, use the projected area method as follows:

- For wind perpendicular to the ridge of a roof with a span of 28 ft (using Table 8-7), 7:12 roof pitch and wall height of 10 ft, Category D as shown in Illustration A, the lateral roof diaphragm load, w_{roof}, can be found by interpolation between the 24 ft and 32 ft roof span w_{roof} values in Table 8.7:

$$w_{roof} = (256 \text{ plf} + 299 \text{ plf})(0.5) = 278 \text{ plf}$$

To adjust w_{roof} for Exposure Category D due to the fact Table 8-7 assumes Exposure Category C:

$$w_{roof} = 1.18(278 \text{ plf}) = 328 \text{ plf}$$

where

   1.18 = Exposure D adjustment factor (33 ft mean roof height)

To adjust w_{roof} for a wall height of 10 ft due to the fact Table 8.7 assumes a wall height of 8 ft:

$$w_{roof} = (328 \text{ plf})\left(\frac{10 \text{ ft}}{8 \text{ ft}}\right) = 410 \text{ plf}$$

- Determine lateral floor diaphragm load, w_{floor} from Table 8-7. Once more, this value needs to be adjusted for Exposure Category D from the assumed Exposure Category C:

$$w_{floor} = 1.18(286 \text{ plf}) = 338 \text{ plf}$$

where

   1.18 = Exposure D adjustment factor (33 ft mean roof height)

EXAMPLE 8.10. LOAD COMBINATION EXAMPLE PROBLEM (continued)

To adjust w_{roof} for a wall height of 10 ft due to the fact Table 8.7 assumes a wall height of 8 ft:

$$w_{floor} = (338 \text{ plf})\left(\frac{10 \text{ ft}}{8 \text{ ft}}\right) = 423 \text{ plf}$$

Finally, adjust this value to account for the reference case in Table 8-7 assuming the lateral floor diaphragm load is from wind pressures on the lower half of the wall above and the upper half of the wall below the floor diaphragm. Because the structure is open below the floor diaphragm level, adjust w_{floor} to account for the presence of only half of the wall area used in the reference case for Table 8-7 (e.g., structure is open below first floor diaphragm):

$$w_{floor} = 0.5(423 \text{ plf}) = 212 \text{ plf}$$

• For building length = 60 ft, total horizontal shear at the top of the foundation is:

$$W_{foundation} = (410 \text{ plf} + 212 \text{ plf})(60 \text{ ft}) = \textbf{37,320 lb}$$

Solution for #2: The horizontal load required for foundation design, can be determined using the following calculations of the load combination equations given in Section 8.10:

• Zone V and Coastal A Zone

5. $D + 0.6W + 1.5F_a$

6a. $D + 0.75L + 0.75(0.6W) + 0.75(L_r \text{ or } S \text{ or } R) + 1.5F_a$

6b. $D + 0.75L + 0.75S + 1.5F_a$

7. $0.6D + 0.6W + 1.5F_a$

Load combination No. 5 produces the maximum shear at the foundation for the loads considered. This load combination includes a wind load factor adjustment of 0.6. Because the value of $W_{foundation}$ from Solution #1 has already been adjusted by 0.6 for ASD, it will not be further adjusted in the calculations that follow.

For flood load, the value of F_a is determined in accordance with Table 8-5. The hydrodynamic load is greater than breaking wave load, therefore, F_a for an individual pile and the foundation as a whole (i.e., global) is calculated as:

$$F_{a(individual)} = F_i + F_{dyn} = 2,440 \text{ lb} + 909 \text{ lb} = 3,349 \text{ lb}$$

$$F_{a(global)} = (1 \text{ pile})F_i + (35 \text{ piles})F_{dyn} = 34,255 \text{ lb}$$

5. Total shear: $37,320 \text{ lb} + 1.5(34,255 \text{ lb}) = \textbf{88,703 lb}$

• Portion of Zone A landward of the LiMWA

5. $D + 0.6W + 0.75F_a$

6a. $D + 0.75L + 0.75(0.6W) + 0.75(L_r \text{ or } S \text{ or } R) + 0.75F_a$

EXAMPLE 8.10. LOAD COMBINATION EXAMPLE PROBLEM (continued)

6b. $D + 0.75L + 0.75S + 0.75F_a$

7. $0.6D + 0.6W + 0.75F_a$

Load combination 5 produces the maximum shear at the foundation for the loads considered.

5.   Total shear: $37,320 + 0.75(34,255 \text{ lb}) = \textbf{63,011 lb}$

Note: Considering seismic force from Example 8.8, ASD shear force at the foundation is determined by load combination No. 8:

8.   Total seismic base shear = $0.7(24,921 \text{ lb}) = \textbf{17,444 lb}$

Solution for #3: To determine the factored global overturning moment due to the factored loads on the building, take the moments about the pivot point in Illustration B.

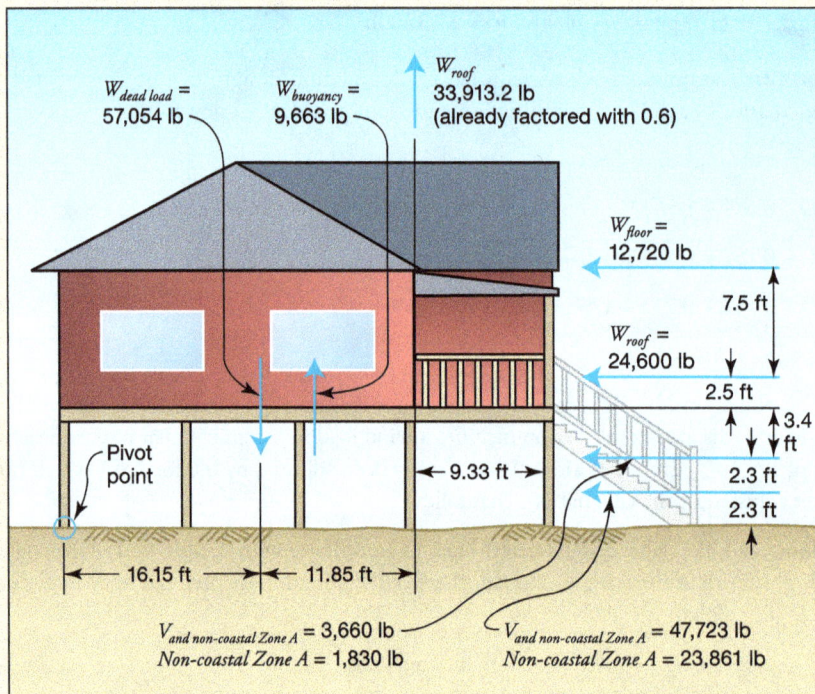

$W_{dead\ load} = $
57,054 lb

$W_{buoyancy} = $
9,663 lb

W_{roof}
33,913.2 lb
(already factored with 0.6)

$W_{floor} = $
12,720 lb

7.5 ft

$W_{roof} = $
24,600 lb

2.5 ft

3.4 ft

2.3 ft

2.3 ft

Pivot
point

9.33 ft

16.15 ft

11.85 ft

$V_{and\ non\text{-}coastal\ Zone\ A} = 3,660\ lb$
Non-coastal Zone A = 1,830 lb

$V_{and\ non\text{-}coastal\ Zone\ A} = 47,723\ lb$
Non-coastal Zone A = 23,861 lb

Illustration B. Loads on building for global overturning moment calculation

Load Combination 7 produces the maximum overturning at the foundation for the loads considered. Factored global overturning moment can be calculated from the factored loads and their location of application as shown in Illustration B.

- Zone V and Coastal A Zone

EXAMPLE 8.10. LOAD COMBINATION EXAMPLE PROBLEM (concluded)

7. $0.6D + 0.6W + 1.5F_a$ gives the appropriate factors to be used in calculating the factored global overturning moment

From Illustration B:

$M_{global} = (0.6)w_{roof}(18\ \text{ft}) + (0.6)w_{floor}(10.5\ \text{ft}) + (1.5)F_i(4.6\ \text{ft}) + (1.5)F_{dyn}(2.3\ \text{ft}) +$
$W_{uplift}(28\ \text{ft}) - (0.6)DL(16.15\ \text{ft}) + (1.5)F_b(19\ \text{ft})$

$M_{global} = (0.6)(24,600\ \text{lb})(18\ \text{ft}) + (0.6)(12,720\ \text{lb})(10.5\ \text{ft}) + (1.5)(2,440\ \text{lb})(4.6\ \text{ft}) +$
$(1.5)(31,815\ \text{lb})(2.3\ \text{ft}) + (33,913\ \text{lb})(28\ \text{ft}) - (0.6)(95,090\ \text{lb})(16.15\ \text{ft}) + (1.5)(9,663\ \text{lb})(19\ \text{ft})$
$= \textbf{776,000 ft-lb}$

• The portion of Zone A landward of the LiMWA

7. $0.6D + 0.6W + 0.75F_a$ gives the appropriate factors to be used in calculating the factored global overturning moment

From Illustration B:

$M_{global} = (0.6)w_{roof}(18\ \text{ft}) + (0.6)w_{floor}(10.5\ \text{ft}) + (0.75)F_i(4.6\ \text{ft}) + (0.75)F_{dyn}(2.3\ \text{ft})$
$+W_{uplift}(28\ \text{ft}) - (0.6)DL(16.15\ \text{ft}) + (.75)F_b(19\ \text{ft})$

$M_{global} = (0.6)(24,600\ \text{lb})(18\ \text{ft}) + (0.6)(12,720\ \text{lb})(10.5\ \text{ft}) + (0.75)(2,440\ \text{lb})(4.6\ \text{ft})$
$+(0.75)(31,815\ \text{lb})(2.3\ \text{ft}) + (33,913\ \text{lb})(28\ \text{ft}) - (0.6)(95,090\ \text{lb})(16.15\ \text{ft})$
$+(0.75)(9,663\ \text{lb})(19\ \text{ft}) = \textbf{575,000 ft-lb}$

Note: In this example, the required uplift capacity to resist overturning is estimated by evaluating the skin friction capacity of the piles. The total pile uplift capacity is approximately 908,000 ft-lb. which exceeds both calculated overturning moments and is based on the horizontal distance to each row of piles from the pivot point and the following assumptions:

• Pile embedment: 19.33 ft

• Pile size: 10 in.

• Coefficient of friction: 0.4 for wood piles

• Density of sand: 128 lb/ft³

• Coefficient of lateral pressure: 09.5

• Critical depth fir sand: 15 ft

• Angle of internal friction: 38°

• Scour depth: 5 ft

• Factor of safety: 2

The following worksheet can be used to facilitate load combination computations.

Worksheet 3. Load Combination Computation

<table>
<tr><td colspan="2" align="center">**Load Combination Computation Worksheet**</td></tr>
<tr><td>OWNER'S NAME: _____</td><td>PREPARED BY: _____</td></tr>
<tr><td>ADDRESS: _____</td><td>DATE: _____</td></tr>
<tr><td colspan="2">PROPERTY LOCATION: _____</td></tr>
</table>

Variables

D (dead load) =

E (earthquake load) =

L (live load) =

F (fluid load) =

F_a (flood load) =

H (lateral soil and water in soil load) =

L_r (roof live load) =

S (snow load) =

R (rain load) =

T (self-straining force) =

W (wind load) =

Summary of Load Combinations:

1.

2.

3.

4.

5.

6a.

6b.

7.

8.

Combination No. 1

$D =$

Combination No. 2

$D + L =$

Combination No. 3

$D + (L_r \text{ or } S \text{ or } R) =$

Worksheet 3. Load Combination Computation (concluded)

Combination No. 4
$D + 0.75L + 0.75(L_r \text{ or } S \text{ or } R) =$
Combination No. 5
$D + (0.6W \text{ or } 0.7E) =$
Combination No. 6a
$D + 0.75L + 0.75(0.6W) + 0.75(L_r \text{ or } S \text{ or } R) =$
Combination No. 6b
$D + 0.75L + 0.75(0.7E) + 0.75S =$
Combination No. 7
$0.6D + 0.6W =$
Combination No. 8
$0.6D + 0.7E =$
When a structure is located in a flood zone, the following load combinations should be considered in addition to the basic combinations:

- In Zone V or Coastal A Zone, $1.5F_a$ should be added to load combinations Nos. 5, 6, and 7, and E should be set equal to zero in Nos. 5 and 6.

- In the portion of Zone A landward of the LiMWA, $0.75F_a$ should be added to load combinations Nos. 5, 6, and 7, and E should be set equal to zero in Nos. 5 and 6.

8.11 References

AF&PA (American Forest & Paper Association). 2012. *Wood Frame Construction Manual for One- and Two-Family Dwellings.* WFCM-12.

AISI (American Institute of Steel Institute). 2007. *Standard for Cold-formed Steel Framing-prescriptive Method for One- and Two-family Dwellings.* AISI S230-07.

AISC (American Institute of Steel Construction) / ICC (International Code Council). 2008. *Standard for the Design and Construction of Storm Shelters.* ANSI/ICC 500-2008.

The American Institute of Architects. 2007. *Architectural Graphic Standards.* A. Pressman, ed. Hoboken, NJ: John Wiley & Sons, Inc.

ASCE (American Society of Civil Engineers). 1995. *Wave Forces on Inclined and Vertical Wall Surfaces.*

ASCE. 2005a. *Flood Resistant Design and Construction.* ASCE Standard ASCE 24-05.

ASCE. 2005b. *Minimum Design Loads for Buildings and Other Structures.* ASCE Standard ASCE 7-05.

ASCE. 2010. *Minimum Design Loads for Buildings and Other Structures.* ASCE Standard ASCE 7-10.

AWC (American Wood Council). 2009. *Prescriptive Residential Wood Deck Construction Guide.* AWC DCA6.

Bea, R.G., T. Xu, J. Stear, and R. Ramos. 1999. "Wave Forces on Decks of Offshore Platforms." *Journal of Waterway, Port, Coastal and Ocean Engineering*, Vol. 125, No. 3, pp. 136–144.

Camfield, F.E. 1980. *Tsunami Engineering.* Vicksburg, MS: Coastal Engineering Research Center.

FEMA (Federal Emergency Management Agency). 2001. *Engineering Principles and Practices for Retrofitting Floodprone Residential Buildings.* FEMA 259.

FEMA. 2006. *Hurricane Katrina in the Gulf Coast.* FEMA 549.

FEMA. 2008a. *Design and Construction Guidance for Community Safe Rooms.* FEMA 361.

FEMA. 2008b. *Guidelines for Design of Structures for Vertical Evacuation from Tsunamis.* FEMA P646.

FEMA. 2008c. *Taking Shelter from the Storm: Building a Safe Room for Your Home or Small Business.* FEMA 320.

FEMA. 2009. *Hurricane Ike in Texas and Louisiana.* FEMA P-757.

Fox, R.W. and A.T. McDonald. 1985. *Introduction to Fluid Mechanics.* New York: John Wiley & Sons, Inc.

ICC (International Code Council). 2008. *Standard for Residential Construction in High-Wind Regions.* ICC 600-2008.

ICC. 2011a. *International Building Code.* 2012 IBC. ICC: Country Club Hills, IL.

ICC. 2011b. *International Residential Code for One-and Two-Family Dwellings.* 2012 IRC. ICC: Country Club Hills, IL.

McConnell, K., W. Allsop, and I. Cruickshank. 2004. *Piers, Jetties and Related Structures Exposed to Waves: Guidelines for Hydraulic Loadings.* London: Thomas Telford Publishing

Sumer, B.M., R.J.S. Whitehouse, and A. Tørum. 2001. "Scour around Coastal Structures: A Summary of Recent Research." *Coastal Engineering*, Vol. 44, Issue 2, pp. 153–190.

USACE (U.S. Army Corps of Engineers). 1984. *Shore Protection Manual, Volume 2.*

USACE. 2006. *Coastal Engineering Manual.*

USACE. 2008. *Coastal Engineering Manual.*

Walton, T.L. Jr., J.P. Ahrens, C.L. Truitt, and R.G. Dean. 1989. *Criteria for Evaluating Coastal Flood-Protection Structures.* Coastal Engineering Research Center Technical Report 89-15. U.S. Army Corps of Engineers Waterways Experiment Station.

Designing the Building

This chapter provides guidance on design considerations for buildings in coastal environments. The topics discussed in this chapter are developing a load path through elements of the building structure, considerations for selecting building materials, requirements for breakaway walls, and considerations for designing appurtenances. Examples of problems for the development of the load path for specific building elements are provided, as well as guidance on requirements for breakaway walls, selection of building materials, and appurtenances.

> **CROSS REFERENCE**
>
> For resources that augment the guidance and other information in this Manual, see the Residential Coastal Construction Web site (http://www.fema.gov/rebuild/mat/fema55.shtm).

9.1   Continuous Load Path

In hazard-resistant construction, the ability of the elements of a building, from the roof to the foundation, to carry or resist loads is critical. Loads include lateral and uplift loads. A critical aspect of hazard-resistant construction is the capability of a building or structure to carry and resist all loads—including lateral and uplift loads—from the roof, walls, and other elements to the foundation and into the ground. The term "continuous load path" refers to the structural condition required to resist loads acting on a building. A load path can be thought of as a chain running through the building. A building may contain hundreds of continuous load paths. The continuous load path starts at the point or surface where loads are applied, moves through the building, continues through the foundation, and terminates where the loads are transferred to the soils that support the building. Because all applied loads must be transferred to the foundation, the load path must connect to the foundation. To be effective, each link in the load path chain must be strong enough to transfer loads without breaking.

Buildings that lack strong and continuous load paths may fail when exposed to forces from coastal hazards, thus causing a breach in the building envelope or the collapse of the building. The ability of a building to resist these forces depends largely on whether the building's construction provides a continuous load path and materials that are appropriate for the harsh coastal environment. The history of storm damage is replete with instances of failures in load paths. Figures 9-1 through 9-5 show instances of load path failure.

Figure 9-1.
Load path failure at gable end, Hurricane Andrew (Dade County, FL, 1992)

Figure 9-2.
Load path failure in connection between home and its foundation, Hurricane Fran (North Carolina, 1996)

Figure 9-3.
Roof framing damage
and loss due to load path
failure at top of wall/roof
structure connection,
Hurricane Charley (Punta
Gorda, FL, 2004)

Figure 9-4.
Load path failure in
connections between roof
decking and roof framing,
Hurricane Charley (Punta
Gorda, FL, 2004)

Most load path failures have been observed to occur at connections as opposed to the failure of an individual structural member (e.g. roof rafter or wall stud). Improvements in codes, design, and materials over the past decade have resulted in improved performance of structural systems. As the structural systems perform better, other issues related to load path—such as building envelope issues—become apparent.

CROSS REFERENCE

For a discussion of building envelope issues, see Chapter 11 of this Manual.

Figure 9-5.
Newer home damaged
from internal
pressurization and
inadequate connections,
Hurricane Katrina (Pass
Christian, MS, 2005)

Load path guidance in this chapter is focused primarily on elements of the building structure, excluding foundation elements. Foundation elements are addressed in Chapter 10. Examples are provided primarily to illustrate how the load path resists wind uplift forces, but a complete building design includes a consideration of numerous other forces on the load path, including those from gravity loads and lateral loads. The examples illustrate important concepts and best practices in accordance with building codes and standards but do not represent an exhaustive collection of load calculation methods. See the applicable building code, standard, or design manual for more detailed guidance.

Figure 9-6 shows a load path for wind uplift beginning with the connection of roof sheathing to roof framing and ending with the resistance of the pile to wind uplift. Links #1 through #8 in the figure show connections that have been observed during investigations after high-wind events to be vulnerable to localized failure. However, the load path does not end with the resistance of the pile to wind uplift. The end of transfer through the load path occurs when the loads from the building are transferred into the soil (see Chapter 10 for information about the interaction of foundations and soils). Adequately sizing and detailing every link is important for overall performance because even a small localized failure can lead to a progressive failure of the building structure. The links shown in Figure 9-6 are discussed in more detail in Sections 9.1.1 through 9.1.8. For additional illustration of the concept of load path, see Fact Sheet 4.1, *Load Paths*, in FEMA P-499 (FEMA 2010b).

9.1.1 Roof Sheathing to Framing Connection (Link #1)

Link #1 is the nailed connection of the roof sheathing to the roof framing (see Figures 9-6 and 9-7). Design considerations include ensuring the connection has adequate strength to resist both the withdrawal of the nail shank from the roof framing and the sheathing's pulling over the head of the fastener (also referred to as "head pull-through"). Because of the potential for head pull-through and the required minimum nailing for diaphragm shear capacity, fastener spacing is typically not increased even where shank withdrawal strength

Figure 9-6.
Example load path for case study building

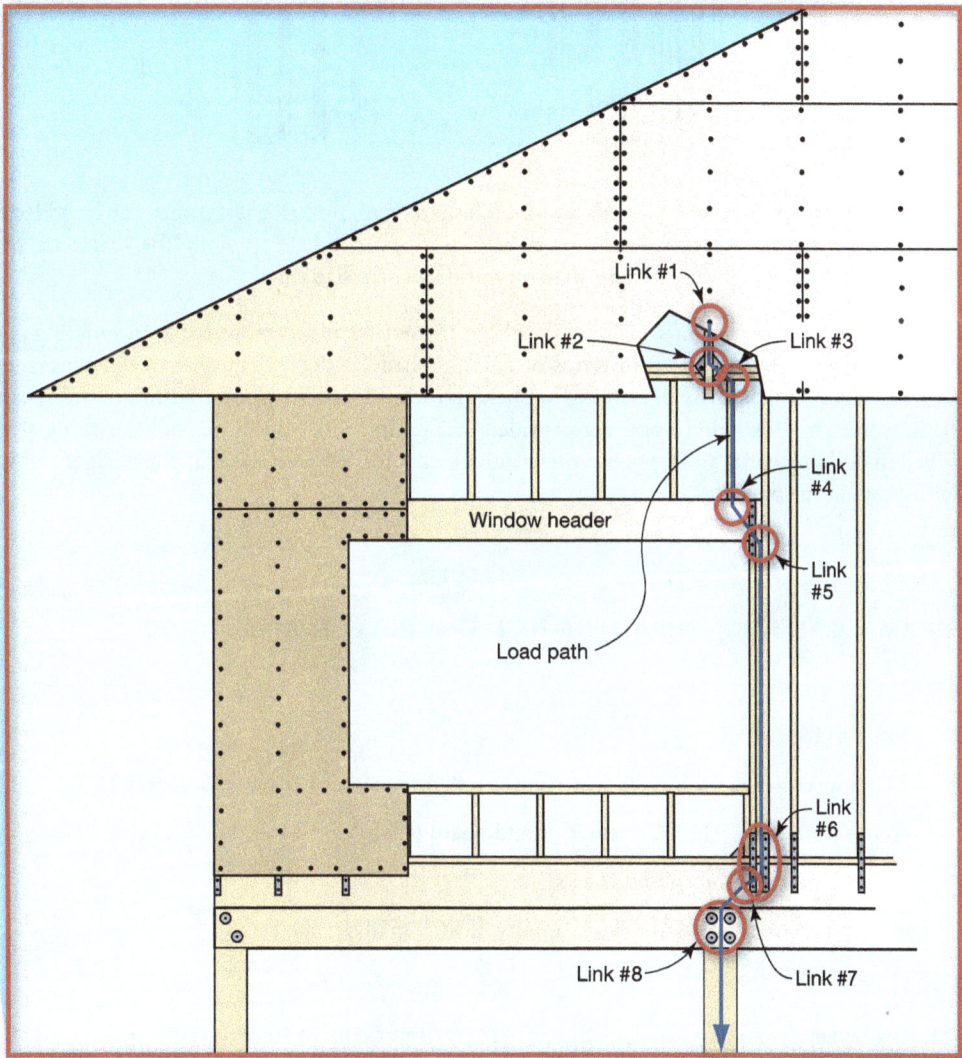

Link #1

Link #2

Link #3

Link #4

Window header

Link #5

Load path

Link #6

Link #8   Link #7

Figure 9-7.
Connection of the roof
sheathing to the roof
framing (Link #1)

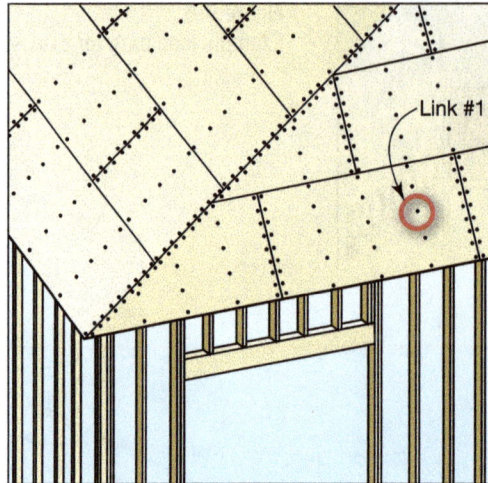

is significantly greater than that provided by a smooth shank nail. Additional strength can be added by using ring shank nails, also called deformed shank nails. The grooves and ridges along the shank act as wedges, giving the nail more withdrawal strength than a typical smooth shank nail.

Fastener attachment requirements for roof sheathing to roof framing are available in building codes and design standards and are presented in terms of nailing schedules dependent on nail diameter and length, framing spacing, specific gravity of framing lumber, and wind speed. Common assumptions for calculating nailing schedules to resist wind uplift are provided in Example 9.1. Minimum roof sheathing attachment prescribed in building codes and reference prescriptive standards is 6 inches o.c. at panel edges and 12 inches o.c. in the field of the panel.

EXAMPLE 9.1. ROOF SHEATHING NAIL SPACING FOR WIND UPLIFT

Given:

- Refer to Figure 9-7

- Wind speed = 150 mph (700-year wind speed, 3-sec gust), Exposure Category D

- Roof sheathing = 7/16-in. oriented strand board (OSB)

- Roof framing specific gravity, G = 0.42

- 8d common nail has withdrawal capacity of 66 lb/nail per the NDS

Find:

1. Nail spacing for the perimeter edge zone for rafter spacing of 16 in. o.c.

2. Nail spacing for the perimeter edge zone for rafter spacing of 24 in. o.c.

EXAMPLE 9.1. ROOF SHEATHING NAIL SPACING FOR WIND UPLIFT (continued)

Solution for #1: The following calculations are used to determine the nail spacing:

- From Table 8-8, the maximum wind suction pressure (based on ASD design) is:

  p = 108.7 psf acting normal to the roof surface (Zone 3 overhang) for Exposure Category C

  The maximum wind suction pressure for Exposure D is:

  p = 108.7 psf (1.18)=128.3 psf

where:

  1.18 = the adjustment factor from Exposure C to Exposure D at 33-ft mean roof height (see Example 8.10)

- The assumed minimum tributary area for calculation of this pressure is 10 ft^2 in accordance with Example 8.7

- For framing at 16 in. o.c., roof suction loads in plf are:

  $$P = 128.3 \text{ psf } \frac{16 \text{ in.}}{12 \text{ in./ft}} = 171.0 \text{ plf}$$

- Nail spacing:

  $$\text{Spacing} = \frac{66 \text{ lb/nail}}{171.0 \text{ plf}} = 0.386 \text{ ft} = 4.6 \text{ in.}$$

  Rounding down to next typical spacing value, specify **4-in. spacing**

Solution for #2: The following calculations are used to determine the nail spacing:

- From Table 8-8, the maximum wind suction pressure is:

  p = 108.7 psf acting normal to the roof surface (Zone 3 overhang) for Exposure Category C

  See Figure 8-18 and Table 8-8.

The maximum wind suction pressure for Exposure D is:

  p = 108.7 psf (1.18)=128.3 psf

where:

  1.18 = adjustment factor from Exposure C to Exposure D at 33-ft mean roof height (see Example 8.10).

- The assumed minimum tributary area for calculation of this pressure is 10 ft^2 in accordance with Example 8.7.

EXAMPLE 9.1. ROOF SHEATHING NAIL SPACING FOR WIND UPLIFT (concluded)

- For framing at 24 in. o.c., roof suction loads on a plf basis is:

$$P = 128.3 \text{ psf} \frac{24 \text{ in.}}{12 \text{ in./ft}} = 256.5 \text{ plf}$$

- Nail spacing:

$$\text{Spacing} = \frac{66 \text{ lb/nail}}{256.5 \text{ plf}} = 0.26 \text{ ft} = 3.09 \text{ in.}$$

Rounding down to next typical spacing value, specify **3-in. spacing**

Note: Edge zone nail spacing associated with Zone 3 OH pressures is conservative for other edge zone locations. Although increased nail spacing may be calculated for an edge zone away from the building corners, it is recommended that the same nailing schedule be used throughout all edge zones.

9.1.2 Roof Framing to Exterior Wall (Link #2)

Link #2 is the connection between the roof framing member (truss or rafter) and the top of the wall below (see Figures 9-6 and 9-8) for resistance to wind uplift. Metal connectors are typically used where uplift forces are large. A variety of metal connectors are available for attaching roof framing to the wall. Manufacturers' literature should be consulted for proper use of the connector and allowable capacities for resistance to uplift. Prescriptive solutions for the connection of the roof framing to the wall top plates are available in building codes and wind design standards. One method of sizing the connection between the roof framing and the exterior wall is provided in Example 9.2.

Figure 9-8.
Connection of roof
framing to exterior wall
(Link #2)

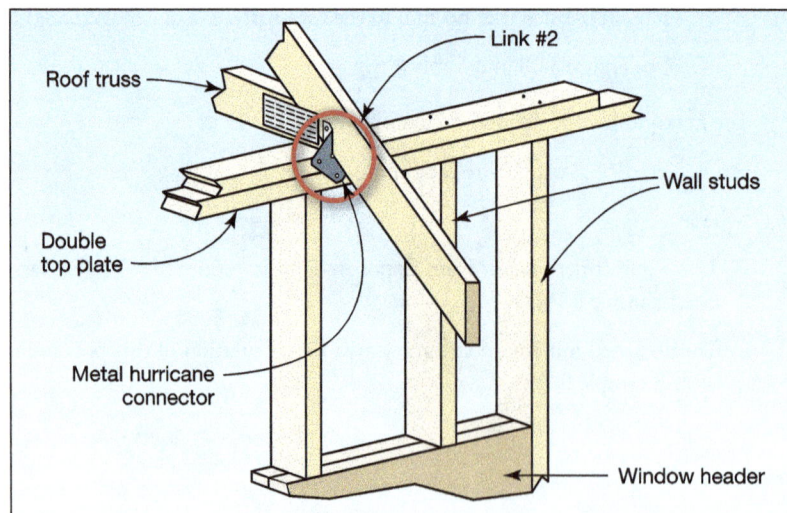

EXAMPLE 9.2. ROOF-TO-WALL CONNECTION FOR UPLIFT

Given:

- Refer to Figure 9-8 and Illustration A

- Wind speed = 150 mph, Exposure D

- Mean roof height = 24 ft

- Rafter spacing = 24 in. o.c.

- Hip rafter span = 14 ft

- Roof pitch = 7:12

- Roof dead load = 10 psf

- Wall height = 10 ft

Illustration A. Location of uplift connection on hip roof

Find:

Determine the required connector size for wind uplift using prescriptive tables for wind uplift loads (i.e., find the uplift and lateral loads for the connector).

Solution: The required connector size using wind uplift prescriptive tables can be determined as follows:

Uplift

- For this example, the maximum hip rafter span = 14 ft

- To use Table 8-6, the uplift strap connector load should be obtained for a 28-ft roof width (e.g., 28 ft is 2 times the 14-ft maximum hip rafter span; see the note at the end of this Example)

- Interpolating between the 24-ft and 32-ft roof span uplift strap connector loads for 150 mph wind speed in Exposure C is:

$$\frac{(424 \text{ plf} + 534 \text{ plf})}{2} = 479 \text{ plf}$$

  Adjust to Exposure Category D by multiplying by 1.18 (see Example 8.10)

  $$1.18(479 \text{ plf}) = 565.2 \text{ plf}$$

- For rafter framing at 2 ft on center, the uplift connector force is:

  $$(565.2 \text{ plf})(2 \text{ ft}) = \textbf{1,131 lb}$$

Lateral

- The lateral load on the connector is = 205 plf (see Table 8-9) for Exposure Category C

- Adjusting for Exposure Category D

EXAMPLE 9.2. ROOF-TO-WALL CONNECTION FOR UPLIFT (concluded)

- 1.18(205 plf) = 241.9 plf for rafter framing at 2 ft o.c., the lateral connector force at each rafter is:

$$(241.9 \text{ plf})(2 \text{ ft}) = \textbf{484 lb}$$

Note: Although the connector forces shown in Table 8-9 assume a gable roof, requirements can be conservatively applied for attaching the hip rafter to the wall. See Table 2.5A, Wood Frame Construction Manual for One- and Two-Family Dwellings (AF&PA 2012). Note that the example roof uses both a gable roof and hip roof framing. For simplicity, the same rafter connection is often used at each connection between the rafter and wall framing. In addition, although smaller forces are developed in shorter hip roof rafter members, the same connector is typically used at all hip rafters.

Figure 9-9 shows truss-to-wood wall connections made with metal connectors. Figure 9-10 shows a rafter-to-masonry wall connector that is embedded into the concrete-filled or grouted masonry cell.

Figure 9-9.
Connection of truss to wood-frame wall

9.1.3   Wall Top Plate to Wall Studs (Link #3)

Link #3 is the connection between the wall top plates and the wall stud over the window header (see Figures 9-6 and 9-11). The connection provides resistance to the same uplift force as used for the roof framing to the exterior connection minus the weight of the top plates. An option for maintaining the uplift load path is the use of metal connectors between the top plates and wall studs or wood structural panel sheathing (see Figure 9-12). The uplift load path can be made with wood structural panel wall sheathing, particularly when the uplift and shear forces in the wall are not very high. Guidance on using wood structural panel wall sheathing for resisting wind uplift is provided in ANSI/AF&PA SDPWS-08. The lateral load path (e.g., out-of-plane wall loads) is maintained by stud-to-top plate nailing.

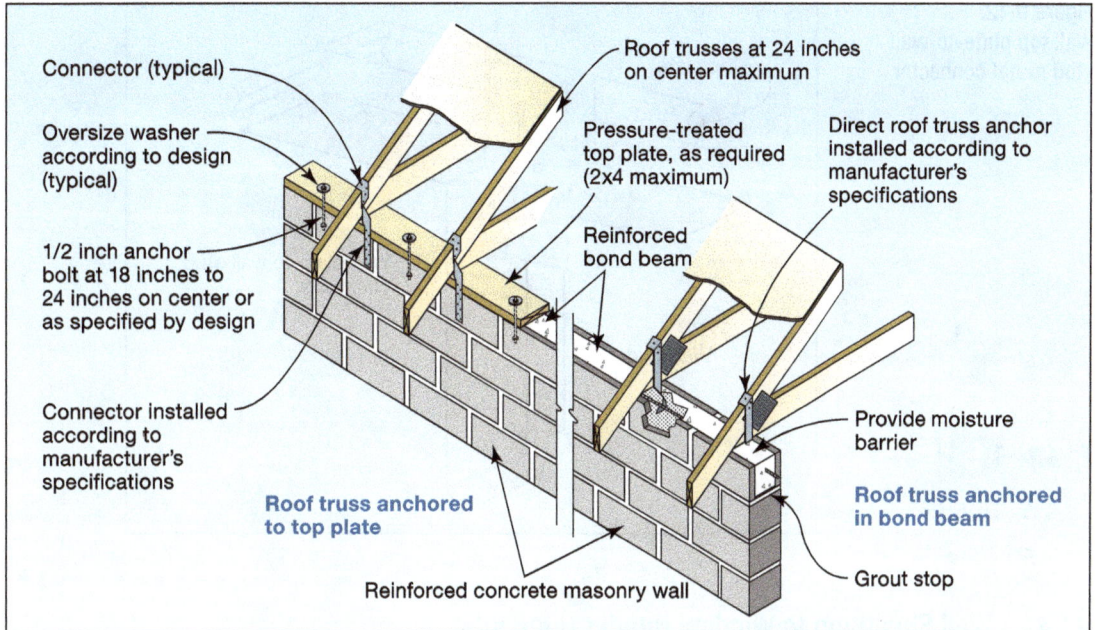

Figure 9-10.
Roof truss-to-masonry wall connectors embedded into concrete-filled or grouted masonry cell (left-hand side image has a top plate installed while the right-hand side does not)

Labels in Figure 9-10:
- Connector (typical)
- Oversize washer according to design (typical)
- 1/2 inch anchor bolt at 18 inches to 24 inches on center or as specified by design
- Connector installed according to manufacturer's specifications
- Roof truss anchored to top plate
- Reinforced concrete masonry wall
- Roof trusses at 24 inches on center maximum
- Pressure-treated top plate, as required (2x4 maximum)
- Reinforced bond beam
- Direct roof truss anchor installed according to manufacturer's specifications
- Provide moisture barrier
- Roof truss anchored in bond beam
- Grout stop

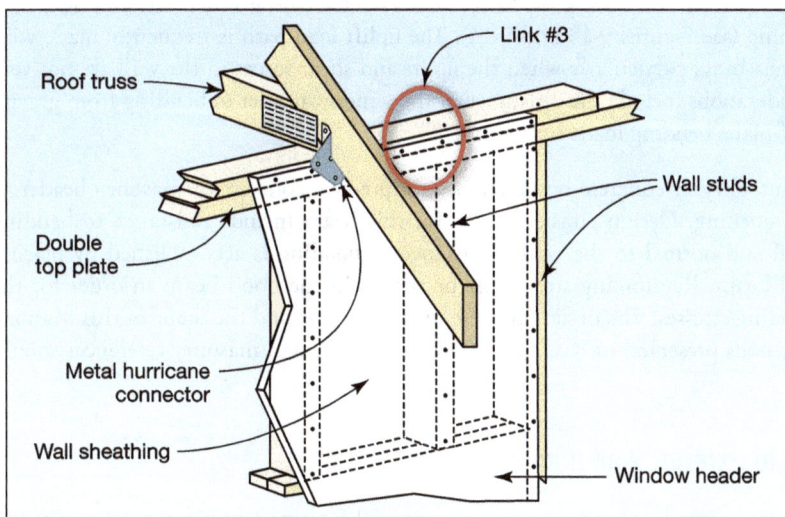

Labels in Figure 9-11:
- Roof truss
- Double top plate
- Metal hurricane connector
- Wall sheathing
- Link #3
- Wall studs
- Window header

Figure 9-11.
Connection of wall top plate-to-wall stud (Link #3)

For masonry or concrete walls, the wood sill plate is typically connected by anchor bolts, cast-in straps, or other approved fasteners capable of maintaining a load path for uplift, lateral, and shear loads. Anchorage spacing varies based on the anchorage resistance to pullout, the resistance of the plate to bending, and strength of the anchorage in shear. Anchorage must be spaced to resist pullout, and the plate must resist bending and splitting. Placing anchor bolts close together assists in reducing the bending stress in the plate.

Figure 9-12.
Wall top plate-to-wall
stud metal connector

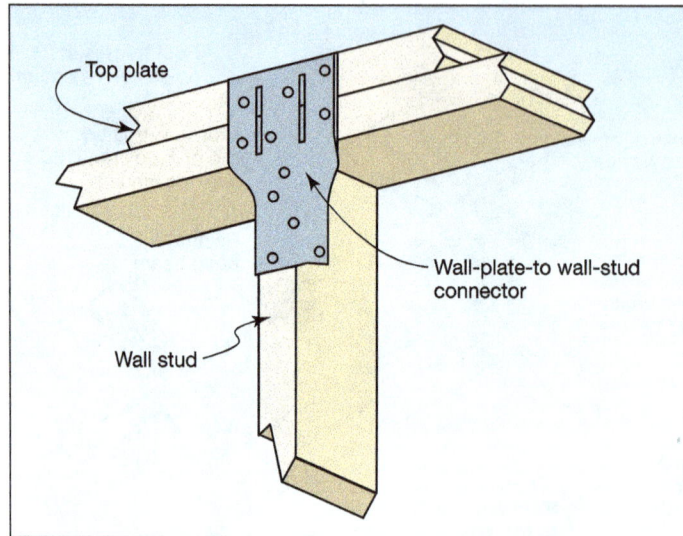

9.1.4   Wall Sheathing to Window Header (Link #4)

Link #4 is the connection between the wood structural panel wall sheathing and the window header (see Figures 9-6 and 9-13). The connection maintains the uplift load path from the wall top plates for the same force as determined for the roof connection to the wall minus additional dead load from the wall. Options for maintaining the uplift load path include using metal connectors between the wall studs and header or wood structural panel sheathing (see Figures 9-13 and 9-14). The uplift load path is frequently made with wood structural panel wall sheathing, particularly when the uplift and shear forces in the wall are not very high. Additional design considerations include the resistance of the window header to bending from gravity loads, wind uplift, and out-of-plane bending loads from wind.

In masonry construction, a masonry or concrete bond beam, or a pre-cast concrete or masonry header, is often used over the window opening. Design considerations for this beam include resistance to bending in both the plane of the wall and normal to the wall. Resistance to bending is accomplished by placing reinforcing steel in the bond beam. Reinforcing steel must be placed in the bond beam in order for the beam to adequately resist bending stresses. The design of these members is beyond the scope of this Manual; therefore, the prescriptive methods presented in ICC 600-2008, or concrete and masonry references should be used.

9.1.5   Window Header to Exterior Wall (Link #5)

Link #5 is the connection from the window header to the adjacent wall framing (see Figures 9-6 and 9-14). Link #5 provides resistance to wind uplift and often consists of a metal strap or end-nailing the stud to the header. The total uplift force is based on the uplift forces tributary to the header. Maintaining the load path for the out-of-plane forces at this location includes consideration of both the positive (inward) and the negative (outward) pressures from wind. This load path is commonly developed by the stud-to-header nailing. One method of sizing the connection between the window header and the exterior wall is provided in Example 9.3.

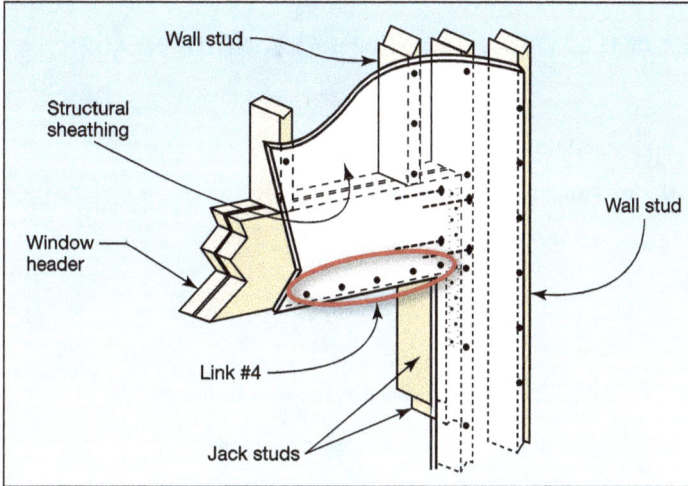

Figure 9-13.
Connection of wall
sheathing to window
header (Link #4)

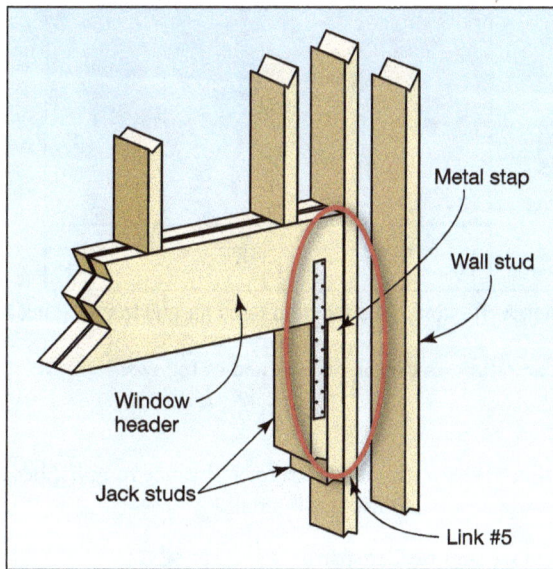

Figure 9-14.
Connection of window
header to exterior wall
(Link #5)

EXAMPLE 9.3. UPLIFT AND LATERAL LOAD PATH AT WINDOW HEADER

Given:

- Refer to Figure 9-14 and Illustration A

- Unit uplift load on window header = 565.2 plf (from Example 9.2)

- Unit lateral load on header = 241.9 plf (from Example 9.2)

- Header span = 14 ft

Find:

- Uplift and lateral load for connection of the header to the wall framing.

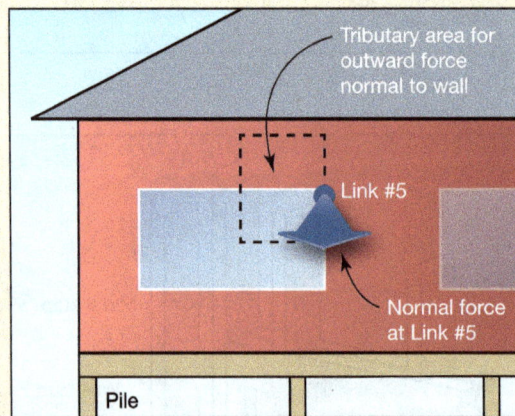

Illustration A. Tributary area for wind force normal to wall (Link #5)

Solution: The uplift and lateral forces can be determined as follows:

Uplift

- Ignore the contribution of the wall's dead load for resistance to uplift because the amount of wall dead load above the header connection is small

$$\text{Uplift load} = \frac{(565.2 \text{ plf})(\text{header span})}{2} = (479 \text{ plf})(7 \text{ ft}) = \mathbf{3,955 \text{ lb}}$$

$$\text{Lateral load} = \frac{(241.9 \text{ plf})(\text{header span})}{2} = (241.9 \text{ plf})(7 \text{ ft}) = \mathbf{1,694 \text{ lb}}$$

9.1.6 Wall to Floor Framing (Link #6)

Link #6 is the connection of the wall framing to the floor framing (see Figures 9-6 and 9-15) for resistance to wind uplift. This connection often includes use of metal connectors between the wall studs and the band joist or wood structural panel sheathing. In addition to uplift, connections between wall and floor framing can be used to maintain the load path for out-of-plane wall forces from positive and negative wind pressures and forces in the plane of the wall from shear. One method of sizing the wind uplift and lateral connections between the wall framing and the floor framing is provided in Example 9.4.

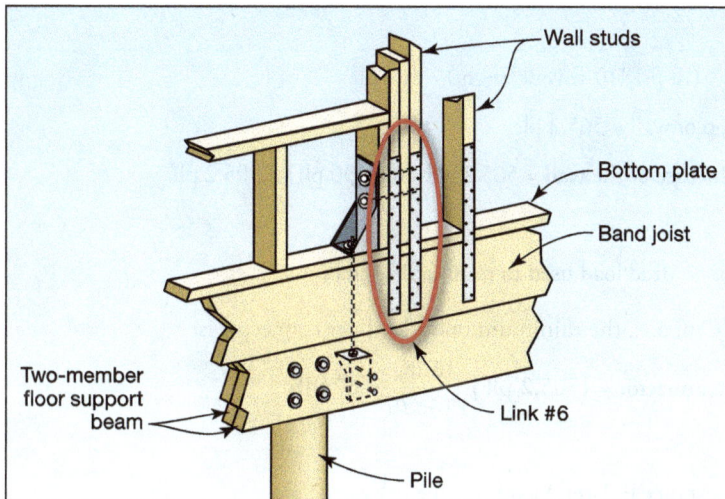

Figure 9-15.
Connection of wall to
floor framing (Link #6)

EXAMPLE 9.4. UPLIFT AND LATERAL LOAD PATH AT WALL-TO-FLOOR FRAMING

Given:

- Refer to Figure 9-15
- Unit uplift load at top of wall 565.2 plf (from Example 9.2)
- Unit lateral load = 241.9 plf (from Example 9.2)
- Wall dead load = 10 psf
- Wall height = 10 ft
- Wood specific gravity, G = 0.42
- Three 16d common stud-to-plate nails per stud to provide resistance to lateral loads
- Two 16d common plate-to-band joist nails per ft to provide resistance to lateral loads

EXAMPLE 9.4. UPLIFT AND LATERAL LOAD PATH AT WALL-TO-FLOOR FRAMING (concluded)

Find:

- Uplift load for wall-to-floor framing connections and if framing connections are adequate to resist the lateral loads.

Solution: Determine the uplift and lateral load for the wall-to-floor framing connections as follows:

Uplift:

Wall dead load = (10 psf)(10-ft wall height) = 100 plf

Uplift load at top of wall = 565.2 plf

Uplift load at the base of the wall = 565.2 plf – 0.6(100 plf) = 505.2 plf

where:

0.6 = load factor on dead load used to resist uplift forces

For connectors spaced at 16 in. o.c., the minimum uplift load per connector is:

$$\text{Uplift load per connector} = \left(505.2 \text{ plf}\right)\frac{16 \text{ in.}}{12 \text{ in./ft}} = \textbf{674 lb}$$

Lateral:

- Stud-to-plate nail resistance to lateral loads can be calculated as:

Lateral resistance = (3 nails/ft)(120 lb/nail)(1.6)(0.67) = 386 lb

where:

1.6 = NDS load duration factor

0.67 = NDS end grain factor

Because studs are at 16 inches o.c., unit lateral load resistance is:

$$\text{Lateral resistance} = \left(386 \text{ lb}\right)\frac{12 \text{ in./ft}}{16 \text{ in.}} = \textbf{289 lb}$$

289 plf > 241.9 plf ✓

- Plate-to-band joist nail resistance to lateral can be calculated as:

Lateral resistance = (2 nails/ft)(120 lb/nail)(1.6) = 384 plf

where:

1.6 = NDS load duration factor

384 plf > 241.9 plf ✓

The wall-to-floor framing connections provide adequate resistance to lateral forces.

9.1.7   Floor Framing to Support Beam (Link #7)

Link #7 is the connection between the floor framing and the floor support beam (see Figures 9-6, 9-16, 9-17, and 9-18). The connection transfers the uplift forces that are calculated in Example 9.4. Options for maintaining the uplift load path for wind uplift include using metal connectors (see Figures 9-16 and 9-17) between the floor joist and the band joist or wood blocking (see Figure 9-18). Connections are also necessary to maintain a load path for lateral and shear forces from the floor and wall framing into the support beam. One method of sizing the wind uplift connections between the floor framing and support beam is provided in Example 9.5.

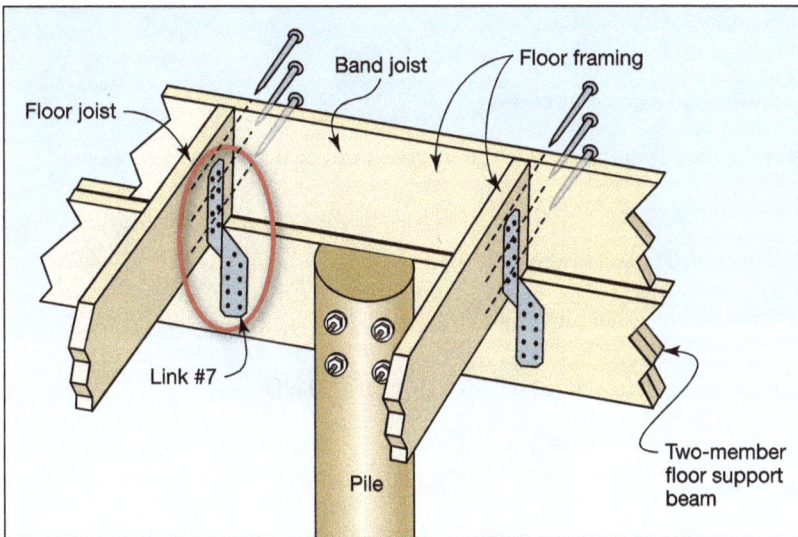

Figure 9-16.
Connection of floor framing to support beam (Link #7) (band joist nailing to the floor joist is adequate to resist uplift forces)

Figure 9-17.
Metal joist-to-beam connector

EXAMPLE 9.5. UPLIFT LOAD PATH AT FLOOR TO SUPPORT BEAM FRAMING

Given:

- Refer to Figure 9-16

- Unit uplift load at top of wall 565.2 plf (from Example 9.2)

- Wall dead load = 10 psf

- Floor dead load = 10 psf

- Wall height = 10 ft

Find:

- Uplift load for floor framing to beam connections

Solution: The uplift load for the floor framing to beam connections can be determined as follows:

Uplift:

$$\text{Wall dead load} = (10 \text{ psf})(10 \text{ ft wall height}) = 100 \text{ plf}$$

$$\text{Floor dead load} = 10 \text{ psf} \frac{14 \text{ ft}}{2} = 70 \text{ plf}$$

$$\text{Uplift load at the base of the floor} = 565.2 \text{ plf} - 0.6 (100 \text{ plf} + 70 \text{ plf}) = 463.2 \text{ plf}$$

where:

0.6 = load factor on dead load used to resist uplift forces

For connectors spaced at 16 in. o.c., the minimum uplift load per connector is:

$$\text{Uplift load per connector} = \left(463.2 \text{ plf}\right) \frac{16 \text{ in.}}{12 \text{ in./ft}} = \mathbf{618\ lb}$$

9.1.8   Floor Support Beam to Foundation (Pile) (Link #8)

Link #8 is the connection of the floor support beam to the top of the pile (see Figures 9-6 and 9-18). Link #8 resists wind uplift forces, and the connection often consist of bolts in the beam-to-pile connection or holddown connectors attached from wall studs above to the pile. One method of sizing the wind uplift connections between the floor support beam and piles is provided in Example 9.6.

The connection of the beam to the pile is also designed to maintain load path for lateral and shear forces. It is typically assumed that lateral and shear forces are transferred through the floor diaphragm and can therefore be distributed to other support beam-to-pile connections. Stiffening of the diaphragm can be achieved by installing braces at each corner pile between the floor support beam in the plane of the floor (see Figure 9-19) or sheathing the underside of the floor framing. Stiffening also reduces pile cap rotation. The load path, however, does not end at Link #8. The load path ends with the transfer of loads from the foundation into the soil. See Chapter 10 for considerations that must be taken into account with regard to the interaction between the foundation members and soil in the load path.

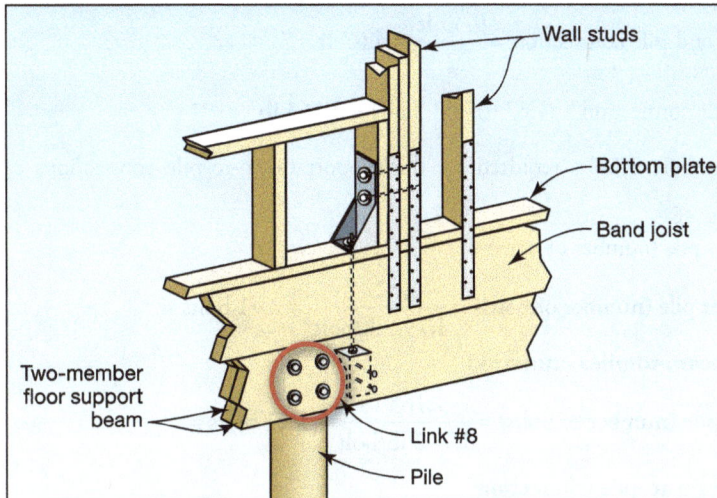

Figure 9-18.
Connection of floor support beam to foundation (Link #8)

EXAMPLE 9.6. UPLIFT LOAD PATH FOR SUPPORT BEAM TO PILE

Given:

- Refer to Figure 9-18

- Unit uplift load at top of floor beams = 463.2 plf (from Example 9.5)

- Pile spacing = 9.33 ft

- Continuous beam of 28-ft length at end wall

- ASD capacity for 1-in. diameter bolt in beam-to-pile connection = 1,792 lb (where wood specific gravity (G) = 0.42, 3.5-in. side member, and 5.25-in. main member

Find:

1. Uplift load for support beam-to-pile connections.

2. Number of bolts required for support beam-to-pile connections for wind uplift.

Solution for #1: The uplift load for the support beam-to-pile connections can be determined as follows:

Uplift:

Tributary length of center pile connection = 9.33 ft

EXAMPLE 9.6. UPLIFT LOAD PATH FOR SUPPORT BEAM-TO-PILE (concluded)

Uplift load at center pile connection = (9.33 ft)(463.2 plf)= **4,322 lb**

$$\text{Tributary length of end pile connection} = \frac{9.33 \text{ ft}}{2} = 4.67 \text{ ft}$$

Uplift load at end pile connection = (4.67 ft)(463.2 plf) = **2,163 lb**

Solution for #2: The number of bolts required for the support beam-to-pile connections can be determined as follows:

Connection at center pile (number of bolts) =

$$\text{Connection at center pile (number of bolts)} = \frac{4,322 \text{ lb}}{1,792 \text{ lb/bolt}} = 2.41 \text{ bolts} =$$

3 bolts at support beam-to-pile connection

$$\text{Connection at end pile (number of bolts)} = \frac{2,163 \text{ lb}}{1,792 \text{ lb/bolt}} = 1.21 \text{ bolts} =$$

2 bolt at support beam-to-pile connection

Figure 9-19.
Diaphragm stiffening and corner pile bracing to reduce pile cap rotation

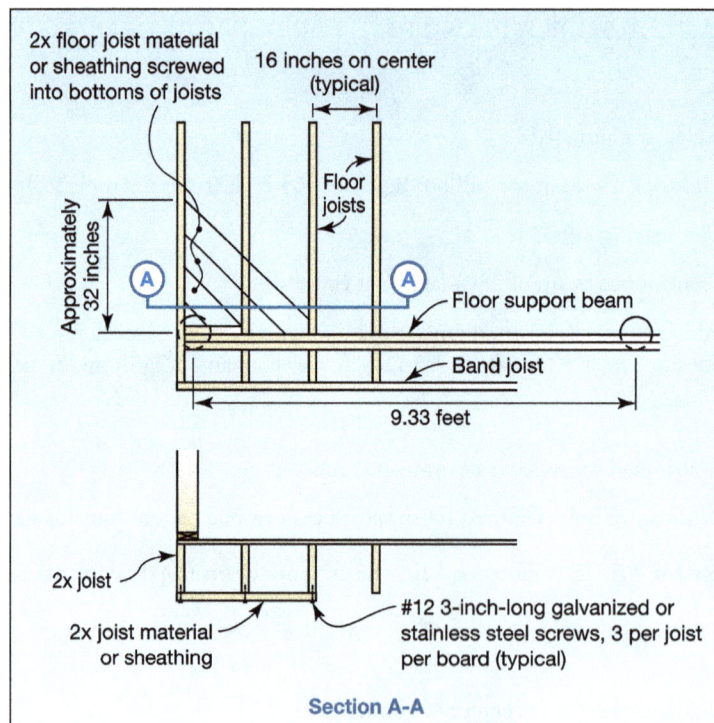

Section A-A

9.2   Other Load Path Considerations

Several additional design considerations must be investigated in order for a design to be complete. The details of these investigations are left to the designer, but they are mentioned here to more thoroughly cover the subject of continuous load paths and to point out that many possible paths require investigation.

Using the example of the building shown in Example 9.3, Illustration A, the following load paths should also be investigated:

- Load paths for shear transfer between shear walls and diaphragms including uplift due to shear wall overturning

- Gable wall support for lateral wind loads

- Uplift of the front porch roof

- Uplift of the main roof section that spans the width of the building

Other factors that influence the building design and its performance are:

- Connection choices

- Building eccentricities

- Framing system

- Roof shape

9.2.1   Uplift Due to Shear Wall Overturning

The shear wall that contains Link #6 includes connections designed to resist overturning forces from wind acting perpendicular to the ridge (see Example 9.7, Illustration A). Calculation of the overturning induced uplift and compressive forces are given in Example 9.7.

EXAMPLE 9.7. UPLIFT AND COMPRESSION DUE TO SHEAR WALL OVERTURNING

Given:

- Refer to Illustration A
- Wind speed = 150 mph, Exposure D
- Mean roof height = 33 ft
- Roof span perpendicular to ridge = 28 ft
- Roof pitch = 7:12
- Wall height = 10 ft

EXAMPLE 9.7. UPLIFT AND COMPRESSION DUE TO SHEAR WALL OVERTURNING
(continued)

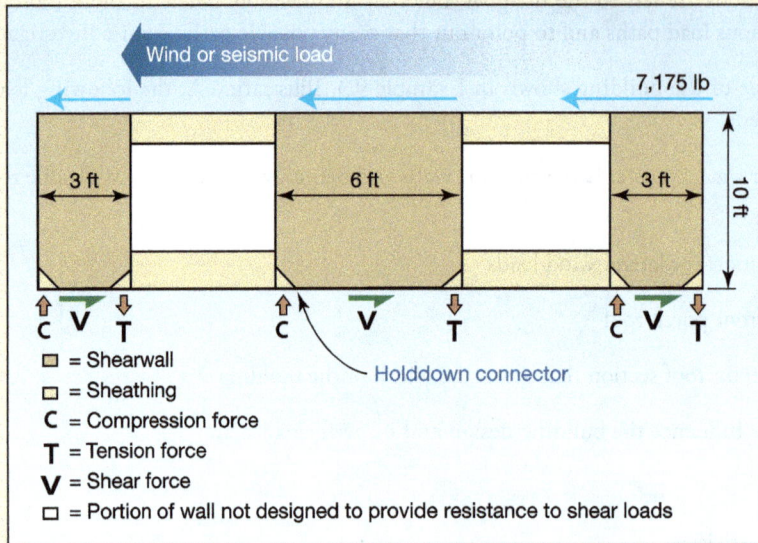

Illustration A. Loads on south shear wall

Find: Uplift and compressive force due to shear wall overturning.

Solution: The uplift and compressive force due to shear wall overturning can be determined as follows:

• The total shear force due to wind acting perpendicular to the ridge is determined for the 28-ft roof span by interpolation from Table 8-7:

Roof diaphragm load for 24-ft roof span = 256 plf

Roof diaphragm load for 32-ft roof span = 299 plf

Roof diaphragm load for 28-ft roof span = $\dfrac{(256 \text{ plf} + 299 \text{ plf})}{2} = 278$ plf

Adjusting the roof diaphragm load to account for the building being located in Exposure Category D:

1.18(278 plf) = 328 plf

To adjust w_{roof} for a wall height of 10 ft because Table 8.7 assumes a wall height of 8 ft

$328 \text{ plf}\left(\dfrac{10 \text{ ft}}{8 \text{ ft}}\right) = 410$ plf

EXAMPLE 9.7. UPLIFT AND COMPRESSION DUE TO SHEAR WALL OVERTURNING
(concluded)

• The total shear load for south wall assuming flexible diaphragm distribution of roof diaphragm load is calculated as follows:

$$\text{Length tributary to shear walls} = \frac{35 \text{ ft}}{2} = 17.5 \text{ ft (see Example 9.3, Illustration A)}$$

Shear load in south shear walls = (17.5 ft)(410 plf) = 7,175 lb

Shear wall segment aspect ratio (see Illustration B):

• Each shear wall segment must meet the requirements for shear wall aspect ratio in order to be considered as a shear resisting element. For wood structural panel shear walls, the maximum ratio of height to length (e.g., aspect ratio, h/L) is 3.5:1.

• The aspect ratio for shear wall segments in Illustration A can be calculated as follows:

$$\text{Aspect ratio of 6-ft long shear wall segment: } \frac{10 \text{ ft}}{6 \text{ ft}} = 1.67 < 3.5 \checkmark$$

$$\text{Aspect ratio of 3-ft long shear wall segment: } \frac{10 \text{ ft}}{3 \text{ ft}} = 3.33 < 3.5 \checkmark$$

$$\text{Unit shear, } v = \text{total shear load/shear wall length} = \frac{7,175 \text{ lb}}{(6 \text{ ft} + 3 \text{ ft} + 3 \text{ ft})} = 598 \text{ plf}$$

Uplift (T) and compressive force (C) at shear wall ends due to overturning
= (598 plf)(10-ft wall height) = **5,980 lb**

Note: As seen in this example, tension and compression forces due to shear wall overturning can be large. Alignment of shear wall end posts with piles below facilitates use of standard connectors and manufacturers' allowable design values. A check of the pile uplift and compressive capacity in soil is needed to ensure an adequate load path for overturning forces.

Because of the magnitude of overturning induced uplift and compression forces, it is desirable to align shear wall ends with piles to provide direct vertical support and to minimize offset of the tension or compression load path from the axis of the pile. Where shear wall end posts are aligned with piles below, detailing that allows connection of the shear wall end post holddown directly to the pile is desirable to minimize forces transferred through other members such as the support beams. Where direct transfer of overturning induced uplift and compression forces into the pile is not possible, minimizing the offset distance reduces bending stresses in the primary support beam (see Figure 9-20). For the holddown connection shown in Figure 9-20, the manufacturers' listed allowable load will be reduced because the bolted connection to the wood beam is loaded perpendicular to grain rather than parallel to grain.

Figure 9-20.
Shear wall holddown
connector with bracket
attached to a wood beam

9.2.2 Gable Wall Support

There are many cases of failures of gable-end frames during high-wind events. The primary failure modes in gable-end frames are as follows:

- A gable wall that is not braced into the structure collapses, and the roof framing falls over (see Figure 9-21)

- An unsupported rake outrigger used for overhangs is lifted off by the wind and takes the roof sheathing with it

- The bottom chord of the truss is pulled outward, twisting the truss and causing an inward collapse

The need for and type of bracing at gable-end frames depend on the method used to construct the gable end. Recommendations for installing rafter outriggers at overhangs for resistance to wind loads are provided in the *Wood Frame Construction Manual* (American Wood Council, 2001). In addition to using the gable-end truss bracing shown in Figure 9-22, installing permanent lateral bracing on all roof truss systems is recommended. Gable-end trusses and conventionally framed gable-end walls should be designed, constructed, and sheathed as individual components to withstand the pressures associated with the established basic wind speed.

9.2.3 Connection Choices

Alternatives for joining building elements include:

- Mechanical connectors such as those available from a variety of manufacturers

- Fasteners such as nails, screws, bolts, and reinforcing steel

CROSS REFERENCE

For recommendations on corrosion-resistant connectors, see Table 1 in NFIP Technical Bulletin 8, *Corrosion Protection for Metal Connectors in Coastal Areas* (FEMA 1996).

- Connectors such as wood blocks

- Alternative materials such as adhesives and strapping

Most commercially available mechanical connectors recognized in product evaluation reports are fabricated metal devices formed into shapes designed to fit snugly around elements such as studs, rafters, and wall plates. To provide their rated load, these devices must be attached as specified by the manufacturer. Mechanical connectors are typically provided with various levels of corrosion resistance such as levels of hot-dip galvanizing and stainless steel. Hot-dip galvanizing may be applied before or after fabrication. Thicker galvanized coatings can consist of 1 to 2 ounces of zinc per square foot. Thicker coatings provide greater protection against corrosion. Welded steel products generally have a hot-dip galvanized zinc coating or are painted for corrosion protection. Stainless steel (A304 and A316) connectors also provide corrosion resistance. Because exposed metal fasteners (even when galvanized) can corrode in coastal areas within a few years of installation, stainless steel is recommended where rapid corrosion is expected. According to FEMA NFIP Technical Bulletin 8-96, the amount of salt spray in the air is greatest near breaking waves and declines with increasing distance away from the shoreline. The decline may be rapid in the first 300 to 3,000 feet. FEMA P-499 recommends using stainless steel within 3,000 feet of the coast (including sounds and back bays).

Metal connectors must be used in accordance with the manufacturer's installation instructions in order for the product to provide the desired strength rating and to ensure that the product is suitable for a particular application. Particular attention should be given to the following information in the installation instructions:

- Preservative treatments used for wood framing

- Level of corrosion protection

- Wood species or lumber type used in framing (e.g., sawn lumber, pre-fabricated wood I-joists, laminated veneer lumber)

Gable-end truss

2x4 Brace

2x Ladder framing

Metal strap

2x4 Blocking

2x4x8 feet
Brace

Braces extend to
fourth truss

No roof sheathing joints parallel to gable-end eave

Roof sheathing
nails at 4 inches
o.c. maximum

2x Ladder framing
at 24 inches o.c.
maximum

Gable-end truss
designed for
end-wall pressure

Engineered wood
roof trusses at
24 inches o.c.
maximum

2x4 continuous
brace

Exterior
structural
sheathing

Metal strap

2x4x8 feet long
brace at 5 feet
4 inches o.c.
maximum

2 – 10d nails at
each truss

2 – 2x4
top plate

Stud wall

2x6 w/10d nails
at 12 inches o.c. to
double top plate
and sheathing

2x4 blocking
between
trusses

Interior finish

Section A-A

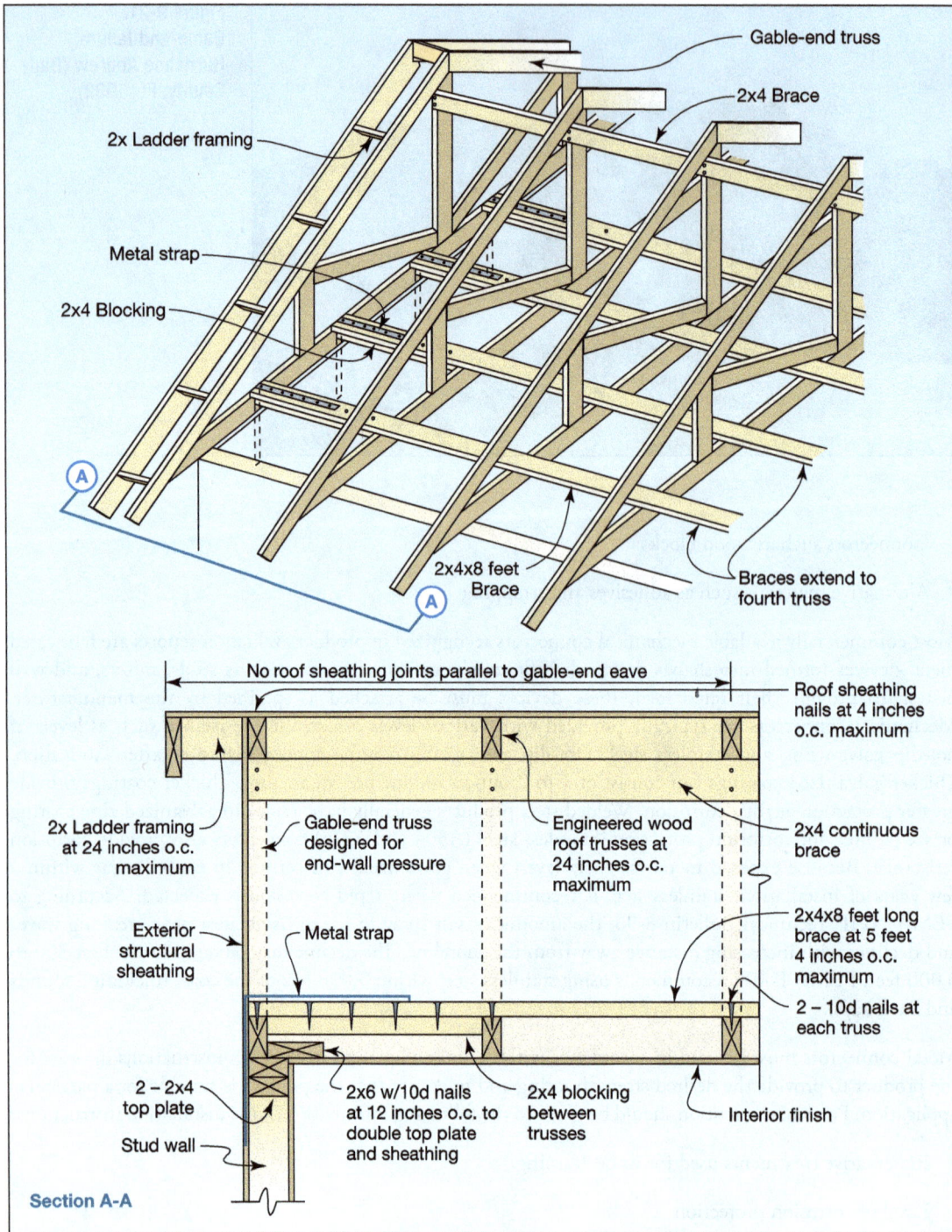

Figure 9-22.
Gable-end bracing detail; nailing schedule, strap specification, brace spacing, and overhang limits should be
adapted for the applicable basic wind speed

- Rated capacity of connector for all modes of failure (e.g., shear, uplift, gravity loading)

- Level of corrosion protection for nails, bolts, and/or screws

- Nail, bolt, and/or screw size and type required to achieve rated loads

9.2.4   Building Eccentricities

The L-shaped building configuration produces stress concentrations in the re-entrant corner of the building structure. Additionally, differences between the center of rotation and the center of mass produce torsional forces that must be transferred by the diaphragms and accounted for in the design of shear walls. Provisions for torsional response are different for wind and seismic hazards. Design methods to account for building eccentricities is beyond the scope of this Manual; therefore, the user is referred to building code requirements and provisions of ASCE 7-10 and applicable material design standards.

9.2.5   Framing System

Methods used for maintaining a load path throughout the structure depend on the framing system or structural system that makes up the building structure. Specifics related to platform framing, concrete/masonry construction, and moment-resistant framing are provided below.

9.2.5.1   Platform Framing

Across the United States, platform framing is by far the most common method of framing a wood-stud or steel-stud residential building. In the platform framing method, a floor assembly consisting of beams, joists, and a subfloor creates a "platform" that supports the exterior and interior walls. The walls are normally laid out and framed flat on top of the floor, tilted up into place, and attached at the bottom to the floor through the wall bottom plate. The walls are attached at the top to the next-level floor framing or in a one-story building to the roof framing. Figure 9-23 is an example of platform framing in a two-story building. This method is commonly used on all types of foundation systems, including walls, piles, piers, and columns consisting of wood, masonry, and concrete materials. Less common framing methods in wood-frame construction are balloon framing in which wall studs are continuous from the foundation to the roof and post-and-beam framing in which a structure of beams and columns is constructed first, including the floors and roof, and then walls are built inside the beam and column structure.

9.2.5.2   Concrete/Masonry

Masonry exterior walls are normally constructed to full height (similar to wood balloon framing), and then wood floors and the roof are framed into the masonry. Fully or partially reinforced and grouted masonry is required in high-wind and seismic hazard areas. Floor framing is normally supported by a ledger board fastened to a bond beam in the masonry, and the roof is anchored to a bond beam at the top of the wall. Connections can be via a top plate as shown in Figure 9-24 or direct embedded truss anchors in the bond beam as shown in

> **NOTE**
>
> Masonry frames typically require continuous footings. However, continuous footings are not allowed in Zone V or Coastal A Zones and are not recommended in Zone A.

Figure 9-23.
Example of two-story
platform framing on a
pile-and-beam foundation

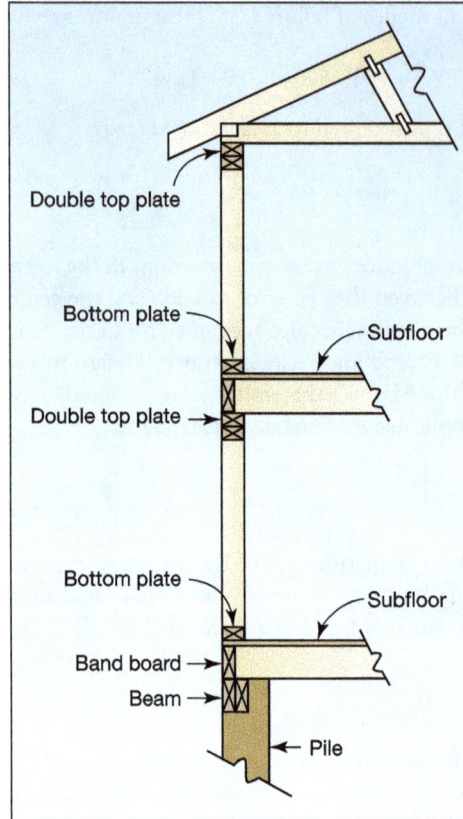

Double top plate

Bottom plate

Subfloor

Double top plate →

Bottom plate

Subfloor

Band board →

Beam →

← Pile

Figure 9-6. Options for end walls are hip roofs, continuous masonry gables, and braced gable frames. Details and design tables for all of the above can be found in ICC 600-2008. Figure 9-24 is an example of masonry wall construction in a two-story building.

9.2.5.3   Moment-Resisting Frames

Over the past few decades, an increasing number of moment-resisting frames have been built and installed in coastal homes (Hamilton 1997). The need for this special design is a result of more buildings in coastal high hazard areas being constructed with large glazed areas on exterior walls, with large open interior areas, and with heights of two to three stories. Figure 9-25 shows a typical steel moment frame.

Large glazed areas pose challenges to the designer because they create:

- Large openings in shear walls

- Large deflection in shear walls

- Difficulties in distributing the shear load to the foundation

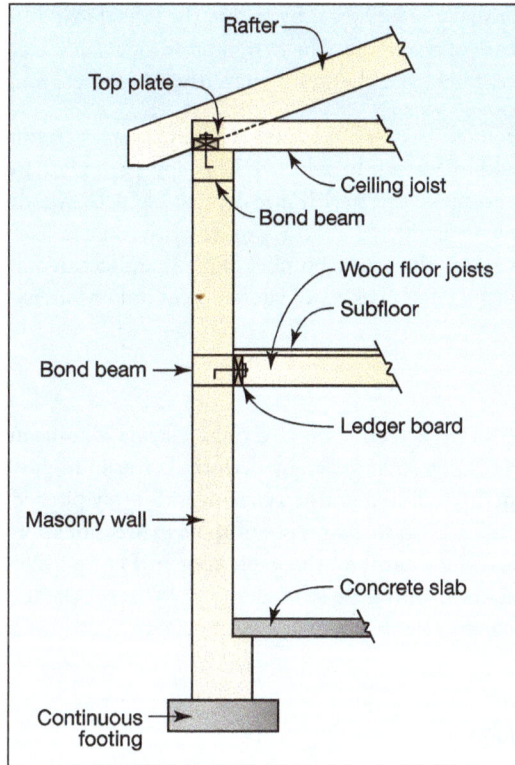

Figure 9-24.
Two-story masonry wall
with wood floor and roof
framing

Figure 9-25.
Steel moment frame with
large opening

• A moment-resisting frame usually resists shear by taking the lateral load into the top of the frame thus creating a moment at the base of the frame. The design professional must design a moment connection at the base between the steel frame and the wood, masonry, or concrete foundation.

In residential construction, moment frames are frequently tubular steel. Tubular steel shapes that are close to the size of nominal framing lumber can be selected. This approach alleviates the need for special, time-consuming methods required to make the steel frame compatible with wood; however, frames made with tubular steel are more difficult to build than frames made with "H" or "WF" flange shapes because all connections in the frame are welded. There are a number of pre-manufactured moment frame products on the market now that have been designed for a variety of lateral forces to fit a variety of wall lengths and heights.

9.2.6 Roof Shape

Roof shape, both the structural aspect and the covering, plays a significant role in roof performance. Compared to other types of roofs, hip roofs generally perform better in high winds because they have fewer sharp corners and fewer distinctive building geometry changes. Steeply pitched roofs usually perform better than flat roofs. Figures 9-26 and 9-27 show two types of roofs in areas of approximately similar terrain that experienced the winds of Hurricane Marilyn. The gable roof in Figure 9-26 failed, while the hip roof in Figure 9-27 survived the same storm with little to no damage. Whether the roof is a gabled roof or hip roof, proper design and construction are necessary for successful performance in high-wind events.

9.3 Breakaway Wall Enclosures

In Zone V and Coastal A Zones, breaking waves are almost certain to occur simultaneously with peak flood conditions. As breaking waves pass an open piling or column foundation, the foundation experiences cyclic fluid impact and drag forces. The flow peaks at the wave crest, just as the wave breaks. Although the flow creates drag on the foundation, most of the flow under the building is undisturbed. This makes open foundations somewhat resistant to wave actions and pile and column foundations a manageable design.

When a breaking wave hits a solid wall, the effect is quite different. When the crest of a breaking wave strikes a vertical surface, a pocket of air is trapped and compressed by the wave. As the air pocket compresses, it exerts a high-pressure burst on the vertical surface, focused at the stillwater level. The pressures can be extreme. For example, a 5-foot wave height can produce a peak force of 4,500 pounds/square foot, roughly 100 times the force caused by a 170-mph wind. These extremely high loads make designing solid foundation walls for small buildings impractical in areas subject to the effects of breaking waves. Prudent design dictates elevating buildings on an open foundation above potential breaking waves. In fact, the 2012 IBC and the 2012 IRC require that new, substantially damaged, and substantially improved buildings in Zone V be elevated above the BFE on an open foundation (e.g., pile, post, column, pier).

The 2012 IBC and 2012 IRC prohibit obstructions below elevated buildings but allow enclosures below the BFE as long as they are constructed with insect screening, lattice, or walls designed and constructed to fail under the loads imposed by floodwaters (termed "breakaway walls"). Because such enclosures fail under flood forces, they do not transfer additional significant loads to the foundation. Regulatory requirements and design criteria concerning enclosures and breakaway walls below elevated buildings in Zone V are discussed in FEMA NFIP Technical Bulletin 9 (FEMA 2008a). Additional guidance is contained in Fact

Figure 9-26.
Gable-end failure caused
by high winds, Hurricane
Marilyn (U.S. Virgin
Islands, 1995)

Figure 9-27.
Hip roof that survived
high winds with little to
no damage, Hurricane
Marilyn (U.S. Virgin
Islands, 1995)

Sheet No. 8.1, *Enclosures and Breakaway Walls* in FEMA P-499. Breakaway walls may be of wood- or metal-frame or masonry construction.

Figure 9-28 shows how a failure begins in a wood-frame breakaway wall. Note the failure of the connection between the bottom plate of the wall and the floor of the enclosed area. Figure 9-29 shows a situation in which utility components placed on and through a breakaway wall prevented it from breaking away cleanly.

To increase the likelihood of collapse as intended, it is recommended that the vertical framing members (such as 2x4s) on which the screen or lattice work is mounted be spaced at least 2 feet apart. Either metal or synthetic screening is acceptable. Wood and plastic lattice is available in 4-foot x 8-foot sheets. The material used to fabricate the lattice should be no thicker than 1/2 inch, and the finished sheet should have an opening ratio of at least 40 percent. Figure 9-30 shows lattice used to enclose an area below an elevated building.

Figure 9-28.
Typical failure mode of breakaway wall beneath an elevated building—failure of the connection between the bottom plate of the wall and the floor of the enclosed area, Hurricane Hugo (South Carolina, 1989)

Figure 9-29.
Breakaway wall panel prevented from breaking away cleanly by utility penetrations, Hurricane Opal (Florida, 1995)

**Figure 9-30.
Lattice beneath an
elevated house in Zone V**

9.4   Building Materials

The choice of materials is influenced by many considerations, including whether the materials will be used above or below the DFE. Below the DFE, design considerations include the risk of inundation by seawater, and the forces to be considered include those from wave action, water velocity, and waterborne debris impact. Materials intermittently wetted by floodwater below the BFE are subject to corrosion and decay.

Above the DFE, building materials also face significant environmental effects. The average wind velocity increases with height above ground. Wind-driven saltwater spray can cause corrosion and moisture intrusion. The evaporation of saltwater leaves crystalline salt that retains water and is corrosive.

Each type of commonly used material (wood, concrete, steel, and masonry) has both characteristics that can be advantageous and that can require special consideration when the materials are used in the coastal environment (see Table 9-1). A coastal residential structure usually has a combination of these materials.

Table 9-1. General Guidance for Selection of Materials

Material	Advantages	Special Considerations
Wood	• Generally available and commonly used • With proper design, can generally be used in most structural applications • Variety of products available • Can be treated to resist decay • Some species are naturally decay-resistant	• Easily over-cut, over-notched, and over-nailed • Requires special treatment and continued maintenance to resist decay and damage from termites and marine borers • Requires protection to resist weathering • Subject to warping and deterioration
Steel	• Used for forces that are larger than wood can resist • Can span long distances • Can be coated to resist corrosion	• Not corrosion-resistant • Heavy and not easily handled and fabricated by carpenters • May require special connections such as welding

Table 9-1. General Guidance for Selection of Materials (concluded)

Material	Advantages	Special Considerations
Reinforced Concrete	• Resistant to corrosion if reinforcing is properly protected • Good material for compressive loads • Can be formed into a variety of shapes • Pre-stressed members have high load capacity	• Saltwater infiltration into concrete cracks causes reinforcing steel corrosion • Pre-stressed members require special handling • Water intrusion and freeze-thaw cause deterioration and spalling
Masonry	• Resistant to corrosion if reinforcing is properly protected • Good material for compressive loads • Commonly used in residential construction	• Not good for beams and girders • Water infiltration into cracks causes reinforcing steel corrosion • Requires reinforcement to resist loads in coastal areas

9.4.1   Materials Below the DFE

The use of flood-resistant materials below the BFE is discussed in FEMA NFIP Technical Bulletin 2 (FEMA 2008b). According to the bulletin, "All construction below the lowest floor is susceptible to flooding and must consist of flood-resistant materials. Uses of enclosed areas below the lowest floor in a residential building are limited to parking, access, and limited storage—areas that can withstand inundation by floodwater without sustaining significant structural damage." The 2012 IBC and 2012 IRC require that all new construction and substantial improvements in the SFHA be constructed with materials that are resistant to flood damage. Compliance with these requirements in coastal areas means that the only building elements below the BFE are:

NOTE

Although NFIP regulations, 2012 IBC, and 2012 IRC specify that flood-resistant materials be used below the BFE, in this Manual, flood-resistant materials below the DFE are recommended.

- Foundations – treated wood; concrete or steel piles; concrete or masonry piers; or concrete, masonry, or treated wood walls

- Breakaway walls

- Enclosures used for parking, building access, or storage below elevated buildings

- Garages in enclosures under elevated buildings or attached to buildings

- Access stairs

CROSS REFERENCE

For NFIP compliance provisions as described in the 2012 IBC and the 2012 IRC, see Chapter 5 of this Manual.

Material choices for these elements are limited to materials that meet the requirements provided in FEMA NFIP Technical Bulletin 2. Even for materials meeting those requirements, characteristics of various materials can be advantageous or may require special consideration when the materials are used for

CROSS REFERENCE

For examples of flood insurance premiums for buildings in which the lowest floor is above the BFE and in which there is an enclosure below the BFE, see Table 7-2 in Chapter 7.

different building elements. Additional information about material selection for various locations and uses in a building is included in "Material Durability in Coastal Environments," available on the Residential Coastal Construction Web site (http://www.fema.gov/rebuild/mat/fema55.shtm).

9.4.2   Materials Above the DFE

Long-term durability, architectural, and structural considerations are normally the most important factors in material selection. Material that will be used in a coastal environment will be subjected to weathering, corrosion, termite damage, and decay from water infiltration, in addition to the stresses induced by loads from natural hazard events. These influences are among the considerations for selecting appropriate materials. "Material Durability in Coastal Environments" contains additional information about a variety of wood products and the considerations that are important in their selection and use.

9.4.3   Material Combinations

Materials are frequently combined in the construction of a single residence. The most common combinations are as follows:

- Masonry or concrete lower structure with wood on upper level

- Wood piles supporting concrete pile caps and columns that support a wood superstructure

- Steel framing with wood sheathing

For the design professional working with of coastal buildings, important design considerations when combining materials include:

1. The compatibility of metals is a design consideration because dissimilar metals that are in contact with each other may corrode in the presence of salt and moisture. "Material Durability in Coastal Environments" addresses a possible problem when galvanized fasteners and hardware are in contact with certain types of treated wood.

2. Connecting the materials together is crucial. Proper embedment of connectors (if into concrete or masonry) and proper placement of connectors are necessary for continuity of the vertical or horizontal load path. Altering a connector location after it has been cast into concrete or grout is a difficult and expensive task.

3. Combining different types of material in the same building adds to construction complexity and necessitates additional skills to construct the project. Figure 9-31 shows a coastal house being constructed with preservative-treated wood piles that support a welded steel frame, resulting in metal coming into direct contact with treated wood.

4. Material properties, such as stiffness of one material relative to another, affect movement or deflection of one material relative to the other.

Figure 9-31.
House being constructed
with a steel frame on
wood piles

9.4.4   Fire Safety Considerations

Designing and constructing townhouses and low-rise multi-family coastal buildings to withstand natural hazards and meet the building code requirements for adequate fire separation presents some challenges. Although fire separation provisions of the 2012 IBC and 2012 IRC differ, they both require that the common walls between living units be constructed of materials that provide a minimum fire resistance rating. The intent is for units to be constructed so that if a fire occurs in one unit, the structural frame of that unit would collapse within itself and not affect either the structure or the fire resistance of adjacent units.

For townhouse-like units, the common framing method is to use the front and rear walls for the exterior load-bearing walls so that firewalls can be placed between the units. Beams that are parallel to the front and rear exterior walls are typically used to provide support for these walls as well as the floor framing. Figure 9-32 illustrates a framing system for a series of townhouses in which floor beams are perpendicular to the primary direction of flood forces. Design issues include the following:

5. The floor support beams are parallel to the shore and perpendicular to the expected flow and may therefore create an obstruction during a greater-than-design flood event.

6. The fire separation between townhouse units limits options for structural connections between units, making the transfer of lateral loads to the foundation more difficult to achieve.

7. The exposed undersides of buildings elevated on an open foundation (e.g., pile, pier, post, column) must be protected with a fire-rated material. Typically, this is accomplished with use of fire-resistant gypsum board; however, gypsum board is not a flood-damage-resistant material. An alternative approach is to use other materials such as cement-fiber board (with appropriate fire rating), which has a greater resistance to damage from floodwaters, and fire retardant treated wood. Other alternative materials or methods of protection that are flood-damage-resistant may be required in order to meet the competing demands of flood- and fire-resistance.

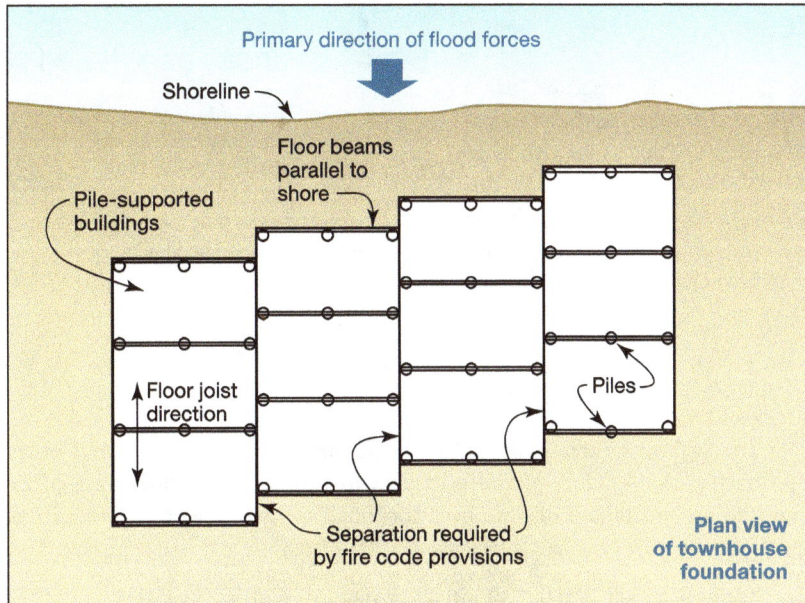

**Figure 9-32.
Townhouse framing
system**

8. The requirement for separation of the foundation elements between townhouse units makes structural rigidity in the direction parallel to the shore more difficult to achieve. If the houses in Figure 9-32 were in a seismic hazard area, the designer could decide to place diagonal bracing parallel to the shore (i.e., perpendicular to the primary flood flow direction) or use more closely spaced and larger piles. Diagonal bracing would provide rigidity but would also create an obstruction below the DFE. The design professional should consult FEMA NFIP Technical Bulletin 5 (FEMA 2008c) for information about the types of construction that constitute an obstruction.

One solution to some of the issues illustrated by Figure 9-32 would be to use two parallel independent walls to provide the required fire separation between units. Each wall could be attached to the framing system of the unit on one side of the separation and supported by a beam running perpendicular to the shore and bearing on the open foundation of that unit.

9.4.5   Corrosion

Modern construction techniques often rely heavily on metal fasteners and connectors to resist the forces of various coastal hazards. To be successful, these products must have lifetimes that are comparable to those of the other materials used for construction. Near saltwater coastlines, corrosion has been found to drastically shorten the lifetime of standard fasteners and connectors. Corrosion is one of the most underestimated

CROSS REFERENCE

For additional information about corrosion of metal connectors in coastal construction, see FEMA NFIP Technical Bulletin 8-96.

hazards affecting the overall strength and lifetime of coastal buildings. To be successful, hazard-resistant buildings must match the corrosion exposure of each element with the proper corrosion-resistant material.

9.5   Appurtenances

The NFIP regulations define "appurtenant structure" as "a structure which is on the same parcel of property as the principal structure to be insured and the use of which is incidental to the use of the principal structure" (44 CFR § 59.1). In this Manual, "appurtenant structure" means any other building or constructed element on the same property as the primary building, such as decks, covered porches, access to elevated buildings, pools, and hot tubs.

> **CROSS REFERENCE**
>
> For additional information about the types of building elements that are allowed below the BFE and for respective site development issues, see FEMA NFIP Technical Bulletin 5.

9.5.1   Decks and Covered Porches Attached to Buildings

Many decks and other exterior attached structures have failed during hurricanes. For decks and other structures without roofs, the primary cause of failure has been inadequate support: the pilings have either not been embedded deep enough to prevent failure or have been too small to carry the large forces from natural hazards.

The following are recommendations for designing decks and other exterior attached structures:

- If a deck is structurally attached to a structure, the bottom of the lowest horizontal supporting member of the deck must be at or above the BFE. Deck supports that extend below the BFE (e.g., pilings, bracing) must comply with Zone V design and construction requirements. The structure must be designed to accommodate any increased loads resulting from the attached deck.

- Some attached decks are located above the BFE but rely on support elements that extend below the BFE. These supports must comply with Zone V design and construction requirements.

- If a deck or patio (not counting its supports) lies in whole or in part below the BFE, it must be structurally independent from the structure and its foundation system.

- If the deck surface is constructed at floor level, the deck surface/floor level joint provides a point of entry for wind-driven rain. This problem can be eliminated by lowering the deck surface below the floor level.

- If deck dimensions can be accommodated with cantilevering from the building, this eliminates the need for piles altogether and should be considered when the deck dimensions can be accommodated with this structural technique. Caution must be exercised with this method to keep water out of the house framing. Chapter 11 discusses construction techniques for flashing cantilever decks that minimize water penetration into the house.

> **WARNING**
>
> Decks should not cantilever over bulkheads or retaining walls where waves can run up the vertical wall and under the deck.

- Exposure to the coastal environment is severe for decks and other exterior appurtenant structures. Wood must be preservative-treated or naturally decay resistant, and fasteners must be corrosion resistant.

9.5.1.1   Handrails

To minimize the effects of wind pressure, flood forces, and wave impacts, deck handrails should be open and have slender vertical or horizontal members spaced in accordance with the locally adopted building code. Many deck designs include solid panels (some made of impact-resistant glazing) between the top of the deck handrail and the deck. These solid panels must be able to resist the design wind and flood loads (below the DFE) or they will become debris.

9.5.1.2   Stairways

Many coastal homes have stairways leading to ground level. During flooding, flood forces often move the stairs and frequently separate them from the point of attachment. When this occurs, the stairs become debris and can cause damage to nearby houses and other buildings. Recommendations for stairs that descend below the BFE include the following:

- To the extent permitted by code, use open-riser stairs to let floodwater through the stair stringers and anchor the stringers to a permanent foundation by using, for example, piles driven to a depth sufficient to prevent failure from scour.

- Extend the bottom of the stair carriages several feet below grade to account for possible scour. Stairs constructed in this fashion are more likely to remain in place during a coastal hazard event and therefore more likely to be usable for access after the event. In addition, by decreasing the likelihood of damage, this approach reduces the likelihood of the stairs becoming debris.

9.5.2   Access to Elevated Buildings

The first floor of buildings in the SFHA is elevated from a few feet to many feet above the exterior grade in order to protect the building and its contents from flood damage. Buildings in Zone A may be only a few feet above grade; buildings in Zone V may be 8 feet to more than 12 feet above grade. Access to these elevated buildings must be provided by one or more of the following:

- Stairs

- Ramps

- Elevator

Stairs must be constructed in accordance with the local building code so that the run and rise of the stairs conform to the requirements. The 2012 IBC and 2012 IRC require a minimum run of 11 inches per stair tread and a maximum rise of 7 inches per tread. An 8-foot elevation difference requires 11 treads or almost 12 feet of horizontal space for the stairs. Local codes also have requirements concerning other stair characteristics, such as stair width and handrail height.

Ramps that comply with regulations for access by persons with disabilities must have a maximum slope of 1:12 with a maximum rise of 30 inches and a maximum run of 30 feet without a level landing. The landing length must be a minimum of 60 inches. As a result, access ramps are generally not practical for buildings elevated more than a few feet above grade and then only when adequate space is available.

Elevators are being installed in many one- to four-family residential structures and provide an easy way to gain access to elevated floors of a building (including the first floor). There must be an elevator entrance on the lowest floor; therefore, in flood hazard areas, some of the elevator equipment may be below the BFE. FEMA's NFIP Technical Bulletin 4 (FEMA 2010a) provides guidance on how to install elevators so that damage to elevator elements is minimized during a flood.

CROSS REFERENCE

For more information about elevator installation in buildings located in SFHAs, see FEMA NFIP Technical Bulletin 4.

9.5.3   Pools and Hot Tubs

Many homes at or near the coast have a swimming pool or hot tub as an accessory. Some of the pools are fiberglass and are installed on a pile-supported structural frame. Others are in-ground concrete pools. The design professional should consider the following when a pool is to be installed at a coastal home:

- Only an in-ground pool may be constructed beneath an elevated Zone V building. In addition, the top of the pool and the accompanying deck or walkway must be flush with the existing grade, and the area below the lowest floor of the building must remain unenclosed.

- Enclosures around pools beneath elevated buildings constitute recreational use and are therefore not allowed, even if constructed to breakaway standards. Lattice and insect screening are allowed because they do not create an enclosure under a community's NFIP-compliant floodplain management ordinance or law.

- A pool adjacent to an elevated Zone V building may be either constructed at grade or elevated. Elevated pools must be constructed on an open foundation and the bottom of the lowest horizontal structural member must be at or above the DFE so that the pool will not act as an obstruction.

- The designer must assure community officials that a pool beneath or adjacent to an elevated Zone V building will not be subject to breaking up or floating out of the ground during a coastal flood

NOTE

Check with local floodplain management officials for information about regulations governing the disturbance of primary frontal dunes. Such regulations can affect various types of coastal construction, including the installation of appurtenant structures such as swimming pools.

NOTE

The construction of pools below or adjacent to buildings in coastal high hazard areas must meet the requirements presented in FEMA NFIP Technical Bulletin 5. In general, pools must be (1) elevated above the BFE on an open foundation or (2) constructed in the ground in such a way as to minimize the effects of scour and the potential for the creation of debris.

and will therefore not increase the potential for damage to the foundations and elevated portions of any nearby buildings. If an in-ground pool is constructed in an area that can be inundated by floodwaters, the elevation of the pool must account for the potential buoyancy of the pool. If a buoyancy check is necessary, it should be made with the pool empty. In addition, the design professional must design and site the pool so that any increased wave or debris impact forces will not affect any nearby buildings.

- Pools and hot tubs have water pumps, piping, heaters, filters, and other equipment that is expensive and that can be damaged by floodwaters and sediment. All such equipment should be placed above the DFE where practical.

- Equipment required for fueling the heater, such as electric meters or gas tanks, should be placed above the DFE. It may also be necessary to anchor the gas tank to prevent a buoyancy failure.

- If buried, tanks must not be susceptible to erosion and scour and thus failure of the anchoring system.

The design intent for concrete pools includes the following:

- Elevation of an in-ground pool should be such that scour will not permit the pool to fail from either normal internal loads of the filled pool or from exterior loads imposed by the flood forces.

- The pool should be located as far landward as possible and should be oriented in such a way that flood forces are minimized. One way to minimize flood forces includes placing the pool with the narrowest dimension facing the direction of flow, orienting the pool so there is little to no angle of attack from floodwater, and installing a pool with rounded instead of square corners. All of these design choices reduce the amount of scour around the pool and improve the chances the pool will survive a storm. These concepts are illustrated in Figure 9-33.

- A concrete pool deck should be frangible so that flood forces create concrete fragments that help reduce scour. The concrete deck should be installed with no reinforcing and should have contraction joints placed at 4-foot squares to "encourage" failure. See Figure 9-34 for details on constructing a frangible concrete pad.

- Pools should not be installed on fill in or near Zone V. Otherwise, a pool failure may result from scour of the fill material.

For concrete pools, buoyancy failure is also possible when floodwaters cover the pool. In addition, flood flows can scour the soil surrounding a buried pool and tear the pool from its anchors. When this happens, the pieces of the pool become large waterborne debris.

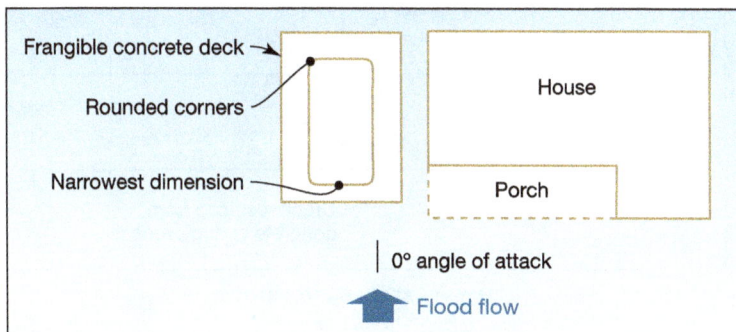

Figure 9-33.
Recommendations for orientation of in-ground pools

Figure 9-34. Recommended contraction joint layout for frangible slab-on-grade below elevated building

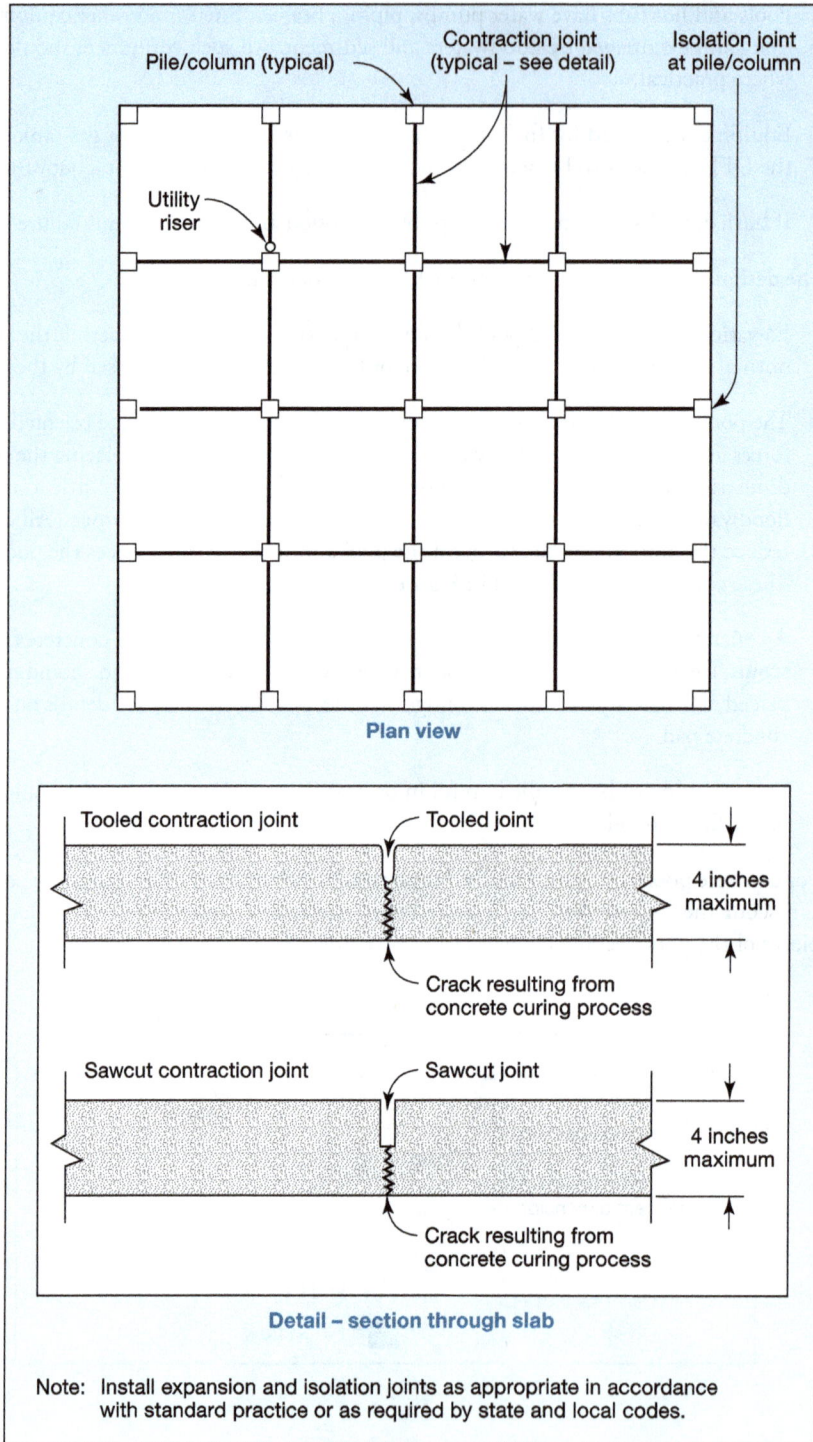

Plan view

Tooled contraction joint — Tooled joint — 4 inches maximum — Crack resulting from concrete curing process

Sawcut contraction joint — Sawcut joint — 4 inches maximum — Crack resulting from concrete curing process

Detail – section through slab

Note: Install expansion and isolation joints as appropriate in accordance with standard practice or as required by state and local codes.

9.6   References

AF&PA (American Forest & Paper Association). 2008. *Special Design Provisions for Wind and Seismic.* ANSI/AF&PA SDPWS-08.

AF&PA. 2012. *Wood Frame Construction Manual for One- and Two-Family Dwellings.* Washington, DC.

ASCE (American Society of Civil Engineers). 2010. *Minimum Design Loads for Buildings and Other Structures,* ASCE Standard ASCE 7-10.

AWC (American Wood Council). 2001. *Wood Frame Construction Manual.*

AWC. 2006. *National Design Specification for Wood Construction (NDS).*

FEMA (Federal Emergency Management Agency). 1996. *Corrosion Protection of Metal Connectors in Coastal Areas for Structures Located in Special Flood Hazard Areas.* NFIP Technical Bulletin 8-96.

FEMA. 2008a. *Design and Construction Guidance for Breakaway Walls Below Elevated Coastal Buildings,* FEMA NFIP Technical Bulletin 9.

FEMA. 2008b. *Flood Damage-Resistant Materials Requirements,* FEMA NFIP Technical Bulletin 2.

FEMA. 2008c. *Free of Obstruction Requirements for Buildings Located in Coastal High Hazard Areas.* NFIP Technical Bulletin 5.

FEMA. 2009. *Local Officials Guide for Coastal Construction.* FEMA 762.

FEMA. 2010a. *Elevator Installation for Buildings Located in Special Flood Hazard Areas in accordance with the National Flood Insurance Program.* NFIP Technical Bulletin 4.

FEMA. 2010b. *Home Builder's Guide to Coastal Construction Technical Fact Sheets.* FEMA P-499.

Hamilton, P. 1997. "Installing a Steel Moment Frame." *Journal of Light Construction.* March.

ICC (International Code Council). 2008. *Standard for Residential Construction in High-Wind Regions.* ICC 600-2008. ICC: Country Club Hills, IL.

ICC. 2011a. *International Building Code.* 2012 IBC. ICC: Country Club Hills, IL.

ICC. 2011b. *International Residential Code for One-and Two-Family Dwellings.* 2012 IRC. ICC: Country Club Hills, IL.

WPPC (Wood Products Promotion Council). 1996. *Guide to Wood Construction in High Wind Areas.*

Designing the Foundation

This chapter provides guidance on designing foundations, including selecting appropriate materials, in coastal areas. It provides general guidance on designing foundations in a coastal environment and is not intended to provide complete guidance on designing foundations in every coastal area. Design professionals should consult other guidance documents, codes, and standards as needed.

CROSS REFERENCE

For resources that augment the guidance and other information in this Manual, see the Residential Coastal Construction Web site (http://www.fema.gov/rebuild/mat/fema55.shtm).

Design considerations for foundations in coastal environments are in many ways similar to those in inland areas. Like all foundations, coastal foundations must support gravity loads, resist uplift and lateral loads, and maintain lateral and vertical load path continuity from the elevated building to the soils below. Foundations in coastal areas are different in that they must generally resist higher winds, function in a corrosive environment, and withstand the environmental aspects that are unique to coastal areas: storm surges, rapidly moving floodwaters, wave action, and scour and erosion. These aspects can make coastal flooding more damaging than inland flooding.

Like many design processes, foundation design is an iterative process. First, the loads on the elevated structure are determined (see Chapter 9). Then a preliminary foundation design is considered, flood loads on the preliminary design are determined, and foundation style is chosen and the respective elements are sized to resist those loads. With information on foundation size, the design professional can accurately determine flood loads on the foundation and can, through iteration, develop an efficient final design.

Because flood loads depend greatly on the foundation design criteria, the discussion of foundation design begins there. The appropriate styles of foundation are then discussed and how the styles can be selected to reduce vulnerability to natural hazards.

The distinction between *code requirements* and *best practices* is described throughout the chapter.

10.1   Foundation Design Criteria

Foundations should be designed in accordance with the latest edition of the 2012 IBC or the 2012 IRC and must address any locally adopted building ordinances. Designers will find that other resources will likely be needed in addition to the building codes in order to properly design a coastal foundation. These resources are listed at the end of this chapter. Properly designed and constructed foundations are expected to:

- Support the elevated building and resist all loads expected to be imposed on the building and its foundation during a design flood, wind, or seismic event

- In SFHAs, prevent flotation, collapse, and lateral movement of the building

- Function after being exposed to scour and erosion

In addition, the foundation must be constructed with flood-resistant materials below the BFE. See Technical Bulletin 2, *Flood Damage-Resistant Materials Requirements* (FEMA 2008a), and Fact Sheet 1.7, *Coastal Building Materials,* in FEMA P-499 (FEMA 2011).

Some coastal areas mapped as Zone A are referred to as "Coastal A Zones." Following Hurricane Katrina (2005), Coastal A Zones have also been referred to as areas with a Limit of Moderate Wave Action (LiMWA). Buildings in Coastal A Zones may be subjected to damaging waves and erosion and, when constructed to minimum NFIP requirements for Zone A, may sustain major damage or be destroyed during the base flood. Therefore, in this Manual, foundations for buildings in Coastal A Zones are strongly recommended to be designed and constructed with foundations that resist the damaging effects of waves.

> **TERMINOLOGY: LiMWA AND COASTAL A ZONE**
>
> Limit of Moderate Wave Action (LiMWA) is an advisory line indicating the limit of the 1.5-foot wave height during the base flood. FEMA requires new flood studies in coastal areas to delineate the LiMWA.

10.2   Foundation Styles

In this Manual, foundations are described as open or closed and shallow or deep. The open and closed descriptions refer to the above-grade portion of the foundation. The shallow and deep descriptions refer to the below-grade portion. Foundations can be open and deep, open and shallow, or closed and shallow. Foundations can also be closed and deep, but these foundations are relatively rare and generally found only in areas where (1) soils near the surface are relatively weak (700 pounds/square foot bearing capacity or less), (2) soils near the surface contain expansive clays (also called shrink/swell soils) that shrink when dry and swell when wet, or (3) other soil conditions exist that necessitate foundations that extend into deep soil strata to provide sufficient strength to resist gravity and lateral loads.

Open, closed, deep, and shallow foundations are described in the following subsections.

10.2.1   Open Foundations

An open foundation allows water to pass through the foundation of an elevated building, reducing the lateral flood loads the foundation must resist. Examples of open foundations are pile, pier, and column foundations. An open foundation is designed and constructed to minimize the amount of vertical surface area that is exposed to damaging flood forces. Open foundations have the added benefit of being less susceptible than closed foundations to damage from flood-borne debris because debris is less likely to be trapped.

Open foundations are required in Zone V and recommended in Coastal A Zone. Table 10-1 shows the recommended practices in Coastal A Zone and Zone V.

Table 10-1. Foundation Styles in Coastal Areas

Foundation Style	Zone V	Coastal A Zone (LiMWA)	Zone A
Open/deep	Acceptable	Acceptable	Acceptable
Open/shallow	Not permitted	Acceptable[(a)]	Acceptable
Closed/shallow	Not permitted	Not recommended	Acceptable
Closed/deep	Not permitted	Not recommended	Acceptable

LiMWA = Limit of Moderate Wave Action

(a) Shallow foundations in Coastal A Zone are acceptable only if the maximum predicted depth of scour and erosion can be accurately predicted and foundations can be constructed to extend below that depth.

10.2.2   Closed Foundations

A closed foundation is typically constructed using continuous perimeter foundation walls. Examples of closed foundations are crawlspace foundations and stem wall foundations,[1] which are usually filled with compacted soil. Slab-on-grade foundations are also considered closed.

A closed foundation does not allow water to pass easily through the foundation elements below an elevated building. Thus, these types of foundations obstruct floodwater flows and present a large surface area upon which waves and flood forces act. Closed foundations are prohibited in Zone V and are not recommended in Coastal A Zones. If perimeter walls enclose space below the DFE, they must be equipped with openings that allow floodwaters to flow in and out of the area enclosed by the walls (see Figure 2-19). The entry and exit of floodwater equalizes the water pressure on both sides of the wall and reduces the likelihood that the wall will fail. See Fact Sheet No. 3.5, *Foundation Walls,* in FEMA P-499, *Home Builder's Guide to Coastal Construction Technical Fact Sheet Series* (FEMA 2010).

Closed foundations also create much larger obstructions to moving floodwaters than open foundations, which significantly increases localized scour. Scour, with and without generalized erosion, can remove soils that support a building and can undermine the foundation and its footings. Once undermined, shallow footings readily fail (see Figure 10-1).

---

1 Stem wall foundations (in some areas, referred to as chain wall foundations) are similar to crawlspace foundations where the area enclosed by the perimeter walls are filled with compacted soil. Most stem wall foundations use a concrete slab-on-grade for the first floor. The NFIP requires flood vents in crawlspace foundations but not in stem wall foundations (see Section 6.1.1.1 and Section 7.6.1.1.5).

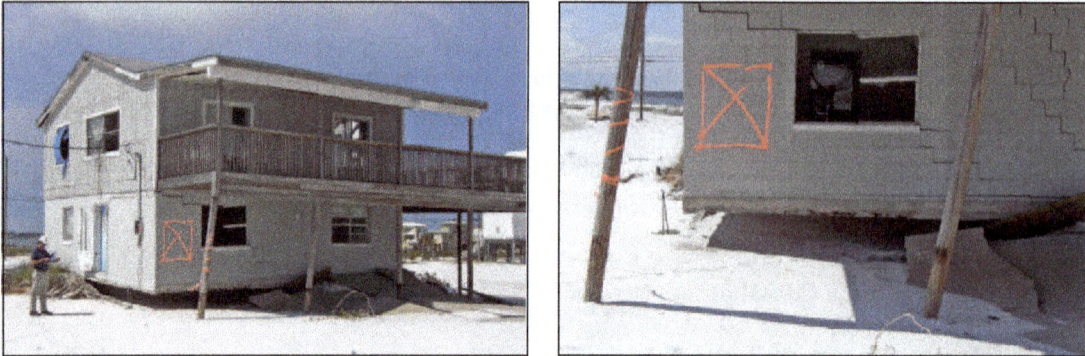

Figure 10-1.
Closed foundation failure due to erosion and scour undermining; photograph on right shows a close-up view of the foundation failure and damaged house wall, Hurricane Dennis (Navarre Beach, FL, 2005)

10.2.3 Deep Foundations

Buildings constructed on deep foundations are supported by soils that are not near grade. Deep foundations include driven timber, concrete or steel piles, and caissons.

Deep foundations are much more resistant to the effects of localized scour and generalized erosion than shallow foundations. Because of that, deep foundations are required in Zone V where scour and erosion effects can be extreme. Open/deep foundations are recommended in Coastal A Zones and in some riverine areas where scour and erosion can undermine foundations.

10.2.4 Shallow Foundations

Buildings constructed on shallow foundations are supported by soils that are relatively close to the ground surface. Shallow foundations include perimeter strip footings, monolithic slabs, discrete pad footings, and some mat foundations. Because of their proximity to grade, shallow foundations are vulnerable to damage from scour and erosion, and because of that, they are not allowed in Zone V and are not recommended in Coastal A Zones unless they extend below the maximum predicted scour and erosion depth.

In colder regions, foundations are typically designed to extend below the frost depth, which can exceed several feet below grade. Extending the foundation below the frost depth is done to prevent the foundation from heaving when water in the soils freeze and to provide adequate protection from scour and erosion. However, scour and erosion depths still need to be investigated to ensure that the foundation is not vulnerable to undermining.

10.3 Foundation Design Requirements and Recommendations

Foundations in coastal areas must elevate the home to satisfy NFIP criteria. NFIP criteria vary for Zone V and Zone A. In Zone V, the NFIP requires that the building be elevated so that the bottom of the lowest horizontal structural member is elevated to the BFE. In Zone A, the NFIP requires that the home be constructed such that the top of the lowest floor is elevated to the BFE.

In addition to elevation, the NFIP contains other requirements regarding foundations. Because of the increased flood, wave, flood-borne debris, and erosion hazards in Zone V, the NFIP requires homes to be elevated on open/deep foundations that are designed to withstand flood forces, wind forces, and forces for flood-borne debris impact. They must also resist scour and erosion.

10.3.1 Foundation Style Selection

Many foundation designs can be used to elevate buildings to the DFE. Table 10-1 shows which foundation styles are acceptable, not recommended, or not permitted in Zone V, Coastal A Zone, and Zone A. Additional information concerning foundation performance can be found in Fact Sheet 3.1, *Foundations in Coastal Areas,* in FEMA P-499.

A best practices approach in the design and construction of coastal foundations is warranted because of the extreme environmental conditions in coastal areas, the vulnerability of shallow foundations to scour and erosion, the fact that the flood loads on open foundations are much lower than those on closed foundations, and foundation failures typically result in extensive damage to or total destruction of the elevated building.

Structural fill can also be used to elevate and support stem wall, crawlspace, solid wall, slab-on-grade, pier, and column foundations in areas not subject to damaging wave action, erosion, and scour. The NFIP precludes the use of structural fill in Zone V. For more information, see FEMA Technical Bulletin 5, *Free-of-Obstruction Requirements* (FEMA 2008b).

10.3.2 Site Considerations

The selected foundation design should be based on the characteristics of the building site. A site characteristic study should include the following:

- **Design flood conditions.** Determine which flood zone the site is located in—Zone V, Coastal A Zone, or Zone A. Flood zones have different hazards and design and construction requirements.

- **Site elevation.** The site elevation and DFE determine how far the foundation needs to extend above grade.

- **Long- and short-term erosion.** Erosion patterns (along with scour) dictate whether a deep foundation is required. Erosion depth affects not only foundation design but also flood loads by virtue of its effect on design stillwater depth (see Section 8.5).

- **Site soils.** A soils investigation report determines the soils that exist on the site and whether certain styles of foundations are acceptable.

10.3.3 Soils Data

Accurate soils data are extremely important in the design of flood-resistant foundations in coastal areas. Although many smaller or less complex commercial buildings and most homes in non-coastal areas are designed without the benefit of specific soils data, all buildings in coastal sites, particularly those in Zone V, should have a thorough investigation of the soils at the construction site. Soils data are available in numerous publications and from onsite soils tests.

10.3.3.1 Sources of Published Soils Data

Numerous sources of soil information are available. Section 12.2 of the *Timber Pile Design and Construction Manual* (Collin 2002) lists the following:

- Topographic maps from the U.S. Geologic Survey (USGS)

- Topographic maps from the Army Map Service

- Topographic maps from the U.S. Coast and Geodetic Survey

- Topographic information from the USACE for some rivers and adjacent shores and for the Great Lakes and their connecting waterways

- Nautical and aeronautical charts from the Hydrographic Office of the Department of the Navy

- Geologic information from State and local governmental agencies, the Association of Engineering Geologists, the Geological Society of America, the Geo-Institute of the American Society of Civil Engineers, and local universities

- Soil survey maps from the Soil Conservation Service of the U.S. Department of Agriculture

10.3.3.2 Soils Data from Site Investigations

Site investigations for soils include surface and subsurface investigations. Surface investigations can identify evidence of landslides, areas affected by erosion or scour, and accessibility for equipment needed for subsurface testing and for equipment needed in construction. Surface investigations can also help identify the suitability or unsuitability of particular foundation styles based on the past performance of existing structures. However, caution should be used when basing the selection of a foundation style solely on the performance of existing structures because the structures may not have experienced a design event.

The 2012 IBC requires that geotechnical investigations be conducted by Registered Design Professionals. Section 1803.2 allows building officials to waive geotechnical investigations where satisfactory data are available from adjacent areas and demonstrate that investigations are not required. The 2012 IRC requires building officials to determine whether soils tests are needed where "quantifiable data created by accepted soil science methodologies indicate expansive, compressible, shifting or other questionable soil characteristics are likely to be present." Because of the hazards in coastal areas, a best practices approach is to follow the 2012 IBC requirements.

Subsurface exploration provides invaluable data on soils at and below grade. The data are both qualitative (e.g., soil classification) and quantitative (e.g., bearing capacity). Although some aspects of subsurface exploration are discussed here, subsurface exploration is too complicated and site-dependent to be covered fully in one document. Consulting with geotechnical engineers familiar with the site is strongly recommended.

Subsurface exploration typically consists of boring or creating test pits, soils sampling, and laboratory tests. The *Timber Pile Design and Construction Manual* (Collin 2002) recommends a minimum of one boring per structure, a minimum of one boring for every 1,000 square feet of building footprint, and a minimum of two borings for structures that are more than 100 feet wide. Areas with varying soil structure and profile dictate more than the minimum number of borings. Again, local geotechnical engineers should be consulted.

The following five types of data from subsurface exploration are discussed in the subsections below: soil classification, bearing capacity, compressive strength, angle of internal friction, and subgrade modulus.

Soil Classification

Soil classification qualifies the types of soils present along the boring depth. ASTM D2487-10 is a consensus standard for soil classification. Soil classification is based on whether soils are cohesive (silts and clays) or non-cohesive (composed of granular soils particles). The degree of cohesiveness affects foundation design. Coupled with other tests such as the plasticity/Atterburg Limits soil classification can identify unsuitable or potentially problematic soils. Table 10-2 contains the soil classifications from ASTM D2487-10. ASTM D2488-09a is a simplified standard for soil classification that may be used when directed by a design professional.

Bearing Capacity

Bearing capacity is a measure of the ability of soil to support gravity loads without soil failure or excessive settlement. Bearing capacity is generally measured in pounds/square foot and occasionally in tons/square foot. Soil bearing capacity typically ranges from 1,000 pounds/square foot (relatively weak soils) to more than 10,000 pounds/square foot (bedrock).

Bearing capacity has a direct effect on the design of shallow foundations. Soils with lower bearing capacities require proportionately larger foundations to effectively distribute gravity loads to the supporting soils. For deep foundations, like piles, bearing capacity has less effect on the ability of the foundation to support gravity loads because most of the resistance to gravity loads is developed by shear forces along the pile.

Presumptive allowable load bearing values of soils are provided in the 2012 IBC and the 2012 IRC. Frequently, designs are initially prepared based on presumed bearing capacities. The builder's responsibility is to verify that the actual site conditions agree with the presumed bearing capacities. As a ***best practices approach***, the actual soil bearing capacity should be determined to allow the building design to properly account for soil capacities and characteristics.

Compressive Strength

Compressive strength is typically determined by Standard Penetration Tests. Compressive strength controls the design of shallow foundations via bearing capacity and deep foundations via the soil's resistance to lateral loads. Compressive strength is also considered when determining the capacity of piles to resist vertical loads.

Compressive strength is determined by advancing a probe, 2 inches in diameter, into the bottom of the boring by dropping a 140-pound slide hammer a height of 30 inches. The number of drops, or blows, required to advance the probe 6 inches is recorded. Blow counts are then correlated to soil properties.

Table 10-2. ASTM D2487-10 Soil Classifications

Major Divisions			Group Symbol	Typical Names	Classification Criteria		
Coarse-grained soils more than 50% retained on No. 200 sieve	Gravels: 50% or more of coarse fraction retained on No. 4 sieve	Clean gravels	GW	Well-graded gravels and gravel-sand mixtures, little or no fines	Classification on basis of percentage of fines: • Less than 5% pass No. 200 sieve: GW, GP, SW, SP • More than 12% pass No. 200 sieve: GM, GC, SM, SC • 5% to 12% pass No. 200 sieve: borderline classification requiring dual symbols	$C_u = \dfrac{D_{60}}{C_{10}}$ greater than 4 $C_Z = \dfrac{(D_{30})^2}{(D_{10})(D_{60})}$ between 1 and 3	
			GP	Poorly graded gravels and gravel-sand mixtures, little or no fines		Not meeting both criteria for GW	
		Gravels with fines	GM	Silty gravels, gravel-sand-silt mixtures		Atterberg limits plot below "A" line or plasticity index less than 4	Atterberg limits plotting in hatched area are borderline classifications requiring use of dual symbols.
			GC	Clayey gravels, gravel-sand-clay mixtures		Atterberg limits plot above "A" line or plasticity index less than 7	
	Sands: More than 50% of coarse fraction passes No. 4 sieve	Clean sands	SW	Well-graded sands and gravelly sands, little or no fines		$C_u = \dfrac{D_{60}}{C_{10}}$ greater than 6 $C_Z = \dfrac{(D_{30})^2}{(D_{10})(D_{60})}$ between 1 and 3	
			SP	Poorly graded sands and gravelly sands, little or no fines		Not meeting both criteria for SW	
		Sands with fines	SM	Silty sands, sand-silt mixtures		Atterberg limits plot below "A" line or plasticity index less than 4	Atterberg limits plotting in hatched area are borderline classifications requiring use of dual symbols.
			SC	Clayey sands, sand-clay mixtures		Atterberg limits plot above "A" line or plasticity index greater than 7	

Table 10-2. ASTM D2487-10 Soil Classifications (concluded)

Major Divisions		Group Symbol	Typical Names	Classification Criteria
Fine-grained soils: 50% or more passes No. 200 sieve	Silts and clay liquid limit 50% or less	ML	Inorganic silts, very fine sands, rock flout, silty or clayey fine sands	
		CL	Inorganic clays of low to medium plasticity, gravelly clays, sandy clays, silty clays, lean clays	
		OL	Organic silts and organic silty clays of low plasticity	
Fine-grained soils: 50% or more passes No. 200 sieve	Silts and clay liquid limit greater than 50%	MH	Inorganic silts, micaceous or diatomaceous fine sands or silts, elastic silts	
		CH	Inorganic clays of high plasticity, fat clays	
		OH	Organic clays of medium to high plasticity	
Highly organic soils		PT	Peat, muck, and other highly organic soils	

Adapted, with permission, from ASTM D2487-10 *Standard Practice for Classification of Soils for Engineering Purposes* (Unified Soil Classification System), copyright ASTM International, 100 Barr Harbor Drive, West Conshohocken, PA 19428. The complete standard is available at ASTM International, http://www.astm.org.

Angle of Internal Friction/Soil Friction Angle

The angle of internal friction is a measure of the soil's ability to resist shear forces without failure. Internal friction depends on soil grain size, grain size distribution, and mineralogy.

The angle of internal friction is used in the design of shallow and deep foundations. It is also used to determine the sliding resistance developed between the bottom of a footing and the foundation at the adjacent soil strata via Equation 10.1.

The following factors should be considered. The normal force includes only the weight of the building (dead load). Live loads should not be considered. Also, ASD load factors in ASCE 7-10 allow only 60 percent of the dead load of a structure to be considered when resisting sliding forces. Foundation materials exert less normal force on a foundation when submerged, so the submerged weight of all foundation materials below the design stillwater depth should be used.

Editions of the IBC contain presumptive coefficients of friction for various soil types (for example, coefficients of friction are contained in Table 1806.2 in the 2009 IBC). Those coefficients can be used in Equation 10.1 by substituting them for the term "tan (φ)."

Subgrade Modulus n_h

The subgrade modulus (n_h) is used primarily in the design of pile foundations. It, along with the pile properties, determines the depth below grade of the point of fixity (point of zero movement and rotation) of a pile under lateral loading.

The inflection point is critical in determining whether piles are strong enough to resist bending moments caused by lateral loads on the foundation and the elevated building. The point of fixity is deep for soft soils (low subgrade modulus) and stiff piles and shallow for stiff soils (high subgrade modulus) and flexible piles.

Subgrade moduli range from 6 to 150 pounds/cubic inch for soft clays to 800 to 1,400 pounds/cubic inch for dense sandy gravel. See Section 10.5.3 for more information on subgrade modulus.

EQUATION 10.1. SLIDING RESISTANCE

$$F = \tan(\varphi)(N)$$

where:

F = resistance to sliding (lb)

φ = angle of internal friction

N = normal force on the footing (lb)

10.4   Design Process

The following are the major steps in foundation analysis and design.

- Determine the flood zone that the building site is in. For a site that spans more than one flood zone (e.g., Zone V and Coastal A Zone, Coastal A Zone and Zone A), design the foundation for the most severe zone (see Chapter 3).

- Determine the design flood elevation and design stillwater elevation (see Chapter 8).

- Determine the projected long- and short-term erosion (see Chapter 8).

- Determine the site elevation and determine design stillwater depths (see Chapter 8).

- Determine flood loads including breaking wave loads, hydrodynamic loads, flood-borne debris loads, and hydrostatic loads. Buoyancy reduces the weight of all submerged materials, so hydrostatic loads need to be considered on all foundations (see Chapter 8).

- Obtain adequate soils data for the site (see Section 10.3.3).

- Determine maximum scour and erosion depths (see Chapter 8).

- Select foundation type (open/deep, open/shallow, closed/deep, or closed/shallow). Use open/deep foundations in Zone V and Coastal A Zone. Use open/shallow foundations in Coastal A Zone only when scour and erosion depths can be accurately predicted and when the foundation can extend beneath the erosion depths. See Sections 10.2 and 10.3.1.

- Determine the basic wind speed, exposure, and wind pressures (see Chapter 8). Determine live and dead loads and calculate all design loads on the elevated building and on the foundation elements (see Chapter 8).

- Determine forces and moments at the top of the foundation elements for all load cases specified in ASCE 7-10. Use load combinations specified in Section 2.3 for strength-based designs or Section 2.4 for stress-based designs. Apply forces and moments to the foundation.

- Design the foundation to resist all design loads and load combinations when exposed to maximum predicted scour and erosion.

10.5   Pile Foundations

Pile foundations are widely used in coastal environments and offer several benefits. Pile foundations are deep and, when properly imbedded, offer resistance to scour and erosion. Piles are often constructed of treated timber, concrete, or steel although other materials are also used.

Treated timber piles are readily available and because they are wood, they can be cut, sawn, and drilled with standard construction tools used for wood framing. ASTM D25-99 contains specifications on round timber piles including quality requirements, straightness, lengths and sizes (circumferences and diameters) as well as limitations on checks, shakes, and knots. The *National Design Specification for Wood Construction* (ANSI/ AF&PA 2005) contains design values for timber piles that meet ASTM D25-99 specifications.

Pre-cast (and typically pre-stressed) concrete piles are not readily available in some areas but offer several benefits over treated timber piles. Generally, they can be fabricated in longer lengths than timber piles. For the same cross section, they are stronger than timber piles and are not vulnerable to rot or wood-destroying insects. The strength of concrete piles can allow them to be used without grade beams. Foundations without grade beams are less vulnerable to scour than foundations that rely on grade beams (See Section 10.5.6).

Steel piles are generally not used in residential construction but are common in commercial construction. Field connections are relatively straightforward, and since steel can be field drilled and welded, steel-to-wood and steel-to-concrete connections can be readily constructed. ASTM A36/A36M-08 contains specifications for mild (36 kip/square inch) steels in cast or rolled shapes. ASTM standards for other shapes and steels include:

- For steel pipe, ASTM A53/A53M-10, *Standard Specification for Pipe, Steel, Black and Hot-Dipped, Zinc-Coated, Welded and Seamless* (ASTM 2010c)

- For structural steel tubing, ASTM A500-10, *Standard Specification for Cold-Formed Welded and Seamless Carbon Steel Structural Tubing in Rounds and Shapes* (ASTM 2010b); and ASTM A501-07, *Standard Specification for Hot-Formed Welded and Seamless Carbon Steel Structural Tubing* (ASTM 2007)

- For welded and seamless steel pipe piles, ASTM 252-10, *Standard Specification for Welded and Seamless Steel Pipe Piles* (ASTM 2010d)

Fiber-reinforced polymer (FRP) piles are becoming more commonplace in transportation and marine infrastructure but are rarely used in residential applications. However, the usage of FRP piles in residential applications is expected to increase. New construction materials can offer many benefits such as sustainability, durability, and longevity but like any new construction material, the appropriateness of FRP piles should be thoroughly investigated before being used in new applications. Although FRP is not discussed in the publication, Technical Fact Sheet 1.8, *Non-Traditional Building Materials and Systems,* in FEMA P-499 provides guidance on using new materials and new systems in coastal environments.

Table 10-3 is a summary of the advantages and special considerations for three of the more common pile materials.

Table 10-3. Advantages and Special Considerations of Three Types of Pile Materials

Material	Advantages	Special Considerations
Wood	• Comparatively low initial cost • Readily available in most areas • Easy to cut, saw and drill • Permanently submerged piles resistant to decay • Relatively easy to drive in soft soil • Suitable for friction and end bearing pile	• Difficult to splice • Subject to eventual decay when in soil or intermittently submerged in water • Vulnerable to damage from driving (splitting) • Comparatively low compressive load • Relatively low allowable bending stress
Concrete	• Available in longer lengths than wood piles • Corrosion resistant • Can be driven through some types of hard material • Suitable for friction and end-bearing piles • Reinforced piles have high bending strength • High bending strength allows taller or more heavily loaded pile foundations to be constructed without grade beams	• High initial cost • Not available in all areas • Difficult to make field adjustments for connections • Because of higher weight, require special consideration in high seismic areas
Steel	• High resistance to bending • Easy to splice • Available in many lengths, sections, and sizes • Can be driven through hard subsurface material • Suitable for friction and end-bearing piles • High bending strength, which allows taller or more heavily loaded pile foundations to be constructed without grade beams	• Vulnerable to corrosion • May be permanently deformed if struck by heavy object • High initial cost • Some difficulty with attaching wood framing

The critical aspects of pile foundations include the pile material and size and pile embedment depth. Pile foundations with inadequate embedment do not have the structural capacity to resist sliding and overturning (see Figure 10-2). Inadequate embedment and improperly sized piles greatly increase the probability for structural collapse. However, when properly sized, installed, and braced with adequate embedment into the soil (with consideration for erosion and scour effects), a building's pile foundation performance allows the building to remain standing and intact following a design flood event (see Figure 10-3).

10.5.1   Compression Capacity of Piles – Resistance to Gravity Loads

The compression capacity of piles determines their ability to resist gravity loads from the elevated structure they support. One source that provides an equation for the compression capacity of piles is the *Foundation and Earth Structures*, Design Manual 7.2 (USDN 1986). The manual contains Equation 10.2 for determining

Figure 10-2.
Near collapse due
to insufficient pile
embedment, Hurricane
Katrina (Dauphin Island,
AL, 2005)

Figure 10-3.
Surviving pile foundation,
Hurricane Katrina
(Dauphin Island, AL,
2005)

the compression capacity of a single pile when placed in granular (non-cohesive) soils. Design Manual 7.2 also contains methods of determining compression capacity of a pile placed in cohesive soils.

The resistance of the pile is the sum of the capacity that results from end bearing and friction. The capacity from end bearing is the first term in Equation 10.2; the capacity from friction is given in the second term.

Equation 10.2 gives the ultimate compression capacity of a pile. The allowable capacity (Q_{allow}) used in ASD depends on a Factor of Safety applied to the ultimate capacity. For ASD, Design Manual 7.2 recommends a Factor of Safety of 3.0; thus, $Q_{allow} = Q_{ult}/3$.

Σ

EQUATION 10.2. ULTIMATE COMPRESSION CAPACITY OF A SINGLE PILE

$$Q_{ult} = P_T N_q A_T + \sum K_{HC} P_0 Ds \tan \delta$$

where:

Q_{ult}	=	ultimate load capacity in compression (lb)
P_T	=	effective vertical stress at pile tip (lb/ft²)
N_q	=	bearing capacity factor (see Table 10-4)
A_T	=	area of pile tip (ft²)
K_{HC}	=	earth pressure in compression (see Table 10-5)
P_0	=	effective vertical stress over the depth of embedment, D (lb/ft²)
δ	=	friction angle between pile and soil (see Table 10-6)
s	=	surface area of pile per unit length (ft)
D	=	depth of embedment (ft)

Table 10-4. Bearing Capacity Factors (N_q)

Parameter	Pile Bearing Capacity Factors													
φ (degrees)[a]	26	28	30	31	32	33	34	35	36	37	38	39	40
N_q (driven pile displacement)	10	15	21	24	29	35	42	50	62	77	86	120	145
N_q (drilled piers)[b]	5	8	10	12	14	17	21	25	30	38	43	60	72

N_q = bearing capacity factor
φ = angle of internal friction

(a) Limit φ to 28° if jetting is used

(b) When a bailer or grab bucket is used below the groundwater table, calculate end bearing based on φ not exceeding 28 degrees. For piers larger than 24 inches in diameter, settlement rather than bearing capacity usually controls the design. For estimating settlement, take 50% of the settlement for an equivalent footing resting on the surface of comparable granular soils.

Table 10-5. Earth Pressure Coefficients

Pile Type	K_{HC}	K_{HT}
Driven single H-pile	0.5 – 1.0	0.3 – 0.5
Driven single displacement pile	1.0 – 1.5	0.6 – 1.0
Driven single displacement tapered pile	1.5 – 2.0	1.0 – 1.3
Driven jetted pile	0.4 – 0.9	0.3 – 0.6
Drilled pile (less than 24-inch diameter)	0.7	0.4

K_{HC} = earth pressure compression coefficient
K_{HT} = earth pressure tension coefficient

Table 10-6. Friction Angle Between Soil and Pile (δ)

Pile Type	δ
Timber	$\frac{3}{4}\varphi$
Concrete	$\frac{3}{4}\varphi$
Steel	20 degrees

φ = angle of internal friction

10.5.2 Tension Capacity of Piles

The tension capacity of piles determines their ability to resist uplift and overturning loads on the elevated structure. One source that provides pile capacity in tension load is the Design Manual 7.2, which is also a reference on compression capacity. Equation 10.3 determines the tension capacity in a single pile.

Σ EQUATION 10.3. ULTIMATE TENSION CAPACITY OF A SINGLE PILE

$$T_{ult} = \sum K_{HT} P_0 Ds \tan \delta$$

where:

T_{ult} = ultimate load capacity in tension (lb)

K_{HT} = earth pressure in tension (see Table 10-5)

P_0 = effective vertical stress over the depth of embedment, D (lb/ft^2)

δ = friction angle between pile and soil (see Table 10-6)

s = surface area of pile per unit length (ft^2/ft or ft)

D = depth of embedment (ft)

Note: *With the recommended Factor of Safety of 3.0, the allowable tension capacity, T_{allow} = T_{ult}/3.*

The Design Manual 7.2 provides tables to identify bearing capacity factors (N_q), earth pressure coefficients (K_{HC} and K_{HT}), and friction angle between pile and soil (δ) based on pile type and the angle of internal friction (φ) of the soil.

Example 10.1 illustrates compression and tension capacity calculations for a single pile not affected by scour or erosion.

Table 10-7 contains example calculations using Equations 10.2 and 10.3 for the allowable compression (gravity loading) and tension (uplift) capacities of wood piles for varying embedments, pile diameters, and installation methods. The table also illustrates the effect of scour around the pile on the allowable compression and tension loads. Scour (and erosion) reduces pile embedment and therefore pile capacity. For this table, a scour depth of twice the pile diameter (2d) with no generalized erosion is considered.

EXAMPLE 10.1. CALCULATION FOR ALLOWABLE CAPACITIES OF WOOD PILES

Given:

- Closed end, driven timber pile

- Diameter (d) = 1 ft

- Depth of embedment (D) = 15 ft

- Soil density (γ) = 65 lb/ft^3

- Angle of internal friction (φ) = 30 K_{HC} = 1.0 (applicable coefficient from Table 10-5)

- Earth pressure in tension (K_{HT}) = 0.6 (applicable coefficient from Table 10-5)

- Bearing capacity factor (N_q) = 21 (applicable coefficient from Table 10-4)

- Factor of Safety = 3.0

Find:

1. Allowable tension and compression capacities of wood piles embedded in soil

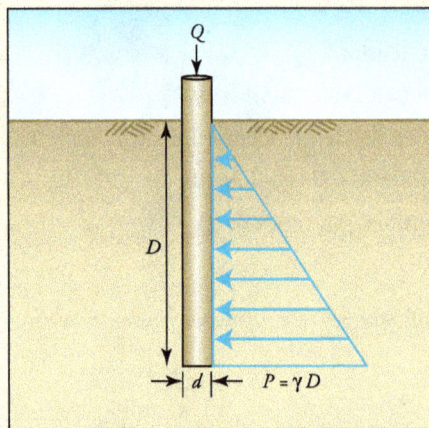

Q = load
D = length of pile
d = diameter of pile
P = pressure
γ = soil density

Illustration A.
Pile schematic and pressure diagram

EXAMPLE 10.1. CALCULATION FOR ALLOWABLE CAPACITIES OF WOOD PILES (concluded)

Solution for #1: Find the allowable tension and compression capacity of the wood pile embedded in soil as follows:

- To determine the resultant pressure from the soil on the pile:

$$\delta = \frac{3}{4}(\varphi) = \frac{3}{4}(30°) = 22.5°$$

$$P_0 = P_t = \gamma D = (65 \text{ lb/ft}^3)(15 \text{ ft}) = 975 \text{ lb/ft}^2$$

- Geometrical properties of the pile surfaces upon which pressure from the soil is applied to the pile are:

$$A_t = (\pi)(\frac{1}{2}d)^2 = (3.14)[(0.5)(1 \text{ ft})]^2 = 0.785 \text{ ft}^2$$

$$P_0 = P_t = \gamma D = (65 \text{ lb/ft}^3)(15 \text{ ft}) = 975 \text{ lb/ft}^2$$

Allowable compression capacity:

$$Q_{ult} = (975 \text{ lb/ft}^2)(21)(0.785 \text{ ft}^2) + (1.0)(975 \text{ lb/ft}^2)(\tan 22.5°)(3.14 \text{ ft}^2/\text{ft})(15 \text{ ft})$$

$$Q_{ult} = 35,095 \text{ lb}$$

$$Q_{all} = \frac{Q_{ult}}{3} = \frac{35,095 \text{ lb}}{3} = \mathbf{11,698 \text{ lb}}$$

Allowable tension capacity:

$$T_{ult} = (0.6)(975 \text{ lb/ft}^2)(\tan 22.5°)(3.14 \text{ ft}^2/\text{ft})(15 \text{ ft}) = 11,413 \text{ lb}$$

$$T_{all} = \frac{T_{ult}}{3} = \frac{11,413 \text{ lb}}{3} = \mathbf{3,804 \text{ lb}}$$

The purpose of Table 10-7 is to illustrate the effects of varying diameters, depths of embedment, and installation methods on allowable capacities. See Section 10.5.4 for information of installation methods. Example calculations used to determine the values in Table 10-7 are used in Example 10.1. The values in Table 10-7 are not intended to be used for design purposes.

Table 10-7. Allowable Compression and Tension of Wood Piles Based on Varying Diameters, Embedments, and Installation Methods

Diameter and Embedment	Installation Method	Compression (pounds)		Tension (pounds)	
		No Scour	2d Scour	No Scour	2d Scour
d = 12 inches *D* = 15 feet	Driven	11,698	9,406	3,804	2,857
	Jetted	7,894	6,548	1,902	1,429
	Augered	6,990	5,545	2,536	1,905
d = 12 inches *D* = 20 feet	Driven	18,416	15,560	6,763	5,478
	Jetted	11,652	10,081	3,382	2,739
	Augered	11,292	9,453	4,509	3,652
d =10 inches *D* = 15 feet	Driven	9,004	7,482	3,170	2,505
	Jetted	5,834	4,977	1,585	1,252
	Augered	5,470	4,497	2,114	1,670

d = diameter
D = depth of embedment

10.5.3   Lateral Capacity of Piles

The lateral capacity of piles is dictated by the piles and the pile/soil interface. The ability of the pile to resist lateral loads depends on the pile size and material, the soil properties, and on presence or absence of pile bracing.

One of the critical aspects of pile design is the distance between the lateral load application point and the point of fixity of the pile. That distance constitutes a moment arm and governs how much bending moment develops when a pile is exposed to lateral loads. For a foundation to perform adequately, that moment must be resisted by the pile without pile failure.

Equation 10.4 determines the distance between the point where the lateral load is applied and the point of fixity for an unbraced pile. Note that in Equation 10.4, "*d*" is the depth below grade of the point of fixity, not the diameter of the pile. Also, see Figure 10-4 for the deflected shape of a laterally loaded pile.

Table 10-8 lists recommended values for n_h, modulus of subgrade reaction, for a variety of soils (Bowles 1996). For wood pilings, the depths to points of fixity range from approximately 1 foot in stiff soils to approximately 5 feet in soft soils.

The ability of site soils to resist lateral loads is a function of the soil characteristics, their location on the site, and their compressive strength. Chapter 7 of the *Timber Pile Design and Construction Manual* (Collin 2002) contains methods of determining the lateral resistance of timber piles for both fixed pile head conditions (i.e., piles used with grade beams or pile caps) and free pile head conditions (i.e., piles free to rotate at their top). The manual also contains methods of approximating lateral capacity and predicting pile capacity when detailed soils data are known.

Σ

EQUATION 10.4. LOAD APPLICATION DISTANCE FOR AN UNBRACED PILE

$$L = H + \frac{d}{12}$$

where:

L = distance between the location where the lateral force in applied and the point of fixity (i.e., moment arm) (ft)

d = depth from grade to inflection point (inches); $d = 1.8 \left(\frac{EI}{n_b} \right)^{\frac{1}{5}}$

H = distance above eroded ground surface (including localized scour) where lateral load is applied (ft)

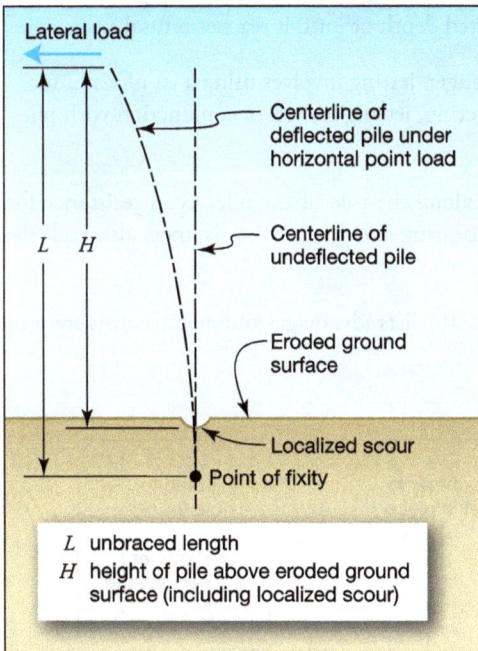

Figure 10-4.
Deflected pile shape for an unbraced pile

Table 10-8. Values of n_b Modulus of Subgrade Reaction

Soil Type	n_b Modulus of Subgrade Reaction (pound/cubic inch)
Dense sandy gravel	800 to 1,400
Medium dense coarse sand	600 to 1,200
Medium sand	400 to 1,000
Fine to silty fine sand	290 to 700
Medium clay (wet)	150 to 500
Soft clay	6 to 150

10.5.4 Pile Installation

Methods for installing piles include driving, augering, and jetting. A combination of methods may also be used. For example, piles may be placed in augered holes and then driven to their final depth. Combining installation methods can increase the achievable embedment depth. With increased depths, a pile's resistance to lateral and vertical loads can be increased, and its vulnerability to scour and erosion will be reduced.

- *Driving* involves hitting the top of the pile with a pile driver or hammer until the pile reaches the desired depth or it is driven to refusal. Piles can be driven with vibratory hammers. Vibratory hammers generate vertical oscillating movements that reduce the soil stress against the pile and which makes the piles easier to drive. Ultimate load resistance is achieved by a combination of end bearing of the pile and frictional resistance between the pile and the soil. A record of the blow counts from the pile driver can be used with a number of empirical equations to determine capacity.

- *Augering* involves placing the pile into a pre-drilled hole typically made with an auger. The augured hole can be the full diameter of the pile or a smaller diameter than the pile. Pre-drilling is completed to a predetermined depth, which often is adjusted for the soils found on the site. After placing the pile into the pre-drilled hole, the pile is then driven to its final desired depth or until it reaches refusal.

- *Jetting* is similar to augering but instead of using a soils auger, jetting involves using a jet of water (or air) to remove soils beneath and around the pile. Like augering, jetting is used in conjunction with pile driving.

Both augering and jetting remove natural, undisturbed soil along the side of the pile. Load resistance for both of these methods is achieved by a combination of end bearing and frictional resistance, although the frictional resistance is much less than that provided by driven piles.

Figure 10-5 illustrates the three pile installation methods. Table 10-9 lists advantages and special considerations for each method.

Figure 10-5.
Pier installation methods

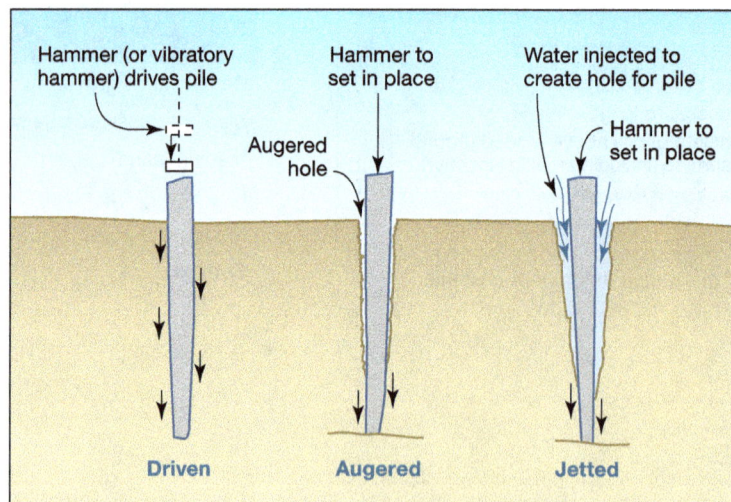

Table 10-9. Advantages and Special Considerations of Pile Installation Methods

Installation Method	Advantages	Special Considerations
Driving	• Well-suited for friction piles • Common construction practice • Pile capacity can be determined empirically	• Requires subsurface investigation • May be difficult to reach terminating soil strata if piles are only driven • Difficult to maintain plumb during driving and thus maintain column lines
Augering	• Economical • Minimal driving vibration to adjacent structures • Well-suited for end bearing • Visual inspection of some soil stratum possible • Convenient for low headroom situations • Easier to maintain column lines	• Requires subsurface investigation • Not suitable for highly compressed material • Disturbs soil adjacent to pile, thus reducing earth pressure coefficients K_{HC} and K_{HT} to 40 percent of that driven for piles • Capacity must be determined by engineering judgment or load test
Jetting	• Minimal driving vibration to adjacent structures • Well-suited for end bearing piles • Easier to maintain column lines	• Requires subsurface investigation • Disturbs soil adjacent to pile, thus reducing earth pressure coefficients K_{HC} and K_{HT} to 40 percent of that driven for piles • Capacity must be determined by engineering judgment or load test

K_{HC} = earth pressure compression coefficient

K_{HT} = earth pressure tension coefficient

10.5.5 Scour and Erosion Effects on Pile Foundations

Coastal homes are often exposed to scour and erosion, and because moving floodwaters cause both scour and erosion, it is rare for an event to produce one and not the other. As Figure 10-6 illustrates, scour and erosion have a cumulative effect on pile foundations. They both reduce piling embedment.

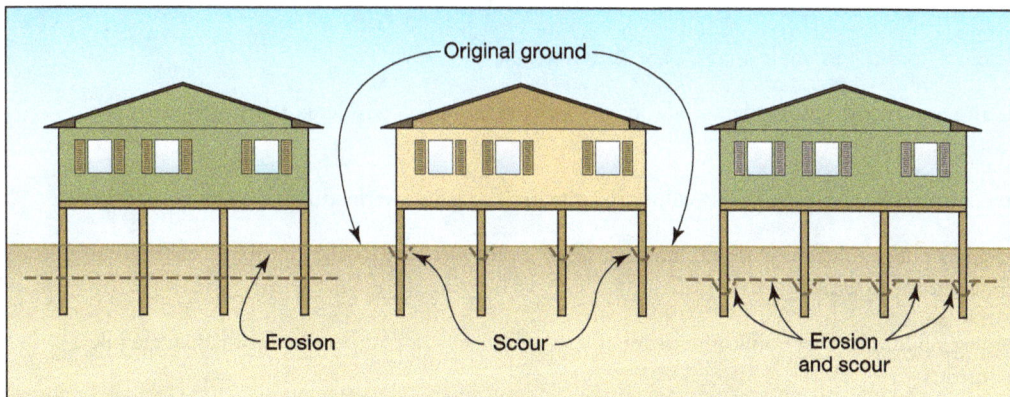

Figure 10-6.
Scour and erosion effects on piling embedment

A properly designed pile foundation must include a consideration of the effects of scour and erosion on the foundation system. Scour washes away soils around the piling, reducing pile embedment, and increases stresses within the pile when the pile is loaded. The reduced embedment can cause the foundation to fail at the pile/soil interface. The increased stresses can cause the pile itself to fracture and fail.

Erosion is even more damaging. In addition to reducing pile embedment depths and increasing stresses on piles, erosion increases the flood forces the foundation must resist by increasing the stillwater depth at the foundation that the flood produces. Pile foundations that are adequate to resist flood and wind forces without being undermined by scour and erosion can fail when exposed to even minor amounts of scour and erosion.

An example analysis of the effects of scour and erosion on a foundation is provided in *Erosion, Scour, and Foundation Design* (FEMA 2009a), published as part of Hurricane Ike Recovery Advisories and available at http://www.fema.gov/library/viewRecord.do?id=3539.

The structure in the example is a two-story house with 10-foot story heights and a 32-foot by 32-foot foundation. The house is away from the shoreline and elevated 8 feet above grade on 25 square timber piles spaced 8 feet apart. Soils are medium dense sands. The house is subjected to a design wind event with a 130-mph (3-second gust) wind speed and a 4-foot stillwater depth above the uneroded grade, with storm surge and broken waves passing under the elevated building.

Lateral wind and flood loads were calculated in accordance with ASCE 7-05. Although the wind loads in ASCE 7-10 vary from ASCE 7-05 somewhat, the results of the analyses do not change significantly. Piles were analyzed under lateral wind and flood loads only; dead, live, and wind uplift loads were neglected. If the neglected loads are included, deeper pile embedment and possibly larger piles than the results of the analysis indicated may be needed. Three timber pile sizes (8-inch square, 10-inch square, and 12-inch square) were evaluated using pre-storm embedment depths of 10 feet, 15 feet, and 20 feet and five erosion and scour conditions (erosion = 0 or 1 foot; scour = 2.0 times the pile diameter to 4.0 times the pile diameter).

The results of the analysis are shown in Table 10-10. A shaded cell indicates that the combination of pile size, pre-storm embedment, and erosion/scour does not provide the bending resistance and/or embedment required to resist lateral loads. The reason for foundation failure is indicated in each shaded cell ("P" for failure due to bending and overstress within the pile and "E" for an embedment failure from the pile/soil interaction). "OK" indicates that the bending and foundation embedment criteria are both satisfied by the particular pile size/pile embedment/erosion-scour combination.

The key points from the example analysis are as follows:

- Scour and erosion can cause pile foundations to fail and must be considered when designing pile foundations.

- Failures can result from either overloading the pile itself or from overloading at the pile/soil interface.

- Increasing a pile's embedment depth does not offset a pile with a cross section that is too small or pile material that is too weak.

- Increasing a pile's cross section (or its material strength) does not compensate for inadequate pile embedment.

Table 10-10. Example Analysis of the Effects of Scour and Erosion on a Foundation

Pile Embedment Before Erosion and Scour	Erosion and Scour Conditions	Pile Diameter (*a*)		
		8 inches	10 inches	12 inches
		Reason for Failure		
10 feet	Erosion = 0, Scour = 0	P, E	E	OK
	Erosion = 1 foot, Scour = 2.0*a*	P, E	E	E
	Erosion = 1 foot, Scour = 2.5*a*	P, E	E	E
	Erosion = 1 foot, Scour = 3.0*a*	P, E	E	E
	Erosion = 1 foot, Scour = 4.0*a*	P, E	P, E	E
15 feet	Erosion = 0, Scour = 0	P	OK	OK
	Erosion = 1 foot, Scour = 2.0*a*	P	OK	OK
	Erosion = 1 foot, Scour = 2.5*a*	P	OK	OK
	Erosion = 1 foot, Scour = 3.0*a*	P	OK	OK
	Erosion = 1 foot, Scour = 4.0*a*	P, E	P, E	E
20 feet	Erosion = 0, Scour = 0	P	OK	OK
	Erosion = 1 foot, Scour = 2.0*a*	P	OK	OK
	Erosion = 1 foot, Scour = 2.5*a*	P	OK	OK
	Erosion = 1 foot, Scour = 3.0*a*	P	OK	OK
	Erosion = 1 foot, Scour = 4.0*a*	P	P	OK

Two-story house supported on square timber piles and located away from the shoreline, storm surge and broken waves passing under the building, 130-mph wind zone, soil = medium dense sand.

a = pile diameter

E = foundation fails to meet embedment requirements

OK = bending and foundation embedment criteria are both satisfied by the particular pile size/pile embedment/erosion-scour combination

P = foundation fails to meet bending

10.5.6   Grade Beams for Pile Foundations

Piles can be used with or without grade beams or pile caps. Grade beams create resistance to rotation (also called "fixity") at the top of the piles and provide a method to accommodate misalignment in piling placement. When used with grade beams, the piles and foundation elements above the grade beams work together to elevate the structure, provide vertical and lateral support for the elevated home, and transfer loads imposed on the elevated home and the foundation to the ground below.

Pile and grade beam foundations should be designed and constructed so that the grade beams act only to provide fixity to the foundation system and not to support the lowest elevated floor. If grade beams support the lowest elevated floor of the home, they become the lowest horizontal structural member and significantly higher flood insurance premiums would result. Grade beams must also be designed to span between adjacent piles, and the piles must be capable of resisting both the weight of the grade beams when undermined by erosion and scour and the loads imposed on them by forces acting on the structure.

Pile foundations with grade beams must be constructed with adequate strength to resist all lateral and vertical loads. Failures during Hurricane Katrina often resulted from inadequate connections between the columns and footings or grade beams below (see Figure 10-7).

If grade beams are used with wood piles, the potential for rot must be considered when designing the connection between the grade beam and the pile. The connection must not encourage water retention. The maximum bending moment in the piles occurs at the grade beams, and decay caused by water retention at critical points in the piles could induce failure under high-wind or flood forces.

While offering some advantages, grade beams can become exposed by moving floodwaters if they are not placed deeply enough. Once exposed, the grade beams create large horizontal obstructions in the flood path that significantly increase scour. Extensive scour was observed after Hurricane Ike in 2008 around scores of homes constructed with grade beams (see Figure 10-8).

Although not possible for all piling materials, foundations should be constructed without grade beams whenever possible. For treated timber piles, this can limit elevations to approximately 8 feet above grade. The actual limit depends greatly on flood forces, number of piles, availability of piles long enough to be driven to the required depth and extend above grade enough to adequately elevate the home, and wind speed and geometry of the elevated structure. For steel and concrete piles, foundations without grade beams are practical in many instances, even for taller foundations. Without grade beams to account for pile placement, additional attention is needed for piling alignment, and soils test are needed for design because pile performance depends on the soils present, and presumptive piling capacities may not adequately predict pile performance.

Figure 10-7.
Column connection failure, Hurricane Katrina (Belle Fontaine Point, Jackson County, MS, 2005)

Figure 10-8.
Scour around grade beam, Hurricane Ike (Galveston Island, TX, 2008)

10.6   Open/Deep Foundations

In this section, some of the more common types of open/deep foundation styles are discussed. Treated timber pile foundations are discussed in Section 10.6.1, and other types of open/deep pile foundations are discussed in Section 10.6.2.

10.6.1   Treated Timber Pile Foundations

In many coastal areas, treated timber piles are the most common type of an open/deep style foundation. Timber piles are the first choice of many builders because they are relatively inexpensive, readily available, and relatively easy to install. The driven timber pile system (see Figure 10-9) is suitable for moderate elevations. Home elevations greater than 10 feet may not be practical because of pile length availability, the pile strength required to resist lateral forces (particularly when considering erosion and scour), and the pile embedment required to resist lateral loads after being undermined by scour and erosion.

When used without grade beams, timber piles typically extend from the pile tip to the lowest floor of the elevated structure. With timber piles and wood floor framing, the connection of the elevated structure to the piling is essentially a pinned connection because moment resisting connections in wood framing are difficult to achieve. Pinned connections do not provide fixity and require stronger piles to resist the same loads as piles that benefit from moment resisting connections at their tops.

Improved performance can be achieved if the piles extend beyond the lowest floor to the roof (or an upper floor level). Doing so provides resistance to rotation where the pile passes through the first floor. This not only reduces stresses within the piles but also increases the stiffness of the pile foundation and reduces movement under lateral forces. Extending piles in this fashion improves survivability of the building.

The timber pile system is vulnerable to flood-borne debris. During a hurricane event, individual piles can be damaged or destroyed by large, floating debris. Two ways of reducing this vulnerability are (1) using piles with diameters that are larger than those called for in the foundation design and (2) using more piles and continuous beams that can redistribute loads around a damaged pile. Using more piles and continuous beams increases structural redundancy and can improve building performance.

Figure 10-9.
Profile of timber pile
foundation type

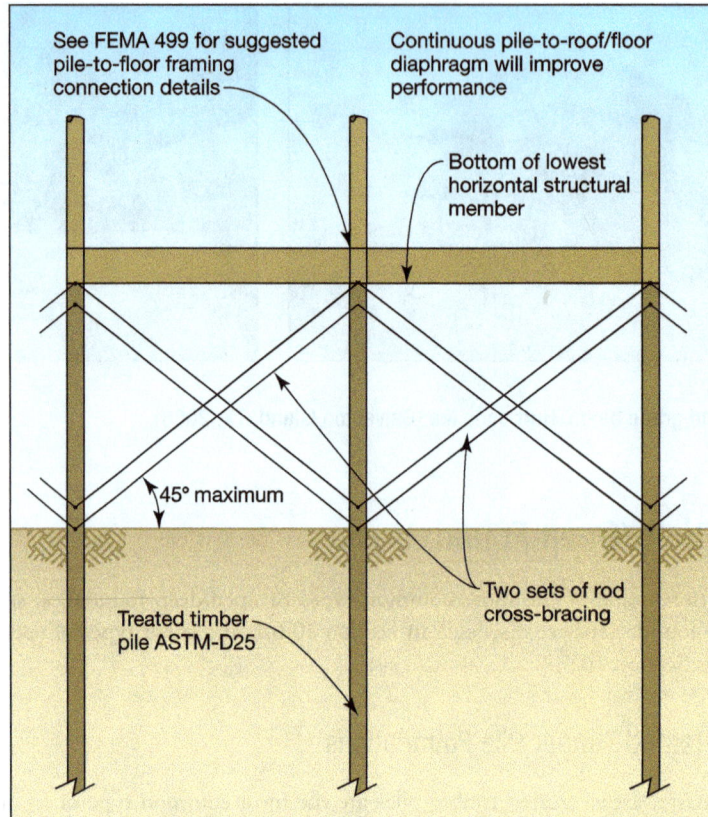

FEMA P-550, *Recommended Residential Construction for Coastal Areas* (FEMA 2006), contains a foundation design using driven timber piles. The foundation design is based on presumptive piling capacities that should be verified prior to construction. Also, the design is intended to support an elevated building with a wide range of widths and roof slopes and as such contains some inherent conservatism in the design. Design professionals who develop foundation designs for specific buildings and have site information on subsurface conditions can augment the FEMA P-550 design to provide more efficient designs that reduce construction costs.

10.6.1.1 Wood Pile-to-Beam Connections

In pile foundations that support wood-framed structures, systems of perimeter and interior beams are needed to support the floors and walls above. Beams must be sized to support gravity loads and, in segmented shear wall construction, resist reactions from shear wall segments. To transfer those loads to the foundation, wood piles are often notched to provide a bearing surface for the beams. Notches should not reduce the pile cross section by more than 50 percent (such information is typically provided by a design professional on contract documents). For proper transfer of gravity loads, beams should bear on the surface of the pile notch.

Although connections play an integral role in the design of structures, they are typically regarded as the weakest link. Guidance for typical wood-pile to wood-girder connections can be found in Fact Sheet 3.3, *Wood Pile to Beam Connections*, in FEMA P-499.

10.6.1.2 Pile Bracing

When timber piles with a sufficiently large cross section are not available, timber piles may require bracing to resist lateral loads. Bracing increases the lateral stiffness of a pile foundation system so that less sway is felt under normal service loads. Bracing also lowers the location where lateral forces are applied to individual piles and reduces bending stresses in the pile. When bracing is used, the forces from moving floodwaters and from flood-borne debris that impacts the braces should be considered.

> **NOTE**
>
> Fact Sheet 3.2, *Pile Installation,* in FEMA P-499 recommends that pile bracing be used only for reducing the structure's sway and vibration for comfort. In other words, bracing should be used to address serviceability issues and not strength issues. The foundation design should consider the piles as being unbraced as the condition that may occur when floating debris removes or damages the bracing. If the pile foundation is not able to provide the desired strength performance without bracing, the designer should consider increasing the pile size.

Bracing is typically provided by diagonal bracing or knee bracing. Diagonal bracing is more effective from a structural standpoint, but because diagonal bracing extends lower into floodwaters, it is more likely to be damaged by flood-borne debris. It can also trap flood-borne debris, and trapped flood-borne debris increases flood forces on the foundation.

Knee bracing does not extend as deeply into floodwaters as cross bracing and is less likely to be affected by flood-borne debris but is less effective at reducing stresses in the pile and also typically requires much stronger connections to achieve similar structural performance as full-length cross bracing.

Diagonal Bracing

Diagonal bracing often consists of dimensional lumber that is nailed or bolted to the wood piles. Steel rod bracing and wire rope (cable) bracing can also be used. Steel rod bracing and cable bracing have the benefit of being able to use tensioning devices, such as turnbuckles, which allow the tension of the bracing to be maintained. Cable bracing has an additional benefit in that the cables can be wrapped around pilings without having to rely on bolted connections, and wrapped connections can transfer greater loads than bolted connections. Figure 10-10 shows an example of diagonal bracing using dimensional lumber.

Diagonal braces tend to be slender, and slender braces are vulnerable to compression buckling. Most bracing is therefore considered tension-only bracing. Because wind and flood loads can act in opposite directions, tension-only bracing must be installed in pairs. One set of braces resists loads from one direction, and the second set resists loads from the opposite direction. Figure 10-11 shows how tension-only bracing pairs resist lateral loads on a home.

The placement of the lower bolted connection of the diagonal brace to the pile requires some judgment. If the connection is too far above grade, the pile length below the connection is not braced and the overall foundation system is less strong and stiff.

Figure 10-10.
Diagonal bracing using
dimensional lumber

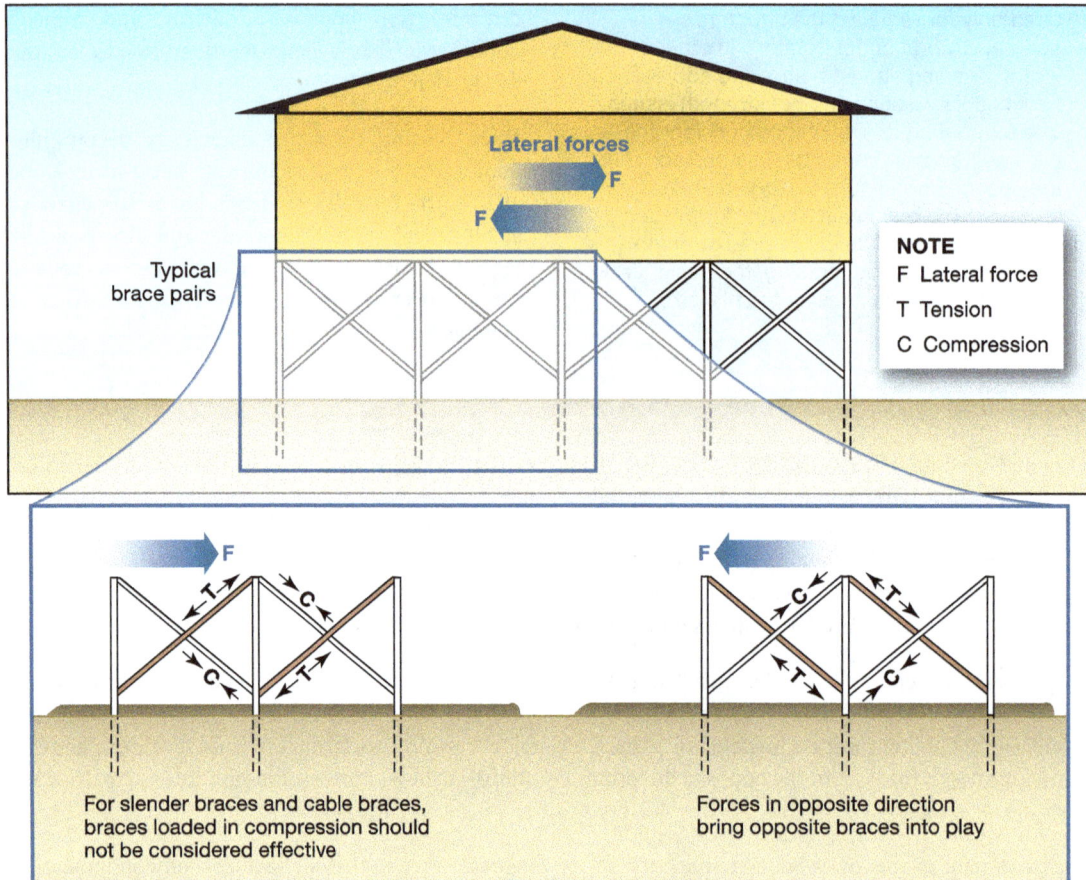

NOTE
F Lateral force
T Tension
C Compression

For slender braces and cable braces,
braces loaded in compression should
not be considered effective

Forces in opposite direction
bring opposite braces into play

Figure 10-11.
Diagonal bracing schematic

EXAMPLE 10.2. DIAGONAL BRACE FORCE

Given:

- Lateral load = 989 lb

- Brace angle = 45

Find:

1. Tension force in the diagonal brace in Illustration A.

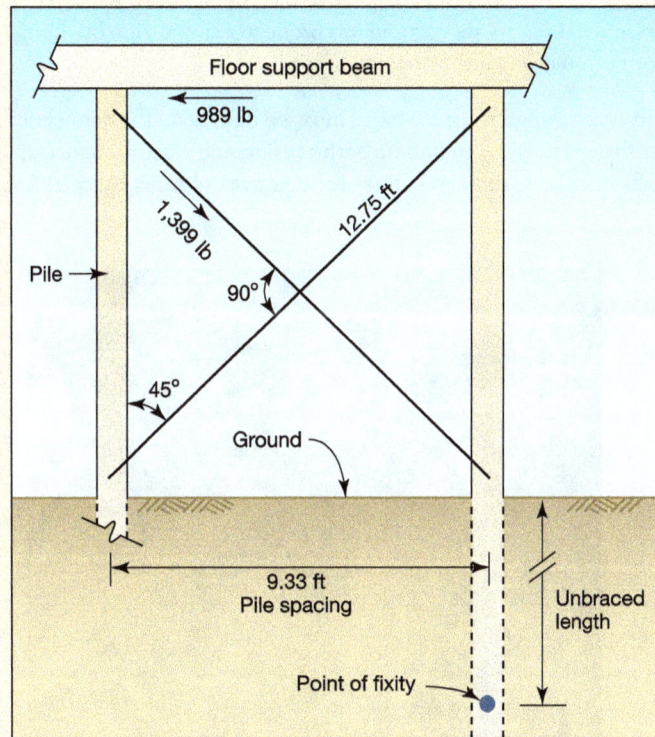

Illustration A. Force diagram for diagonal bracing

Solution for #1: The tension force in the diagonal brace can be found as follows:

Rod bracing is used and assumed to act in tension only because of the rigidity of the rod brace in tension and lack of stiffness of the rod in compression.

- The tension brace force is calculated as follows:

$$T_{diagonal} = \frac{989 \text{ lb}}{\cos 45°} = \textbf{1,399 lb}$$

Interaction of the soil and the pile should be checked to ensure that the uplift component of the brace force can be resisted.

For timber piles, if the connection is too close to grade, the bolt hole is more likely to be flooded and subject to decay or termite infestation, which can weaken the pile at a vulnerable location. All bolt holes should be treated with preservative after drilling and prior to bolt placement.

NOTE

Bolt holes in timber piles should be field-treated (see Chapter 11).

Knee Bracing

Knee braces involve installing short diagonal braces between the upper portions of the pilings and the floor system of the elevated structure (see Figure 10-12). The braces increase the stiffness of an elevated pile foundation and can contribute to resisting lateral forces. Although knee braces do not stiffen a foundation as much as diagonal bracing, they offer some advantages over diagonal braces. For example, knee braces present less obstruction to waves and debris, are shorter and less prone to compression buckling than diagonal braces, and may be designed for both tension and compression loads.

The entire load path into and through the knee brace must be designed. The connections at each end of each knee brace must have sufficient capacity to handle both tension and compression and to resist axial loads in the brace. The brace itself must have sufficient cross-sectional area to resist compression and tensile loads.

Figure 10-12. Knee bracing

The feasibility of knee bracing is often governed by the ability to construct strong connections in the braces that connect the wood piles to the elevated structure.

10.6.1.3 Timber Pile Treatment

Although timber piles are chemically treated to resist rot and damage from insects, they can be vulnerable to wood-destroying organisms such as fungi and insects if the piles are subject to both wetted and dry conditions. If the piles are constantly submerged, fungal growth and insect colonies cannot be sustained; if only periodically submerged, conditions exist that are sufficient to sustain wood-destroying organisms. Local design professionals familiar with the performance of driven, treated timber piles can help quantify the risk. Grade beams can be constructed at greater depths or alternative pile materials can be selected if damage from wood-destroying organisms is a major concern.

Cutting, drilling, and notching treated timber piles disturb portions of the piles that have been treated for rot and insect damage. Because pressure-preservative-treated piles, timbers, and lumber are used for many purposes in coastal construction, the interior, untreated parts of the wood can be exposed to possible decay and infestation. Although treatments applied in the field are much less effective than factory treatments, the potential for decay can be minimized with field treatments. AWPA M4-06 describes field treatment procedures and field cutting restrictions for poles, piles, and sawn lumber.

Field application of preservatives should be done in accordance with the instructions on the label, but if instructions are not provided, dip soaking for at least 3 minutes is considered effective. When dip soaking for 3 minutes is impractical, treatment can be accomplished by thoroughly brushing or spraying the exposed area. The preservative is absorbed better at the end of a member or end grains than on the sides or side grains. To safeguard against decay in bored holes, the preservative should be poured into the holes. If the hole passes through a check (such as a shrinkage crack caused by drying), the hole should be brushed; otherwise, the preservative will run into the check instead of saturating the hole.

Copper naphthenate is the most widely used preservative for field treatment. Its color (deep green) may be objectionable aesthetically, but the wood can be painted with alkyd paints after extended drying. Zinc naphthenate is a clear alternative to copper naphthenate but is not as effective in preventing insect infestation and should not be painted with latex paints. Tributyltin oxide is available but should not be used in or near marine environments because the leachates are toxic to aquatic organisms. Sodium borate is also available, but it does not readily penetrate dry wood and rapidly leaches out when water is present. Sodium borate is therefore not recommended. Waterborne arsenicals, pentachlorophenol, and creosote are unacceptable for field applications.

10.6.2   Other Open/Deep Pile Foundation Styles

Several other styles of pile foundations, in addition to treated timber piles, are used although their use often varies geographically depending on the availability of materials and trained contractors.

FEMA P-550 contains foundation designs that use deep, driven steel and treated timber piles and grade beams that support a system of concrete columns. The second edition of FEMA P-550 (FEMA 2006) added a new design for treated timber piles that incorporates elevated reinforced beams constructed on the concrete columns. In the new design, the elevated beams work with the columns and grade beams to create reinforced concrete portal frames that assist in resisting lateral loads. The elevated beams also create a suitable platform that can support a home designed to a prescriptive standard such as *Wood Framed Construction Manual for One- and Two-Family Dwellings* (AF&PA 2012) or ICC 600-2008.

Figure 10-13 shows one of the deep pile foundation systems that uses treated timber piles and grade beams. The steel pipe pile and grade beam foundation system contained in FEMA P-550 is similar but requires fewer piles because the higher presumptive strength of the steel piles compared to the timber piles. Figure 10-14 shows the foundation system added in the Second Edition of FEMA P-550 (FEMA 2009b), which incorporates an elevated concrete beam.

Figure 10-13.
Section view of a steel pipe pile with concrete column and grade beam foundation type
DEVELOPED FROM FEMA P-550, CASE B

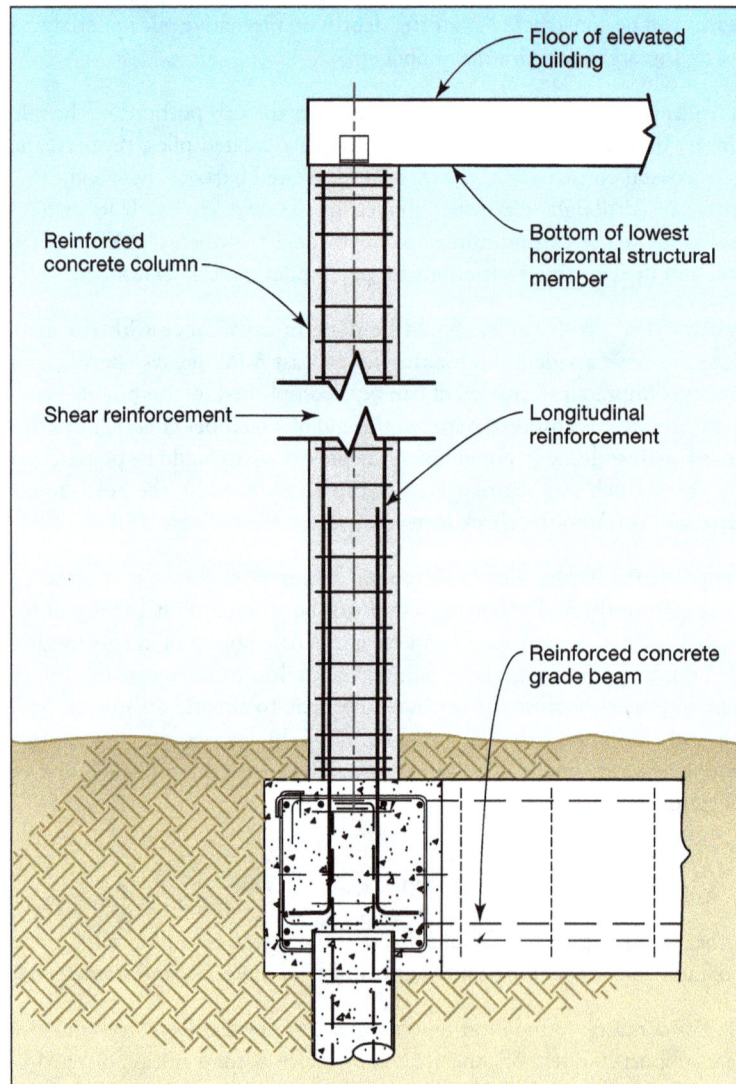

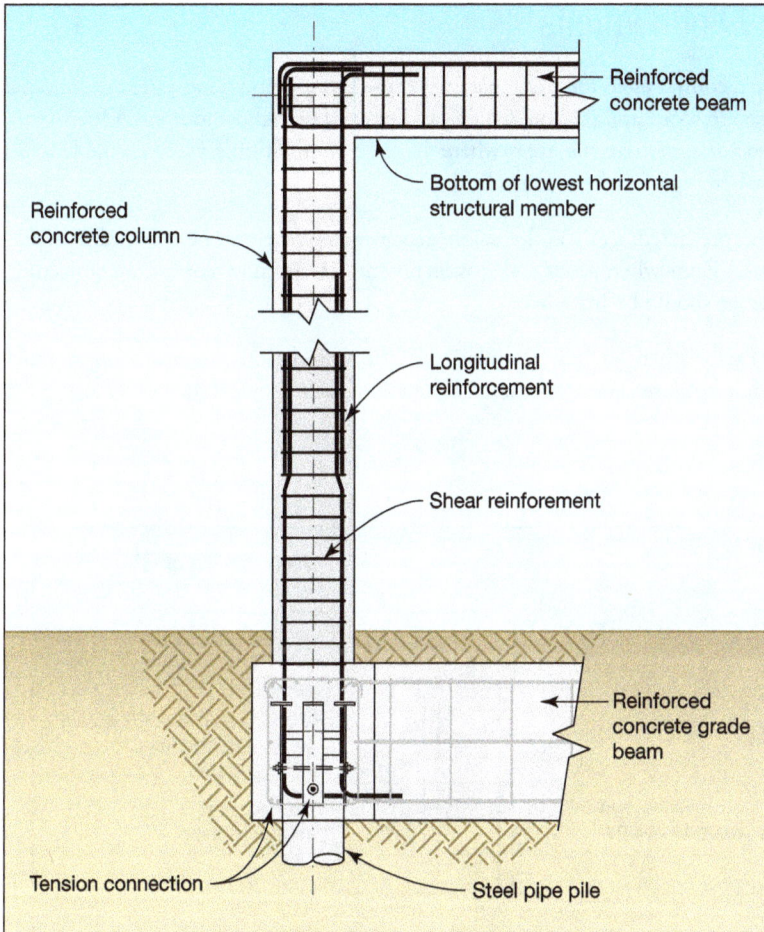

Figure 10-14.
Section view of a foundation constructed with reinforced concrete beams and columns to create portal frames
SOURCE: ADAPTED FROM FEMA P-550, SECOND EDITION, CASE H

The grade beams that are shown in Figures 10-13 and 10-14 should not be used as structural support for a concrete slab that is below an elevated building in Zone V. Although a concrete slab may serve as the floor of a ground-level enclosure (usable only for parking, storage, or building access), the slab must be independent of the building foundation. If a grade beam is used to support the slab, the slab becomes the lowest floor of the building, the beam becomes the lowest horizontal structural member supporting the lowest floor, and the bottom of the beam becomes the reference elevation for flood insurance purposes. For buildings in Zone V, the NFIP, IBC and IRC require that the lowest floor elevated to or above the BFE be supported by the bottom of the lowest horizontal structural member. Keeping the slab from being considered the lowest floor requires keeping the slab and grade beams separate, which means the slab and grade beams cannot be monolithic or connected by reinforcing steel or other means.

Like the driven, treated pile foundation discussed in Section 10.6.1, the foundation designs discussed in this section are based on presumptive piling capacities that should be verified prior to construction. Also, design professionals who develop foundations designs for specific buildings and have site information on subsurface conditions can augment the FEMA P-550 design to provide more efficient designs that reduce construction costs.

10.7   Open/Shallow Foundations

Open/shallow foundations are recommended for areas that are exposed to moving floodwaters and moderate wave actions but are not exposed to scour and erosion, which can undermine shallow footings. Open/shallow foundations are recommended for some riverine areas where an open foundation style is desirable and for buildings in Coastal A Zone where scour and erosion is limited.

In Coastal A Zones where the predicted scour and erosion depths extend below the achievable depth of shallow footings and in Coastal A Zone where scour and erosion potential is unknown or cannot be accurately predicted, open/deep foundations should be installed.

FEMA P-550 contains designs for open shallow foundations. The foundations are resistant to moving floodwaters and wave action, but because they are founded on shallow soils, they can be vulnerable to scour and erosion.

Figure 10-15.
Profile of an open/
shallow foundation
SOURCE: ADAPTED FROM
FEMA P-550, CASE D

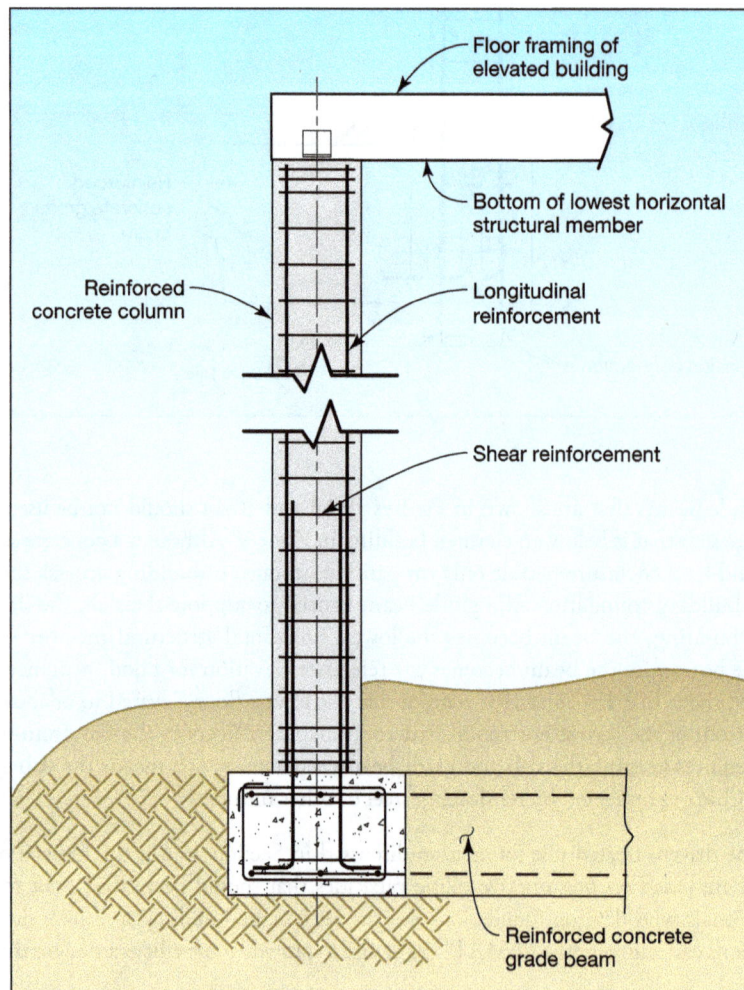

The FEMA P-550 designs make use of a rigid mat to resist lateral forces and overturning moment. Frictional resistance between the grade beams and the supporting soils resist lateral loads. The weight of the foundation and the elevated structure resist uplift forces. Because the foundation lacks the uplift resistance provided by piles, foundation elements often need to be relatively large to provide sufficient dead load to resist uplift, particularly when they are submerged. Grade beams need to be continuous because, as is shown in Section 10.9, discrete foundations that have sufficient capacity to resist lateral and uplift forces without overturning are difficult to design.

FEMA P-550 contains two types of open/shallow foundations. The foundation type shown in Figure 10-15 uses a matrix of grade beams and concrete columns to elevate the building. The grade beam shown in Figure 10-15 should not be used as structural support for a concrete slab that is below an elevated building in Zone V. If the grade beam is used to support the slab, the slab will be considered the lowest floor of the building, which will lead to the insurance ramifications described in Section 10.6.2.

When used to support wood framing, the columns of open/shallow foundations are typically designed as cantilevered beam/columns subjected to lateral forces, gravity forces and uplift forces from the elevated structure and flood forces on the foundation columns. Because of the inherent difficulty of creating moment connections with wood framing, the connections between the top of the columns and the bottom of the elevated structure are typically considered pinned. Maximum shear and moment occurs at the bottom of the columns, and proper reinforcement and detailing is needed in these areas. Also, because there are typically construction joints between the tops of the grade beams and the bases of the columns where salt-laden water can seep into the joints, special detailing is needed to prevent corrosion.

Designing an open/shallow foundation that uses concrete columns and elevated concrete beams can create a frame action that increases the foundation's ability to resist lateral loads. This design accomplishes two things. First, the frame action reduces the size of the columns and in turn reduces flood loads on them, and second, when properly designed, the elevated beams act like the tops of a perimeter foundation wall. Homes constructed to one of the designs contained in prescriptive codes can be attached to the elevated concrete beams with minimal custom design.

Unlike deep, driven-pile foundations, both types of open/shallow foundations can be undermined by erosion and scour. Neither foundation type should be used where erosion or scour is anticipated to expose the grade beam.

10.8   Closed/Shallow Foundations

Closed/shallow foundations are similar to the foundations that are used in non-coastal areas where flood forces are limited to slowly rising floods with no wave action and only limited flood velocities. In those areas, conventional foundation designs, many of which are included for residential construction in prescriptive codes and standards such as the 2012 IRC and ICC 600-2008, may be used. However, these codes and standards do not take into account forces from moving floodwaters and short breaking waves that can exist inland of Coastal A Zones. Therefore, caution should be used when using prescribed foundation designs in areas exposed to moving floodwaters and breaking waves.

FEMA P-550 contains two foundation designs for closed/shallow foundations: a stem wall foundation and a crawlspace foundation. Crawlspace foundation walls in SFHAs must be equipped with flood vents to equalize hydrostatic pressures on either side of the wall. See FEMA Technical Bulletin 1, *Openings in Foundation Walls and Walls of Enclosures* (FEMA 2008c). However, the flood vents do not significantly reduce hydrodynamic loads or breaking wave loads, and even with flood vents, flood forces in Coastal A Zones can damage or destroy these foundation styles.

Both closed/shallow foundations contained in FEMA P-550 are similar to foundations found in prescriptive codes but contain the additional reinforcement requirement to resist moving floodwaters and short (approximately 1.5-foot) breaking waves. Figure 10-16 shows the stem wall foundation design in FEMA P-550.

Figure 10-16.
Stem wall foundation design
SOURCE: ADAPTED FROM
FEMA P-550, CASE F

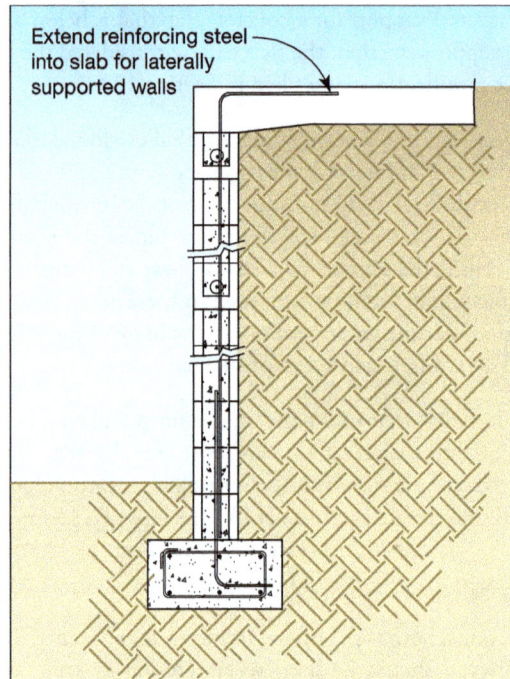

Extend reinforcing steel into slab for laterally supported walls

10.9   Pier Foundations

Properly designed pier foundations offer the following benefits: (1) their open nature reduces the loads they must resist from moving floodwaters, (2) taller piers can often be constructed to provide additional protection without requiring a lot more reinforcement, and (3) the piers can be constructed with reinforced concrete and masonry materials commonly used in residential construction.

Pier foundations, however, can have drawbacks. If not properly designed and constructed, pier foundations lack the required strength and stability to resist loads from flood, wind or seismic events. Many pier foundation failures occurred when Hurricane Katrina struck the Gulf Coast in 2005.

The type of footing used in pier foundations greatly affects the foundation's performance (see Figure 10-17). When exposed to lateral loads, discrete footings can rotate so piers placed on discrete footings are suitable

Figure 10-17.
Performance comparison
of pier foundations: piers
on discrete footings
(foreground) failed by
rotating and overturning
while piers on more
substantial footings (in
this case a concrete
mat) survived Hurricane
Katrina (Pass Christian,
MS, 2005)

only when wind and flood loads are relatively low. Piers placed on continuous concrete grade beams or concrete footings provide much greater resistance to lateral loads and are much less prone to failure. Footings and grade beams must be reinforced to resist the moment forces that develop at the base of the piers from the lateral loads on the foundation and the elevated home.

Like other open/shallow foundations, pier foundations are appropriate only where there is limited potential for erosion or scour. The maximum estimated depth for long- and short-term erosion and localized scour should not extend below the bottom of the footing or grade beam. In addition, adequate resistance to lateral loads is often difficult to achieve for common pier sizes on continuous footings. Even for relatively small lateral loads, larger piers designed as shear walls are often necessary to provide adequate resistance.

The following section provides an analysis of a pier foundation on discrete concrete footings. The analysis shows that discrete pier footings that must resist lateral loads are typically not practical.

10.9.1   Pier Foundation Design Examples

The following three examples discuss pier foundation design. Example 10.3 provides an analysis of the pier footing under gravity loads only (see Figure 10-18) and the footing size required to ensure that the allowable soil bearing pressure is not exceeded. Example 10.4 provides a consideration of uplift forces that many footings (see Figure 10-19) must resist to prevent failure during a design wind event. The analysis in Example 10.4 assumes that other foundation elements are in place to resist the lateral loads that must accompany uplift forces. Example 10.5 adds lateral loads to the pier and footing (see Figure 10-20) to model buildings that lack continuous foundation walls or other lateral load resisting features. The lateral loads can result from wind, seismic or moving floodwaters.

Figure 10-17.
Performance comparison
of pier foundations: piers
on discrete footings
(foreground) failed by
rotating and overturning
while piers on more
substantial footings (in
this case a concrete
mat) survived Hurricane
Katrina (Pass Christian,
MS, 2005)

only when wind and flood loads are relatively low. Piers placed on continuous concrete grade beams or concrete footings provide much greater resistance to lateral loads and are much less prone to failure. Footings and grade beams must be reinforced to resist the moment forces that develop at the base of the piers from the lateral loads on the foundation and the elevated home.

Like other open/shallow foundations, pier foundations are appropriate only where there is limited potential for erosion or scour. The maximum estimated depth for long- and short-term erosion and localized scour should not extend below the bottom of the footing or grade beam. In addition, adequate resistance to lateral loads is often difficult to achieve for common pier sizes on continuous footings. Even for relatively small lateral loads, larger piers designed as shear walls are often necessary to provide adequate resistance.

The following section provides an analysis of a pier foundation on discrete concrete footings. The analysis shows that discrete pier footings that must resist lateral loads are typically not practical.

10.9.1 Pier Foundation Design Examples

The following three examples discuss pier foundation design. Example 10.3 provides an analysis of the pier footing under gravity loads only (see Figure 10-18) and the footing size required to ensure that the allowable soil bearing pressure is not exceeded. Example 10.4 provides a consideration of uplift forces that many footings (see Figure 10-19) must resist to prevent failure during a design wind event. The analysis in Example 10.4 assumes that other foundation elements are in place to resist the lateral loads that must accompany uplift forces. Example 10.5 adds lateral loads to the pier and footing (see Figure 10-20) to model buildings that lack continuous foundation walls or other lateral load resisting features. The lateral loads can result from wind, seismic or moving floodwaters.

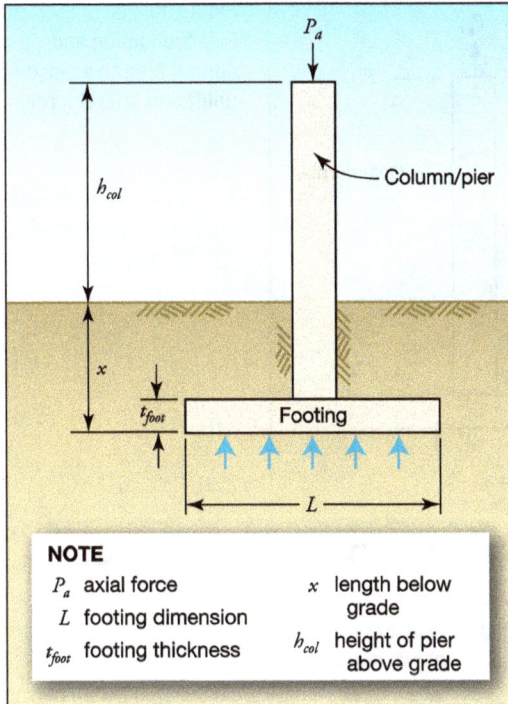

Figure 10-18.
Pier foundation and
spread footing under
gravity loading

NOTE

P_a	axial force	x	length below grade
L	footing dimension		
t_{foot}	footing thickness	h_{col}	height of pier above grade

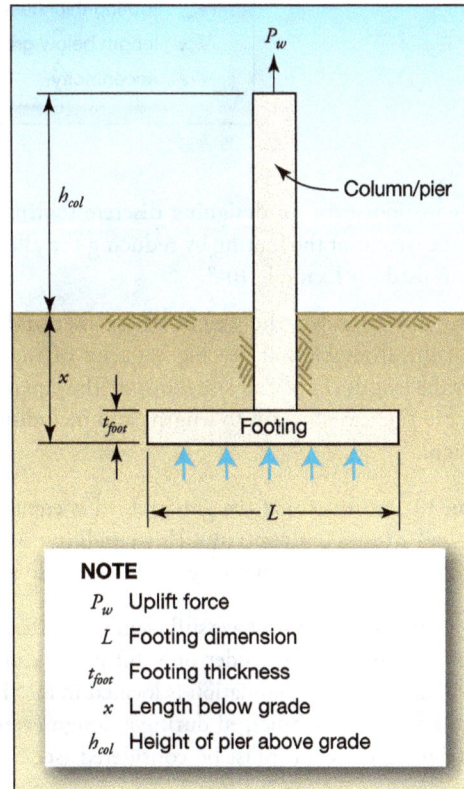

NOTE

P_w	Uplift force
L	Footing dimension
t_{foot}	Footing thickness
x	Length below grade
h_{col}	Height of pier above grade

Figure 10-19.
Pier foundation and
spread footing exposed to
uplift forces

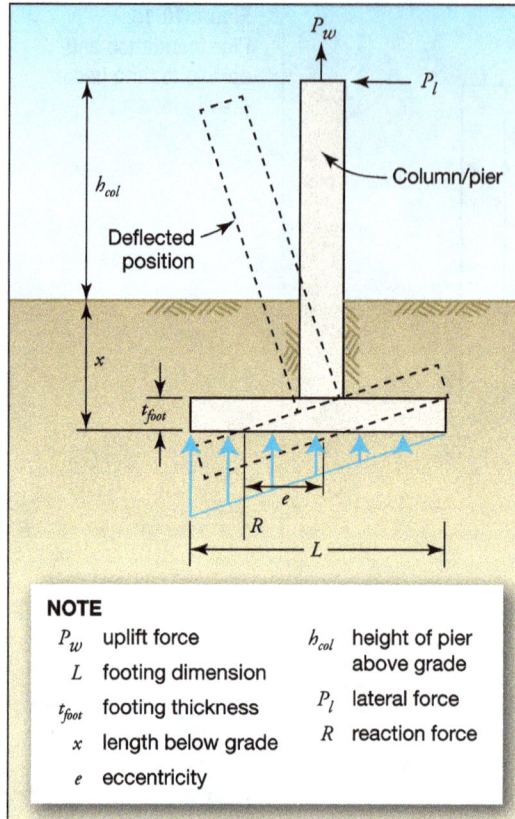

Figure 10-20.
Pier foundation and
spread footing exposed to
uplift and lateral forces

NOTE

P_w	uplift force	h_{col}	height of pier above grade
L	footing dimension		
t_{foot}	footing thickness	P_l	lateral force
x	length below grade	R	reaction force
e	eccentricity		

Several equations exist for designing discrete footings exposed to gravity loads only. Equation 10.5, which models the weight of the footing by reducing the allowable bearing capacity of the soils by the weight of the footing, is used for Example 10-3.

Equation 10.5 considers the weight of the pier and footing, the gravity load imposed on the top of the pier, and the allowable soil bearing capacity of the soils to determine footing dimensions. The equation provides the length (L) of a square footing. The equation can be modified for rectangular footings of a given aspect ratio β (ratio of width to length) and including β in the denominator of the term to the right of the equals sign.

Equation 10.5 assumes that the gravity load is equally distributed across the bottom surface of the footing and the soil stresses are constant. This condition is appropriate when the gravity loads are applied at the center of the pier (and the pier is centered on the footing) and when no lateral loads are applied.

The foundation system must have sufficient weight to prevent failure when uplift loads are applied. ASCE 7-10 requires the designer to consider only 60 percent of the dead load when designing for uplift (ASD load combination #7). If the foundation is located in an SFHA, portions of it will be located below the stillwater elevation and will be submerged during a design event. The dead load of a material is less when submerged so the submerged weight must be considered (see Section 8.5.7). In Example 10.4, it is assumed that the stillwater depth at the site is 2 feet.

EQUATION 10.5. DETERMINATION OF SQUARE FOOTING SIZE FOR GRAVITY LOADS

$$L = \left[\frac{P_a + (h_{col} + x - t_{foot})W_{col}t_{col}w_c}{q - t_{foot}w_c} \right]^{0.5}$$

where:

L = square footing dimension (ft)

P_a = gravity load on pier (lb)

h_{col} = height of pier above grade (ft)

x = distance from grade to bottom of footing (ft)

W_{col} = column width (ft)

t_{col} = column thickness (ft)

w_c = unit weight of column and footing material (lb/ft³)

q = soil bearing pressure (psf)

t_{foot} = footing thickness (ft)

EXAMPLE 10.3. PIER FOOTING UNDER GRAVITY LOAD

Given:

- Figure 10-18
- Gravity load on pier (P_a) = 2,880 lb (includes roof live load, live load, and dead load)
- Height of pier above grade (h_{col}) = 4 ft
- Distance from grade to bottom of footing (x) = 2 ft
- Column width (W_{col}) = 1.33 ft
- Column thickness (t_{col}) = 1.33 ft
- Unit weight of column and footing material (w_c) = 150 lb/ft³
- Soil bearing pressure (q) = 2,000 psf
- Footing thickness (t_{foot}) = 1 ft
- Home is 24 ft x 30 ft consisting of a matrix of 30 16-in. square piers (see Illustration A)
- Piers spaced 6 ft o.c. (see Illustration A)

EXAMPLE 10.3. PIER FOOTING UNDER GRAVITY LOAD (concluded)

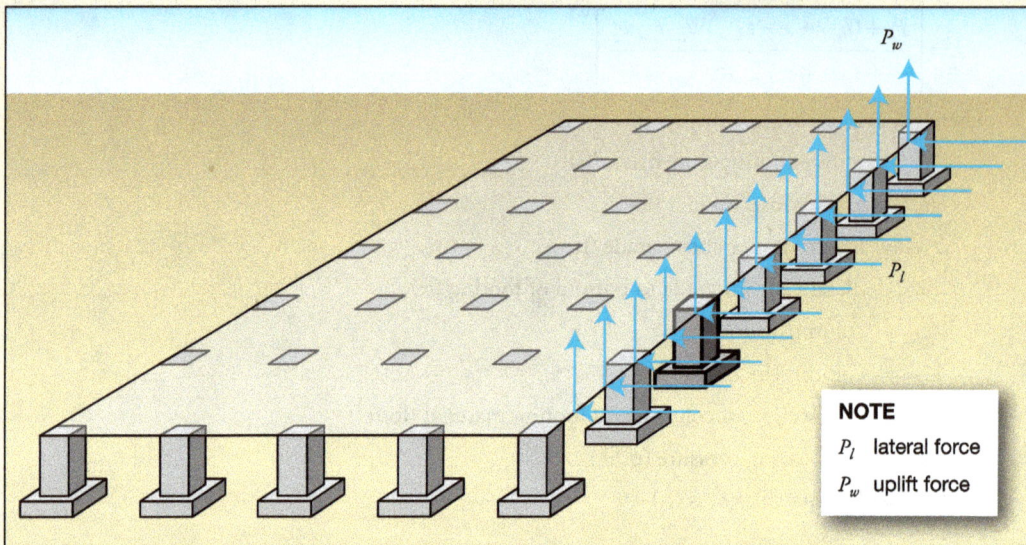

Illustration A. Site layout

NOTE
P_l lateral force
P_w uplift force

Find: The appropriate square footing size for the given gravity load.

Solution: The square footing size can be found using Equation 10.5:

$$L = \left[\frac{P_a + (h_{col} + x - t_{foot})W_{col}t_{col}w_c}{q - t_{foot}w_c} \right]^{0.5}$$

$$L = \left[\frac{2,880 \text{ lb} + (4 \text{ ft} + 2 \text{ ft} - 1 \text{ ft})(1.33 \text{ ft})(1.33 \text{ ft})(150 \text{ lb/ft}^3)}{2,000 \text{ psf} - (1 \text{ ft})(150 \text{ lb/ft}^3)} \right]^{0.5}$$

L = 1.5 ft

The IRC requires a minimum of 2-in. projection for spread footing. Moving to the next minimum standard footing size, a **24-in. x 24-in. x 12-in. square footing** to resist the gravity loads should be used.

Example 10.3 and Example 10.4 model the conditions where the pier and footing only resist axial loads that create no moment on the footing. In those states, the soils are equally loaded across the footing. When a pier and footing foundation must resist lateral loads (or must resist gravity loads applied at some distance Δ from the centroid of the pier), the footing must resist applied moments, and soils below the footing are no longer stressed equally. Soils on one side of the footing experience compressive stresses that are greater than the average compressive stress; soils on the opposite side of the footing experience stresses lower than the average.

EXAMPLE 10.4. PIER FOOTING UNDER UPLIFT LOAD

Given:

- Figure 10-19

- Stillwater flood depth (d_s) = 2 ft

- Density of water (ρ_{water}) = 64 lb/ft³

- Uplift load on pier (P_w) = 2,514 lb

- Height of pier above grade (h_{col}) = 4 ft

- Distance from grade to bottom of footing (x) = 2 ft

- Column width (W_{col}) = 1.33 ft

- Column thickness (t_{col}) = 1.33 ft

- Unit weight of column and footing material (w_c) = 150 lb/ft³

- Soil bearing pressure (q) = 2,000 psf

- Footing thickness (t_{foot}) = 1 ft

- Home is 24 ft x 30 ft consisting of a matrix of 30 16-in. square piers (see Example 10.4, Illustration A)

- Piers spaced 6 ft. on center (see Illustration A)

Find: The appropriate square footing size for the given uplift loads.

Solution: The square footing size can be found as follows:

First consider the dead load of submerged portion of column

$$DL_{submerged} = (w_c - \rho_{water})(x + d_s - t_{foot})(W_{col})(t_{col})$$

$$DL_{submerged} = (150 \text{ lb/ft}^3 - 64 \text{ lb/ft}^3)(2 \text{ ft} + 2 \text{ ft} - 1 \text{ ft})(1.33 \text{ ft})(1.33 \text{ ft}) = 459 \text{ lb}$$

Then consider the dead load of portion of column above the stillwater level

$$DL_{above} = (w_c)(h_{col} - d_x)(W_{col})(t_{col})$$

$$DL_{above} = (150 \text{ lb/ft}^3)(4 \text{ ft} - 2 \text{ ft})(1.33 \text{ ft})(1.33 \text{ ft}) = 533 \text{ lb}$$

Total column dead load can then be found

$$DL_{Total} = DL_{submerged} + DL_{above} = 992 \text{ lb}$$

The footing, when submerged, must provide sufficient weight to resist the deficit of the column dead load. The submerged footing dead load required is given by the following equation:

EXAMPLE 10.4. PIER FOOTING UNDER UPLIFT LOAD (concluded)

Submerged footing dead load =

$$\frac{1}{0.6}\left[P_W - 0.6(DL_{Total})\right] = \frac{1}{0.6}\left[2,514\ \text{lb} - 0.6(992\ \text{lb})\right] = 3,198\ \text{lb}$$

Footing volume required =

$$\frac{3,198\ \text{lb}}{(150\ \text{lb/ft}^3 - 64\ \text{lb/ft}^3)} = 37.0\ \text{ft}^3$$

For a 12-inch-thick footing, the footing area = 37 ft²

The analysis shows that **a square, 6 ft by 6 ft by 12 in., submerged concrete footing** and a 5-ft tall, 16-in. square, partially submerged concrete column are required to resist 2,514 lb of uplift. Increasing the footing thickness to 2 ft would allow the footing dimensions to be reduced to 4 ft 6 in.

At some value of lateral load or eccentricity, the compressive stresses on one side of the footing go to zero. Because there are no tensile connections between the footing and the supporting soils, the footing becomes unstable at that point and can fail by rotation. Failure can also occur when the bearing strength on the other side of the footing is exceeded.

Equation 10.6 relates soil bearing pressure to axial load, lateral load, and footing dimension. For a given axial load, lateral load, and footing dimension, the equation can be used to solve for the maximum and minimum soil bearing pressures, q on each edge of the footing. The maximum can be compared to the allowable soil bearing pressure to determine whether the soils will be overstressed. The minimum stress determines whether instability occurs. Both maximum and minimum stresses are used to determine footing size. Alternatively, for a given allowable soil bearing pressure, axial load, and lateral load, the equation can be solved for the minimum footing size.

Σ

EQUATION 10.6. DETERMINATION OF SOIL PRESSURE

$$q = \frac{P_t}{L^2} \pm 6\frac{M}{L^3}$$

where:

q = minimum and maximum soil bearing pressures at the edges of the footing (lb/ft2)

P_t = total vertical load for the load combination being analyzed

M = applied moment $P_l\,(h_{col} + x)$ (ft lbs) where x and h_{col} are as defined previously and P_l is the lateral load applied at the top of the column

When designing a pier and footing, P_t and P_l depend on the load combination being analyzed.

EXAMPLE 10.5. PIER FOOTING UNDER UPLIFT AND LATERAL LOADS

Given:

- Figure 10-20
- Stillwater flood depth (d_s) = 2 ft
- Lateral load on pier (P_l) = 246 lb (from design example in Chapter 9: (205 plf)/6 ft times 5 piers assumed to be resisting this force)
- Uplift load on pier (P_w) = 2,514 lb (derived from 419 psf from Chapter 9 times 6 ft)
- Height of pier above grade (h_{col}) = 4 ft
- Distance from grade to bottom of footing (x) = 2 ft
- Column width (W_{col}) = 1.33 ft
- Column thickness (t_{col}) = 1.33 ft
- Unit weight of column and footing material (w_c) = 150 lb/ft^2
- Soil bearing pressure (q) = 2,000 psf
- Footing thickness (t_{foot}) = 1 ft
- Home is 24 ft x 30 ft consisting of a matrix of 30 16-in. square piers (see Example 10.3, Illustration A)
- Piers spaced 6 ft o.c. (see Illustration A)

Find: The appropriate square footing size for the given uplift and lateral loads.

Solution: The square footing size can be found using Equation 10.6:

For simplicity, this example assumes the pier is partially submerged and exposed to uplift forces (as in Example 10.4) but that there are no loads from moving floodwaters or wave action. In an actual design, those forces would need to be considered. Also, if the vertical load is applied at an eccentricity "Δ", the moment $P_t\Delta$ must be combined with $P_l(H + x)$ (by vector addition) to determine the total moment applied to the footing.[2]

The total induced moment at the footing can be modeled by considering an effective reaction R numerically equal to the total vertical load P_t but applied at an eccentricity e from the centroid of the footing. The lateral load is modeled at the centroid of the footing where it contributes only to sliding. The equivalent eccentricity e is given by the following formula:

---

2 Unless the eccentricity from the lateral loads is collinear with the eccentricity from the vertical loads, the footing will be exposed to biaxial bending. For biaxial bending, soil stresses must be checked in both directions.

EXAMPLE 10.5. PIER FOOTING UNDER UPLIFT AND LATERAL LOADS (concluded)

EQUATION A

$$e = \frac{M}{P_t} \text{ (see Figure 10-20)}$$

where:

e = eccentricity

P_t = total vertical load for the load combination being analyzed

M = applied moment $P_l(H + x)$ (ft-lbs) where x and H are as defined previously

P_l is the lateral load applied at the top of the column. For equilibrium, R must be applied within the "kern" of the footing (for a square footing, the kern is a square with dimension of $L/3$ centered about the centroid of the footing). Mathematically, e cannot exceed $L/6$. Ensuring that the reaction R is applied within the kern of the footing prevents tensile stresses from forming on the edge of the footing.

Calculating the minimum soils stress for various footing widths (using a recursive solution) shows that the **footing would need to be 11 ft 4 in. wide to prevent overturning**. Increasing the footing thickness to 2 feet would allow the footing size to be reduced to approximately 8 ft 9 in. Either design is not practical to construct.

10.9.2 Pier Foundation Summary

These analyses indicate that piers with discrete footings are practical to construct when they are required to resist gravity loads only but are not practical when they must resist uplift forces or lateral loads. Although prescriptive designs for pier foundations are available in some codes and standards, users of the codes and standards should ensure that the designs take into account all of the loads the foundations must resist. Prescriptive designs should only be used to resist lateral and uplift loads after they have been confirmed to be adequate.

Constructing piers on continuous footings makes pier foundations much more resistant to coastal hazards, but prescriptive designs for piers on continuous footings are not present in widely adopted codes such as the IRC and IBC. Until prescriptive designs using piers are developed, these styles of foundations should be engineered. Continuous footings are discussed in Section 11.1.5 of FEMA 549, *Hurricane Katrina in the Gulf Coast* (FEMA 2006), and continuous footing designs that can be used for the basis of engineered foundations are contained in FEMA P-550.

10.10 References

ACI (American Concrete Institute). 2008. *Building Code Requirements for Structural Concrete and Commentary*, ACI 318-08.

ACI ASCE (American Society of Civil Engineer / TMS (The Masonry Society). 2008. *Building Code Requirements and Specifications for Masonry Structures and Related Commentaries*, ACI 530-08.

AF&PA (American Forest & Paper Association). 2012. *Wood Frame Construction Manual for One- and Two-Family Dwellings*. Washington, D.C.

ANSI (American National Standards Institute) / AF&PA. 2005. *National Design Specification for Wood Construction*.

ASTM (American Society for Testing and Materials). 2005. *Standard Specification for Round Timber Piles*, ASTM D25-99.

ASTM. 2007. *Standard Specification for Hot-Formed Welded and Seamless Carbon Steel Structural Tubing*. ASTM A501-07.

ASTM. 2008. *Standard Specification for Carbon Structural Steel*. ASTM A36/A36M-08.

ASTM. 2009. *Standard Practice for Description and Identification of Soils* (Visual-Manual Procedure). ASTM D2488-09a.

ASTM. 2010a. *Standard Practice for Classification of Soils for Engineering Purposes (Unified Soil Classification System)*. ASTM D2487-10.

ASTM. 2010b. *Standard Specification for Cold-Formed Welded and Seamless Carbon Steel Structural Tubing in Rounds and Shapes*. ASTM A500-10.

ASTM. 2010c. *Standard Specification for Pipe, Steel, Black and Hot-Dipped, Zinc-Coated, Welded and Seamless*. ASTM A53/A53M-10.

ASTM. 2010d. *Standard Specification for Welded and Seamless Steel Pipe Piles*. ASTM 252-10.

ASCE (American Society of Civil Engineers). 2010 *Minimum Design Loads for Buildings and Other Structures*. ASCE Standard ASCE 7-10.

AWPA (American Wood Protection Association). 2006. *Standard for the Care of Preservative-Treated Wood Products*, AWPA M4-06.

Bowles, J.E. 1996. *Foundation Analysis and Design*, 5th Ed. New York: McGraw-Hill.

Collin, J.G. 2002. *Timber Pile Design and Construction Manual*. The Timber Piling Council of the American Wood Preservers Institute.

FEMA (Federal Emergency Management Agency). 2006. *Recommended Residential Construction for the Gulf Coast*. FEMA P-550.

FEMA. 2008a. *Flood Damage-Resistant Materials Requirements*. Technical Bulletin 2.

FEMA. 2008b. *Free-of-Obstruction Requirements*. Technical Bulletin 5.

FEMA. 2008c. *Openings in Foundation Walls and Walls of Enclosures*. Technical Bulletin 1.

FEMA. 2009a. *Erosion, Scour, and Foundation Design*. Available at http://www.fema.gov/library/viewRecord.do?id=3539. Accessed on June 12, 2011.

FEMA. 2009b. *Recommended Residential Construction for Coastal Areas: Building on Strong and Safe Foundations*. FEMA P-550, Second Edition.

FEMA. 2010. *Home Builder's Guide to Coastal Construction Technical Fact Sheet Series*, FEMA P-499.

ICC (International Code Council). 2008. *Standard for Residential Construction in High-Wind Regions*, ICC 600-2008. ICC: Country Club Hills, IL.

ICC. 2011a. *International Building Code*. 2012 IBC. ICC: Country Club Hills, IL.

ICC. 2011b. *International Residential Code for One and Two Family Residences*. 2012 IRC. ICC: Country Club Hills, IL.

TMS (The Masonry Society). 2007. *Masonry Designers' Guide,* Fifth Edition, MDG-5.

USDN (U.S. Department of the Navy). 1982. *Foundation and Earth Structures*, Design Manual 7.2.

Designing the Building Envelope

This chapter provides guidance on the design of the building envelope in the coastal environment.[1] The building envelope comprises exterior doors, windows, skylights, exterior wall coverings, soffits, roof systems, and attic vents. In buildings elevated on open foundations, the floor is also considered a part of the envelope.

High wind is the predominant natural hazard in the coastal environment that can cause damage to the building envelope. Other natural hazards also exist in some localities. These may include wind-driven rain, salt-laden air, seismic events, hail, and wildfire. The vulnerabilities of the building envelope to these hazards are discussed in this chapter, and recommendations on mitigating them are provided.

Good structural system performance is critical to avoiding injury and minimizing damage to a building and its contents during natural hazard events but does not ensure occupant or building protection. Good

CROSS REFERENCE

For resources that augment the guidance and other information in this Manual, see the Residential Coastal Construction Web site (http://www.fema.gov/rebuild/mat/fema55.shtm).

---

1 The guidance in this chapter is based on a literature review and field investigations of a large number of houses that were struck by hurricanes, tornadoes, or straight-line winds. Some of the houses were exposed to extremely high wind speeds while others experienced moderately high wind speeds. Notable investigations include Hurricane Hugo (South Carolina, 1989) (McDonald and Smith, 1990); Hurricane Andrew (Florida, 1992) (FEMA FIA 22; Smith, 1994); Hurricane Iniki (Hawaii, 1992) (FEMA FIA 23); Hurricane Marilyn (U.S. Virgin Islands, 1995) (FEMA unpublished); Typhoon Paka (Guam, 1997) (FEMA-1193-DR-GU); Hurricane Georges (Puerto Rico, 1998) (FEMA 339); Hurricane Charley (Florida, 2004) (FEMA 488); Hurricane Ivan (Alabama and Florida, 2004) (FEMA 489); Hurricane Katrina (Louisiana and Mississippi, 2005) (FEMA 549); and Hurricane Ike (Texas, 2008) (FEMA P-757).

performance of the building envelope is also necessary. Good building envelope performance is critical for buildings exposed to high winds and wildfire.

Good performance depends on good design, materials, installation, maintenance, and repair. A significant shortcoming in any of these five elements could jeopardize the performance of the building. Good design, however, is the key element to achieving good performance. Good design can compensate to some extent for inadequacies in the other elements, but the other elements frequently cannot compensate for inadequacies in design.

The predominant cause of damage to buildings and their contents during high-wind events has been shown to be breaching of the building envelope, as shown in Figure 11-1, and subsequent water infiltration. Breaching includes catastrophic failure (e.g., loss of the roof covering or windows) and is often followed by wind-driven water infiltration through small openings at doors, windows, and walls. The loss of roof and wall coverings and soffits on the house in Figure 11-1 resulted in significant interior water damage. Recommendations for avoiding breaching are provided in this chapter.

For buildings that are in a Special Wind Region (see Figure 3-7) or in an area where the basic (design) wind speed is greater than 115 mph,[2] it is particularly important to consider the building envelope design and construction recommendations in this chapter in order to avoid wind and wind-driven water damage. In wind-borne debris regions (as defined in ASCE 7), building envelope elements from damaged buildings are often the predominant source of wind-borne debris. The wall shown in Figure 11-2 has numerous wind-borne debris scars. Asphalt shingles from nearby residences were the primary source of debris. Following the design and construction recommendations in this chapter will minimize the generation of wind-borne debris from residences.

Figure 11-1.
Good structural system performance but the loss of shingles, underlayment, siding, housewrap, and soffits resulted in significant interior water damage. Estimated wind speed: 125 mph.[3] Hurricane Katrina (Louisiana, 2005)

---

2   The 115-mph basic wind speed is based on ASCE 7-10, Risk Category II buildings. If ASCE 7-05, or an earlier version is used, the equivalent wind speed trigger is 90 mph.

3   The estimated wind speeds given in this chapter are for a 3-second gust at a 33-foot elevation for Exposure C (as defined in ASCE 7). Most of the buildings for which estimated speeds are given in this chapter are located in Exposure B, and some are in Exposure D. For buildings in Exposure B, the actual wind speed is less than the wind speed for Exposure C conditions. For example, a 130-mph Exposure C speed is equivalent to 110 mph in Exposure B.

Building integrity in earthquakes is partly dependent on the performance of the building envelope. Residential building envelopes have historically performed well during seismic events because most envelope elements are relatively lightweight. Exceptions have been inadequately attached heavy elements such as roof tile. This chapter provides recommendations for envelope elements that are susceptible to damage in earthquakes.

A building's susceptibility to wildfire depends largely on the presence of nearby vegetation and the characteristics of the building envelope, as illustrated in Figure 11-3. See FEMA P-737, *Home Builder's Guide to Construction in Wildfire Zones* (FEMA 2008), for guidance on materials and construction techniques to reduce risks associated with wildfire.

Figure 11-2. Numerous wind-borne debris scars on the wall of this house and several missing asphalt shingles. Estimated wind speed: 140 to 150 mph. Hurricane Charley (Florida, 2004)

Figure 11-3. House that survived a wildfire due in part to fire-resistant walls and roof while surrounding houses were destroyed

SOURCE: DECRA ROOFING SYSTEMS, USED WITH PERMISSION

This chapter does not address basic design issues or the general good practices that are applicable to residential design. Rather, the chapter builds on the basics by addressing the special design and construction considerations of the building envelope for buildings that are susceptible to natural hazards in the coastal environment. Flooding effects on the building envelope are not addressed because of the assumption that the envelope will not be inundated by floodwater, but envelope resistance to wind-driven rain is addressed. The recommended measures for protection against wind-driven rain should also be adequate to protect against wave spray.

11.1   Floors in Elevated Buildings

Sheathing is commonly applied to the underside of the bottom floor framing of a building that is elevated on an open foundation. The sheathing provides the following protection: (1) it protects insulation between joists or trusses from wave spray, (2) it helps minimize corrosion of framing connectors and fasteners, and (3) it protects the floor framing from being knocked out of alignment by flood-borne debris passing under the building.

A variety of sheathing materials have been used to sheath the framing, including cement-fiber panels, gypsum board, metal panels, plywood, and vinyl siding. Damage investigations have revealed that plywood offers the most reliable performance in high winds. However, as shown in Figure 11-4, even though plywood has been used, a sufficient number of fasteners are needed to avoid blow-off. Since ASCE 7 does not provide guidance for load determination, professional judgment in specifying the attachment schedule is needed. As a conservative approach, loads can be calculated by using the C&C coefficients for a roof with the slope of 7 degrees or less. However, the roof corner load is likely overly conservative for the underside of elevated floors. Applying the perimeter load to the corner area is likely sufficiently conservative.

To achieve good long-term performance, exterior grade plywood attached with stainless steel or hot-dip galvanized nails or screws is recommended (see the corroded nails in Figure 11-4).

11.2   Exterior Doors

This section addresses exterior personnel doors and garage doors. The most common problems are entrance of wind-driven rain and breakage of glass vision panels and sliding glass doors by wind-borne debris. Blow-off of personnel doors is uncommon but as shown in Figure 11-5, it can occur. Personnel door blow-off is typically caused by inadequate attachment of the door frame to the wall. Garage door failure via negative (suction) or positive pressure was common before doors with high-wind resistance became available (see Figure 11-6). Garage door failure is typically caused by the use of door and track assemblies that have insufficient wind resistance or by inadequate attachment of the tracks to nailers or to the wall. Failures such as those shown in Figures 11-5 and 11-6 can result in a substantial increase in internal pressure and can allow entrance of a significant amount of wind-driven rain.

CROSS REFERENCE

For information regarding garage doors in breakaway walls, see Fact Sheet 8.1, *Enclosures and Breakaway Walls*, in FEMA P-499, *Home Builder's Guide to Coastal Construction Technical Fact Sheet Series* (FEMA 2010b).

Figure 11-4.
Plywood panels on the underside of a house that blew away because of excessive nail spacing. Note the corroded nails (inset). Estimated wind speed: 105 to 115 mph. Hurricane Ivan (Alabama, 2004)

Figure 11-5.
Sliding glass doors pulled out of their tracks by wind suction. Estimated wind speed: 140 to 160 mph. Hurricane Charley (Florida, 2004)

Figure 11-6.
Garage door blown from its track as a result of positive pressure. Note the damage to the adhesive-set tiles (left arrow; see Section 11.5.4.1). This house was equipped with roll-up shutters (right arrow; see Section 11.3.1.2). Estimated wind speed: 140 to 160 mph. Hurricane Charley (Florida, 2004)

11.2.1   High Winds

Exterior door assemblies (i.e., door, hardware, frame, and frame attachment to the wall) should be designed to resist high winds and wind-driven rain.

11.2.1.1   Loads and Resistance

The IBC and IRC require door assemblies to have sufficient strength to resist the positive and negative design wind pressure. Personnel doors are normally specified to comply with AAMA/WDMA/CSA 101/I.S.2/A440, which references ASTM E330 for wind load testing. However, where the basic wind speed is greater than 150 mph,[4] it is recommended that design professionals specify that personnel doors comply with wind load testing in accordance with ASTM E1233. ASTM E1233 is the recommended test method in high-wind areas because it is a cyclic test method, whereas ASTM E330 is a static test. The cyclical test method is more representative of loading conditions in high-wind areas than ASTM E330. Design professionals should also specify the attachment of the door frame to the wall (e.g., type, size, spacing, edge distance of frame fasteners).

It is recommended that design professionals specify that garage doors comply with wind load testing in accordance with ANSI/DASMA 108. For garage doors attached to wood nailers, design professionals should also specify the attachment of the nailer to the wall.

CROSS REFERENCE

For design guidance on the attachment of door frames, see AAMA TIR-A-14.

For a methodology to confirm an anchorage system provides load resistance with an appropriate safety factor to meet project requirements, see AAMA 2501.

Both documents are available for purchase from the American Architectural Manufacturers Association (http://aamanet.org).

CROSS REFERENCE

For design guidance on the attachment of garage door frames, see Technical Data Sheet #161, *Connecting Garage Door Jambs to Building Framing* (DASMA 2010). Available at http://www.dasma.com/PubTechData.asp.

---

4   The 150-mph basic wind speed is based on ASCE 7-10, Risk Category II buildings. If ASCE 7-05 or an earlier version is used, the equivalent wind speed trigger is 120 mph.

11.2.1.2 Wind-Borne Debris

If a solid door is hit with wind-borne debris, the debris may penetrate the door, but in most cases, the debris opening will not be large enough to result in significant water infiltration or in a substantial increase in internal pressure. Therefore, in wind-borne debris regions, except for glazed vision panels and glass doors, ASCE 7, IBC, and IRC do not require doors to resist wind-borne debris. However, the 2007 FBC requires all exterior doors in the High-Velocity Hurricane Zone (as defined in the FBC) to be tested for wind-borne debris resistance.

It is possible for wind-borne debris to cause door latch or hinge failure, resulting in the door being pushed open, an increase in internal pressure, and potentially the entrance of a significant amount of wind-driven rain. As a conservative measure in wind-borne debris regions, solid personnel door assemblies could be specified that resist the test missile load specified in ASTM E1996. Test Missile C is applicable where the basic wind speed is less than 164 mph. Test Missile D is applicable where the basic wind speed is 164 mph or greater.[5] See Section 11.3.1.2 regarding wind-borne debris testing. If wind-borne debris-resistant garage doors are desired, the designer should specify testing in accordance with ANSI/DASMA 115.

CROSS REFERENCE

For more information about wind-borne debris and glazing in doors, see Section 11.3.1.2.

11.2.1.3 Durability

For door assemblies to achieve good wind performance, it is necessary to avoid strength degradation caused by corrosion and termites. To avoid corrosion problems with metal doors or frames, anodized aluminum or galvanized doors and frames and stainless steel frame anchors and hardware are recommended for buildings within 3,000 feet of an ocean shoreline (including sounds and back bays). Galvanized steel doors and frames should be painted for additional protection. Fiberglass doors may also be used with wood frames.

In areas with severe termite problems, metal door assemblies are recommended. If concrete, masonry, or metal wall construction is used to eliminate termite problems, it is recommended that wood not be specified for blocking or nailers. If wood is specified, see "Material Durability in Coastal Environments," a resource document available on the Residential Coastal Construction Web site, for information on wood treatment methods.

11.2.1.4 Water Infiltration

Heavy rain that accompanies high winds can cause significant wind-driven water infiltration. The magnitude of the problem increases with the wind speed. Leakage can occur between the door and its frame, the frame and the wall, and the threshold and the door. When wind speeds approach 150 mph, some leakage should be anticipated because of the high-wind pressures and numerous opportunities for leakage path development.[6]

---

5 The 164-mph basic wind speed is based on ASCE 7-10, Risk Category II buildings. If ASCE 7-05 or an earlier version is used, the equivalent wind speed trigger is 130 mph.

6 The 150-mph basic wind speed is based on ASCE 7-10, Risk Category II buildings. If ASCE 7-05 or an earlier version is used, the equivalent wind speed trigger is 120 mph.

The following elements can minimize infiltration around exterior doors:

- **Vestibule.** Adding a vestibule allows both the inner and outer doors to be equipped with weatherstripping. The vestibule can be designed with water-resistant finishes (e.g., tile), and the floor can be equipped with a drain. In addition, installing exterior threshold trench drains can be helpful (openings must be small enough to avoid trapping high-heeled shoes). Trench drains do not eliminate the problem because water can penetrate at door edges.

- **Door swing.** Out-swinging doors have weatherstripping on the interior side where it is less susceptible to degradation, which is an advantage to in-swinging doors. Some interlocking weatherstripping assemblies are available for out-swinging doors.

- **Pan flashing.** Adding flashing under the door threshold helps prevent penetration of water into the subflooring, a common place for water entry and subsequent wood decay. More information is available in Fact Sheet 6.1, *Window and Door Installation*, in FEMA P-499, *Home Builder's Guide to Coastal Construction Technical Fact Sheet Series* (FEMA 2010b).

- **Door/wall integration.** Successfully integrating the door frame and wall is a special challenge when designing and installing doors to resist wind-driven rain. More information is available in Fact Sheet 6.1 in FEMA P-499.

- **Weatherstripping.** A variety of pre-manufactured weatherstripping elements are available, including drips, door shoes and bottoms, thresholds, and jamb/ head weatherstripping. More information is available in Fact Sheet 6.1 in FEMA P-499.

Figure 11-7 shows a pair of doors that successfully resisted winds that were estimated at between 140 and 160 mph. However, as shown in the inset, a gap of about 3/8 inch between the threshold and the bottom of the door allowed a significant amount of water to be blown into the house. The weatherstripping and thresholds shown in Fact Sheet 6.1 in FEMA P-499 can minimize water entry.

Figure 11-7.
A 3/8-inch gap between the threshold and door (illustrated by the spatula handle), which allowed wind-driven rain to enter the house. Estimated wind speed: 140 to 160 mph. Hurricane Charley (Florida, 2004)

11.3   Windows and Sklylights

This section addresses exterior windows (including door vision panels) and skylights. The most common problems in the coastal environment are entrance of wind-driven rain and glazing breakage by wind-borne debris. It is uncommon for windows to be blown-in or blown-out, but it does occur (see Figure 11-8). The type of damage shown in Figure 11-8 is typically caused by inadequate attachment of the window frame to the wall, but occasionally the glazing itself is blown out of the frame. Breakage of glazing from over-pressurization sometimes occurs with windows that were manufactured before windows with high-wind resistance became available. Strong seismic events can also damage windows although it is uncommon in residential construction. Hail can cause significant damage to skylights and occasionally cause window breakage.

11.3.1   High Winds

Window and skylight assemblies (i.e., glazing, hardware for operable units, frame, and frame attachment to the wall or roof curb) should be designed to resist high winds and wind-driven rain. In wind-borne debris regions, the assemblies should also be designed to resist wind-borne debris or be equipped with shutters, as discussed below.

11.3.1.1   Loads and Resistance

The IBC and IRC require that window and skylight assemblies have sufficient strength to resist the positive and negative design wind pressures. Windows and skylights are normally specified to comply with AAMA/WDMA/CSA 101/I.S.2/A440, which references ASTM E330 for wind load testing. However, where the basic wind speed is greater than 150 mph,[7] it is recommended that design professionals specify that

Figure 11-8.
Window frame pulled out of the wall because of inadequate window frame attachment. Hurricane Georges (Puerto Rico, 1998)

---

7   The 150-mph basic wind speed is based on ASCE 7-10, Risk Category II buildings. If ASCE 7-05 or an earlier version is used, the equivalent wind speed trigger is 120 mph.

windows and skylights comply with wind load testing in accordance with ASTM E1233. ASTM E1233 is the recommended test method in high-wind areas because it is a cyclic test method, whereas ASTM E330 is a static test. The cyclical test method is more representative of loading conditions in high-wind areas than ASTM E330. Design professionals should also specify the attachment of the window and skylight frames to the wall and roof curb (e.g., type, size, spacing, edge distance of frame fasteners). Curb attachment to the roof deck should also be specified.

For design guidance on the attachment of frames, see AAMA TIR-A14 and AAMA 2501.

11.3.1.2 Wind-Borne Debris

When wind-borne debris penetrates most materials, only a small opening results, but when debris penetrates most glazing materials, a very large opening can result. Exterior glazing that is not impact-resistant (such as annealed, heat-strengthened, or tempered glass) or not protected by shutters is extremely susceptible to breaking if struck by debris. Even small, low-momentum debris can easily break glazing that is not protected. Broken windows can allow a substantial amount of water to be blown into a building and the internal air pressure to increase greatly, both of which can damage interior partitions and ceilings.

In windstorms other than hurricanes and tornadoes, the probability of a window or skylight being struck by debris is extremely low, but in hurricane-prone regions, the probability is higher. Although the debris issue was recognized decades ago, as illustrated by Figure 11-9, wind-borne debris protection was not incorporated into U.S. codes and standards until the 1990s. In order to minimize interior damage, the IBC and IRC, through ASCE 7, prescribe that exterior glazing in wind-borne debris regions be impact-resistant (i.e., laminated glass or polycarbonate) or protected with an impact-resistant covering (shutters). ASCE 7 refers to ASTM E1996 for missile (debris) loads and to ASTM E1886 for the test method to be used to demonstrate compliance with the ASTM E1996 load criteria. Regardless of whether the glazing is laminated glass, polycarbonate, or protected by shutters, glazing is required to meet the positive and negative design air pressures.

Figure 11-9.
Very old building
with robust shutters
constructed of
2x4 lumber, bolted
connections, and heavy
metal hinges. Hurricane
Marilyn (U.S. Virgin
Islands, 1995)

Wind-borne debris also occurs in the portions of hurricane-prone regions that are inland of wind-borne debris regions, but the quantity and momentum of debris are typically lower outside the wind-borne debris region. As a conservative measure, impact-resistant glazing or shutters could be specified inland of the wind-borne debris region. If the building is located where the basic wind is 125 mph[8] or greater and is within a few hundred feet of a building with an aggregate surface roof or other buildings that have limited wind resistance, it is prudent to consider impact-resistant glazing or shutters.

With the advent of building codes requiring glazing protection in wind-borne debris regions, a variety of shutter designs have entered the market. Shutters typically have a lower initial cost than laminated glass. However, unless the shutter is permanently anchored to the building (e.g., accordion shutter, roll-up shutter), storage space is needed. Also, when a hurricane is forecast, the shutters need to be deployed. The difficulty of shutter deployment and demobilization on upper-level glazing can be avoided by using motorized shutters, although laminated glass may be a more economical solution.

Because hurricane winds can approach from any direction, when debris protection is specified, it is important to specify that all exterior glazing be protected, including glazing that faces open water. At the house shown in Figure 11-10, all of the windows were protected with roll-up shutters except for those in the cupola. One of the cupola windows was broken. Although the window opening was relatively small, a substantial amount of interior water damage likely occurred.

Figure 11-10. Unprotected cupola window that was broken. Estimated wind speed: 110 mph. Hurricane Ike (Texas, 2008)

The FBC requires exterior windows and sliding glass doors to have a permanent label or marking, indicating information such as the positive and negative design pressure rating and impact-resistant rating (if applicable). Impact-resistant shutters are also required to be labeled. Figure 11-11 is an example of a permanent label on a window assembly. This label provides the positive and negative design pressure rating, test missile rating,

8   The 125-mph basic wind speed is based on ASCE 7-10, Risk Category II buildings. If ASCE 7-05 or an earlier version is used, the equivalent wind speed trigger is 100 mph.

and test standards that were used to evaluate the pressure and impact resistance. Without a label, ascertaining whether a window or shutter has sufficient strength to meet pressure and wind-borne debris loads is difficult (see Figure 11-12). It is therefore recommended that design professionals specify that windows and shutters have permanently mounted labels that contain the type of information shown in Figure 11-11.

Figure 11-11.
Design pressure and impact-resistance information in a permanent window label. Hurricane Ike (Texas, 2008)

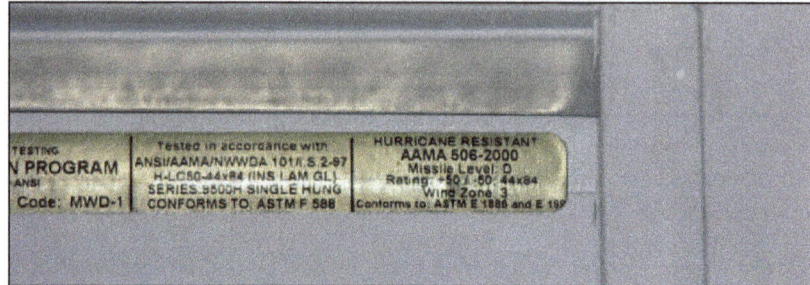

Figure 11-12.
Roll-up shutter slats that detached from the tracks. The lack of a label makes it unclear whether the shutter was tested in accordance with a recognized method. Estimated wind speed: 110 mph. Hurricane Katrina (Louisiana, 2005)

Glazing Protection from Tile Debris

Residential glazing in wind-borne debris regions is required to resist the test missile C or D, depending on the basic wind speed. However, field investigations have shown that roof tile can penetrate shutters that comply with test missile D (see Figure 11-13). Laboratory research conducted at the University of Florida indicates that test missile D compliant shutters do not provide adequate protection against tile debris (Fernandez et al. 2010). Accordingly, if tile roofs occur within 100 to 200 feet (depending on basic wind speed), it is recommended that shutters complying with test missile E be specified.

CROSS REFERENCE

More information, including a discussion of various types of shutters and recommendations pertaining to them, is available in Fact Sheet 6.2, *Protection of Openings – Shutters and Glazing*, in FEMA P-499.

Figure 11-13.
Shutter punctured by
roof tile. Estimated
wind speed: 140 to 160
mph. Hurricane Charley
(Florida, 2004)

Jalousie Louvers

In tropical climates such as Puerto Rico, some houses have metal jalousie louvers in lieu of glazed window openings (see Figure 11-14). Metal jalousies have the appearance of a debris-resistant shutter, but they typically offer little debris resistance. Neither the UBC nor IRC require openings equipped with metal jalousie louvers to be debris resistant because glazing does not occur. However, the louvers are required to meet the design wind pressure.

Because the louvers are not tightly sealed, the building should be evaluated to determine whether it is enclosed or partially enclosed (which depends on the distribution and size of the jalousie windows). Jalousie louvers are susceptible to significant water infiltration during high winds.

11.3.1.3 Durability

Achieving good wind performance in window assemblies requires avoiding strength degradation caused by corrosion and termites. To avoid corrosion, wood or vinyl frames are recommended for buildings within 3,000 feet of an ocean shoreline (including sounds and back bays). Stainless steel frame anchors and hardware are also recommended in these areas.

In areas with severe termite problems, wood frames should either be treated or not used. If concrete, masonry, or metal wall construction is used to eliminate termite problems, it is recommended that wood not be specified for blocking or nailers. If wood is specified, see "Material Durability in Coastal Environments," a resource document available on the Residential Coastal Construction Web site, for information on wood treatment methods.

Figure 11-14.
House in Puerto Rico with
metal jalousie louvers

11.3.1.4 Water Infiltration

Heavy rain accompanied by high winds can cause wind-driven water infiltration. The magnitude of the problem increases with wind speed. Leakage can occur at the glazing/frame interface, the frame itself, or between the frame and wall. When the basic wind speed is greater than 150 mph,[9] because of the very high design wind pressures and numerous opportunities for leakage path development, some leakage should be anticipated when the design wind speed conditions are approached.

A design option that partially addresses this problem is to specify a strip of water-resistant material, such as tile, along walls that have a large amount of glazing instead of extending the carpeting to the wall. During a storm, towels can be placed along the strip to absorb water infiltration. These actions can help protect carpets from water damage.

It is recommended that design professionals specify that window and skylight assemblies comply with AAMA 520. AAMA 520 has 10 performance levels. The level that is commensurate with the project location should be specified.

The successful integration of windows into exterior walls to protect against water infiltration is a challenge. To the extent possible, when detailing the interface between the wall and

NOTE

Laboratory research at the University of Florida indicates that windows with compression seals (i.e., awning and casement windows) are generally more resistant to wind-driven water infiltration than windows with sliding seals (i.e., hung and horizontal sliding windows) (Lopez et al. 2011).

CROSS REFERENCE

For guidance on window installation, see:

- FMA/AAMA 100
- FMA/AAMA 200

---

9   The 150-mph basic wind speed is based on ASCE 7-10, Risk Category II buildings. If ASCE 7-05 or an earlier version is used, the equivalent wind speed trigger is 120 mph.

the window, design professionals should rely on sealants as the secondary line of defense against water infiltration rather than making the sealant the primary protection. If a sealant joint is the first line of defense, a second line of defense should be designed to intercept and drain water that drives past the sealant joint.

CROSS REFERENCE

For a comparison of wind-driven rain resistance as a function of window installation in accordance with ASTM E2112 (as referenced in Fact Sheet 6.1 in FEMA P-499), FMA/AAMA 100, and FMA/AAMA 200, see Salzano et al. (2010).

When designing joints between walls and windows, the design professional should consider the shape of the sealant joint (i.e., , hour-glass shape with a width-to-depth ratio of at least 2:1) and the type of sealant to be specified. The sealant joint should be designed to enable the sealant to bond on only two opposing surfaces (i.e., a backer rod or bond-breaker tape should be specified). Butyl is recommended as a sealant for concealed joints and polyurethane for exposed joints. During installation, cleanliness of the sealant substrate is important, particularly if polyurethane or silicone sealants are specified, as is the tooling of the sealant.

Sealant joints can be protected with a removable stop (as illustrated in Figure 2 of Fact Sheet 6.1 of FEMA P-499). The stop protects the sealant from direct exposure to the weather and reduces the possibility of wind-driven rain penetration.

Where water infiltration protection is particularly demanding and important, onsite water infiltration testing in accordance with AAMA 502 can be specified. AAMA 502 provides pass/fail criteria based on testing in accordance with either of two ASTM water infiltration test methods. ASTM E1105 is the recommended test method.

11.3.2 Seismic

Glass breakage due to in-plane wall deflection is unlikely, but special consideration should be given to walls with a high percentage of windows and limited shear capacity. In these cases, it is important to analyze the in-plane wall deflection to verify that it does not exceed the limits prescribed in the building code.

11.3.3 Hail

A test method has not been developed for testing skylights for hail resistance, but ASTM E822 for testing hail resistance of solar collectors could be used for assessing the hail resistance of skylights.

11.4 Non-Load-Bearing Walls, Wall Coverings, and Soffits

This section addresses exterior non-load-bearing walls, wall coverings, and soffits. The most common problems in the coastal environment are soffit blow-off with subsequent entrance of wind-driven rain into attics and wall covering blow-off with subsequent entrance of wind-driven rain into wall cavities. Seismic events can also damage heavy wall systems including coverings. Although hail can damage walls, significant damage is not common.

A variety of exterior wall systems can be used in the coastal environment. The following wall coverings are commonly used over wood-frame construction: aluminum siding, brick veneer, fiber cement siding, exterior insulation finish systems (EIFS), stucco, vinyl siding, and wood siding (boards, panels, or shakes). Concrete or concrete masonry unit (CMU) wall construction can also be used, with or without a wall covering.

11.4.1 High Winds

Exterior non-load-bearing walls, wall coverings, and soffits should be designed to resist high winds and wind-driven rain. The IBC and IRC require that exterior non-load-bearing walls, wall coverings, and soffits have sufficient strength to resist the positive and negative design wind pressures.

> **NOTE**
>
> ASCE 7, IBC, and IRC do not require exterior walls or soffits to resist wind-borne debris. However, the FBC requires exterior wall assemblies in the High-Velocity Hurricane Zone (as defined in the FBC) to be tested for wind-borne debris or to be deemed to comply with the wind-borne debris provisions that are stipulated in the FBC.

11.4.1.1 Exterior Walls

It is recommended that the exterior face of studs be fully clad with plywood or oriented strand board (OSB) sheathing so the sheathing can withstand design wind pressures that produce both in-plane and out-of-plane loads because a house that is fully sheathed with plywood or OSB is more resistant to wind-borne debris and water infiltration if the wall cladding is lost.[10] The disadvantage of not fully cladding the studs with plywood or OSB is illustrated by Figure 11-15. At this residence, OSB was installed at the corner areas to provide shear resistance, but foam insulation was used in lieu of OSB in the field of the wall. In some wall areas, the vinyl siding and foam insulation on the exterior side of the studs and the gypsum board on the interior side of the studs were blown off. Also, although required by building codes, this wall system did not have a moisture barrier between the siding and OSB/foam sheathing. In addition to the wall covering damage, OSB roof sheathing was also blown off.

Wood siding and panels (e.g., textured plywood) and stucco over CMU or concrete typically perform well during high winds. However, blow-off of stucco applied directly to concrete walls (i.e., wire mesh is not applied over the concrete) has occurred during high winds. This problem can be avoided by leaving the concrete exposed or by painting it. More blow-off problems have been experienced with vinyl siding than with

> **NOTE**
>
> Almost all wall coverings permit the passage of some water past the exterior surface of the covering, particularly when the rain is wind-driven. For this reason, most wall coverings should be considered water-shedding rather than waterproofing. A secondary line of protection with a moisture barrier is recommended to avoid moisture-related problems. Asphalt-saturated felt is the traditional moisture barrier, but housewrap is now the predominate moisture barrier. Housewrap is more resistant to air flow than asphalt-saturated felt and therefore offers improved energy performance.
>
> Fact Sheet 1.9, *Moisture Barrier Systems,* and Fact Sheet 5.1, *Housewrap,* in FEMA P-499 address key issues regarding selecting and installing moisture barriers as secondary protection in exterior walls.

---

10 This recommendation is based on FEMA P-757, *Mitigation Assessment Team Report: Hurricane Ike in Texas and Louisiana* (FEMA 2009).

other siding or panel materials (see Figure 11-15). Problems with aluminum and fiber cement siding have also occurred (see Figure 11-16).

Siding

A key to the successful performance of siding and panel systems is attachment with a sufficient number of proper fasteners (based on design loads and tested resistance) that are correctly located. Fact Sheet 5.3, *Siding Installation and Connectors*, in FEMA P-499 provides guidance on specifying and installing vinyl, wood siding, and fiber cement siding in high-wind regions.

> **NOTE**
>
> In areas that experience frequent wind-driven rain and in areas that are susceptible to high winds, a pressure-equalized rain screen design should be considered when specifying wood or fiber cement siding. A rain screen design is accomplished by installing suitable vertical furring strips between the moisture barrier and siding material. The cavity facilitates drainage of water from the space between the moisture barrier and backside of the siding and facilitates drying of the siding and moisture barrier.
>
> For more information, see Fact Sheet 5.3, *Siding Installation in High-Wind Regions,* in FEMA P-499.

Figure 11-15. Blown-off vinyl siding and foam sheathing; some blow-off of interior gypsum board (circle). Estimated wind speed: 130 mph. Hurricane Katrina (Mississippi, 2006)

Brick Veneer

Blow-off of brick veneer has occurred often during high winds. Common failure modes include tie (anchor corrosion), tie fastener pull-out, failure of masons to embed ties into the mortar, and poor bonding between ties and mortar, and poor-quality mortar. Four of these failure modes occurred at the house shown in Figure 11-17. The lower bricks were attached to CMU and the upper bricks were attached to wood studs. In addition to the wall covering damage, roof sheathing was blown off along the eave.

Figure 11-16.
Blown-off fiber cement
siding; broken window
(arrow). Estimated
wind speed: 125 mph.
Hurricane Katrina
(Mississippi, 2006)

Figure 11-17.
Four brick veneer failure modes; five corrugated ties that were not embedded in the mortar joints (inset).
Hurricane Ivan (Florida, 2004)

A key to the successful performance of brick veneer is attachment with a sufficient number of properly located ties and proper tie fasteners (based on design loads and tested resistance). Fact Sheet 5.4, *Attachment of Brick Veneer in High-Wind Regions*, in FEMA P-499 provides guidance on specifying and installing brick veneer in high-wind regions.

Exterior Insulating Finishing System

EIFS can be applied over steel-frame, wood-frame, concrete, or CMU construction. An EIFS assembly is composed of several types of materials, as illustrated in Figure 11-18. Some of the layers are adhered to one another, and one or more of the layers is typically mechanically attached to the wall. If mechanical fasteners are used, they need to be correctly located, of the proper type and size, and of sufficient number (based on design loads and tested resistance). Most EIFS failures are caused by an inadequate number of fasteners or an inadequate amount of adhesive.

> **NOTE**
>
> When a window or door assembly is installed in an EIFS wall assembly, sealant between the window or door frame and the EIFS should be applied to the EIFS base coat. After sealant application, the top coat is then applied. The top coat is somewhat porous; if sealant is applied to it, water can migrate between the top and base coats and escape past the sealant.

At the residence shown in Figure 11-19, the synthetic stucco was installed over molded expanded polystyrene (MEPS) insulation that was adhered to gypsum board that was mechanically attached to wood studs. Essentially all of the gypsum board blew off (the boards typically pulled over the fasteners). The failure was initiated by detachment of the gypsum board or by stud blow off. Some of the gypsum board on the interior side of the studs was also blown off. Also, two windows were broken by debris.

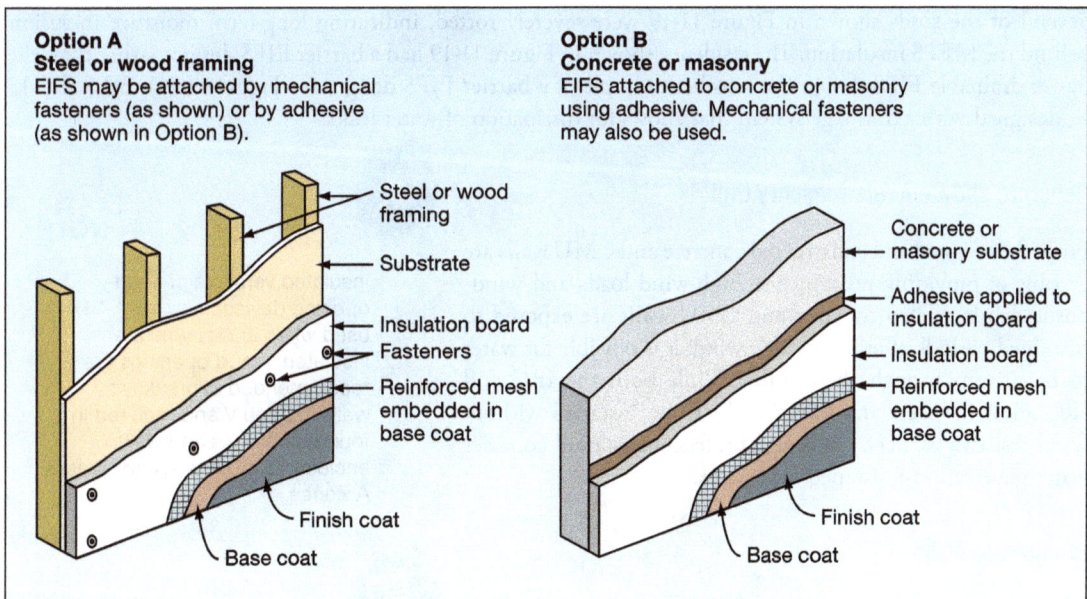

Option A
Steel or wood framing
EIFS may be attached by mechanical fasteners (as shown) or by adhesive (as shown in Option B).

- Steel or wood framing
- Substrate
- Insulation board
- Fasteners
- Reinforced mesh embedded in base coat
- Finish coat
- Base coat

Option B
Concrete or masonry
EIFS attached to concrete or masonry using adhesive. Mechanical fasteners may also be used.

- Concrete or masonry substrate
- Adhesive applied to insulation board
- Insulation board
- Reinforced mesh embedded in base coat
- Finish coat
- Base coat

Figure 11-18.
Typical EIFS assemblies

Figure 11-19.
Blown-off EIFS, resulting
in extensive interior water
damage; detachment
of the gypsum board or
stud blow off (circle);
two windows broken by
debris (arrow). Estimated
wind speed: 105 to 115
mph. Hurricane Ivan
(Florida, 2004)

Several of the studs shown in Figure 11-19 were severely rotted, indicating long-term moisture intrusion behind the MEPS insulation. The residence shown in Figure 11-19 had a barrier EIFS design, rather than the newer drainable EIFS design (for another example of a barrier EIFS design, see Figure 11-21). EIFS should be designed with a drainage system that allows for dissipation of water leaks.

Concrete and Concrete Masonry Unit

Properly designed and constructed concrete and CMU walls are capable of providing resistance to high-wind loads and wind-borne debris. When concrete and CMU walls are exposed to sustained periods of rain and high wind, it is possible for water to be driven through these walls. While both the IBC and IRC allow concrete and CMU walls to be installed without water-resistive barriers, the design professional should consider water-penetration-resistance treatments.

Breakaway Walls

Breakaway walls (enclosures) are designed to fail under base flood conditions without jeopardizing the elevated building. Breakaway walls should also be designed and constructed so that when they break away, they do so without damaging the wall above the line of separation.

NOTE

Insulated versions of flood-opening devices can be used when enclosures are insulated. Flood openings are recommended in breakaway walls in Zone V and required in foundation walls and walls of enclosures in Zone A and Coastal A Zones.

CROSS REFERENCE

For information on breakaway walls, see Fact Sheet 8.1, *Enclosures and Breakaway Walls*, in FEMA P-499.

At the house shown in Figure 11-20, floodwater collapsed the breakaway wall and initiated progressive peeling of the EIFS wall covering. A suitable flashing at the top of the breakaway wall would have avoided the progressive failure. When a wall covering progressively fails above the top of a breakaway wall, wave spray and/or wind-driven water may cause interior damage.

Figure 11-20.
Collapse of the breakaway wall, resulting in EIFS peeling. A suitable transition detail at the top of breakaway walls avoids the type of peeling damage shown by the arrows. Estimated wind speed: 105 to 115 mph. Hurricane Ivan (Alabama, 2004)

11.4.1.2 Flashings

Water infiltration at wall openings and wall transitions due to poor flashing design and/or installation is a common problem in many coastal homes (see Figure 11-21). In areas that experience frequent wind-driven rain and areas susceptible to high winds, enhanced flashing details and attention to their execution are recommended. Enhancements include flashings that have extra-long flanges, use of sealant, and use of self-adhering modified bitumen tape.

When designing flashing, the design professional should recognize that wind-driven rain can be pushed vertically. The height to which water can be pushed increases with wind speed. Water can also migrate vertically and horizontally by capillary action between layers of materials (e.g., between a flashing flange and housewrap) unless there is sealant between the layers.

NOTE

Some housewrap manufacturers have comprehensive, illustrated installation guides that address integrating housewrap and flashings at openings.

A key to successful water diversion is installing layers of building materials correctly to avoid water getting behind any one layer and leaking into the building. General guidance is offered below, design professionals should also attempt to determine the type of flashing details that have been used successfully in the area.

Figure 11-21.
EIFS with a barrier design: blown-off roof decking (top circle); severely rotted OSB due to leakage at windows (inset). Hurricane Ivan (2004)

Door and Window Flashings

An important aspect of flashing design and application is the integration of the door and window flashings with the moisture barrier. See the recommendations in FMA/AAMA 100, FMA/AAMA 200, and Salzano et al. (2010), as described in Section 11.3.1.4, regarding installation of doors and windows, as well as the recommendations given in Fact Sheet 5.1, *Housewrap*, in FEMA P-499. Applying self-adhering modified bitumen flashing tape at doors and windows is also recommended.

Roof-to-Wall and Deck-to-Wall Flashing

Where enhanced protection at roof-to-wall intersections is desired, step flashing with a vertical leg that is 2 to 4 inches longer than normal is recommended. For a more conservative design, in addition to the long leg, the top of the vertical flashing can be taped to the wall sheathing with 4-inch-wide self-adhering modified bitumen tape (approximately 1 inch of tape on the metal flashing and 3 inches on the sheathing). The housewrap should be extended over the flashing in the normal fashion. The housewrap should not be sealed to the flashing—if water reaches the backside of the housewrap farther up the wall, it needs to be able to drain out at the bottom of the wall. This detail and a deck-to-wall flashing detail are illustrated in Fact Sheet No. 5.2, *Roof-to-Wall and Deck-to-Wall Flashing*, in FEMA P-499.

11.4.1.3 Soffits

Depending on the wind direction, soffits can be subjected to either positive or negative pressure. Failed soffits may provide a convenient path for wind-driven rain to enter the building, as illustrated by Figure 11-22. This house had a steep-slope roof with a ventilated attic space. The exterior CMU/stucco wall stopped just above the vinyl soffit. Wind-driven rain entered the attic space where the soffit had blown away. This example and other storm-damage research have shown that water blown into attic spaces after the loss of soffits can cause significant damage and the collapse of ceilings. Even when soffits remain in place, water can penetrate through soffit vents and cause damage (see Section 11.6).

**Figure 11-22.
Blown-away soffit
(arrow), which allowed
wind-driven rain to enter
the attic. Estimated
wind speed: 140 to 160
mph. Hurricane Charley
(Florida, 2004)**

Loading criteria for soffits were added in ASCE 7-10. At this time, the only known test standard pertaining to soffit wind and wind-driven rain resistance is the FBC *Testing Application Standard (TAS) No. 100(A)-95* (ICC 2008). Wind-pressure testing is conducted to a maximum test speed of 140 mph, and wind-driven rain testing is conducted to a maximum test speed of 110 mph. Laboratory research has shown the need for an improved test method to evaluate the wind pressure and wind-driven rain resistance of soffits.

Plywood or wood soffits are generally adequately anchored to wood framing attached to the roof structure or walls. However, it has been common practice for vinyl and aluminum soffit panels to be installed in tracks that are frequently poorly connected to the walls and fascia at the edge of the roof overhang. Properly installed vinyl and aluminum soffit panels should be fastened to the building structure or to nailing strips placed at intervals specified by the manufacturer. Key elements of soffit installation are illustrated in Fact Sheet 7.5, *Minimizing Water Intrusion Through Roof Vents in High-Wind Regions*, in FEMA P-499.

11.4.1.4 Durability

For buildings within 3,000 feet of an ocean shoreline (including sounds and back bays), stainless steel fasteners are recommended for wall and soffit systems. For other components (e.g., furring, blocking, struts, hangers), nonferrous components (such as wood), stainless steel, or steel with a minimum of G-90 hot-dipped galvanized coating are recommended. Additionally, access panels are recommended so components within soffit cavities can be inspected periodically for corrosion or wood decay.

See "Material Durability in Coastal Environments," a resource document located on the Residential Coastal Construction Web site, for information on wood treatment if wood is specified in areas with severe termite problems.

11.4.2   Seismic

Concrete and CMU walls need to be designed for the seismic load. When a heavy covering such as brick veneer or stucco is specified, the seismic design should account for the added weight of the covering. Inadequate connection of veneer material to the base substrate has been a problem in earthquakes and can result in a life-safety hazard. For more information on the seismic design of brick veneer, see Fact Sheet 5.4, *Attachment of Brick Veneer in High-Wind Regions*, in FEMA P-499.

Some non-ductile coverings such as stucco can be cracked or spalled during seismic events. If these coverings are specified in areas prone to large ground-motion accelerations, the structure should be designed with additional stiffness to minimize damage to the wall covering.

11.5   Roof Systems

This section addresses roof systems. High winds, seismic events, and hail are the natural hazards that can cause the greatest damage to roof systems in the coastal environment. When high winds damage the roof covering, water infiltration commonly occurs and can cause significant damage to the interior of the building and its contents. Water infiltration may also occur after very large hail impact. During seismic events, heavy roof coverings such as tile or slate may be dislodged and fall from the roof and present a hazard. A roof system that is not highly resistant to fire exposure can result in the destruction of the building during a wildfire.

> **NOTE**
>
> When reroofing in high-wind areas, the existing roof covering should be removed rather than re-covered so that the roof deck can be checked for deterioration and adequate attachment. See Figure 11-23. Also see Chapter 14 in this Manual.

Residential buildings typically have steep-slope roofs (i.e., a slope greater than 3:12), but some have low-slope roofs. Low-slope roof systems are discussed in Section 11.5.8.

> **NOTE**
>
> Historically, damage to roof systems has been the leading cause of building performance problems during high winds.

A variety of products can be used for coverings on steep-slope roofs. The following commonly used products are discussed in this section: asphalt shingles, cement-fiber shingles, liquid-applied membranes, tiles, metal panels, metal shingles, slate, and wood shingles and shakes. The liquid-applied membrane and metal panel systems are air-impermeable, and the other systems are air-permeable.[11]

At the residence shown in Figure 11-23, new asphalt shingles had been installed on top of old shingles. Several of the newer shingles blew off. Re-covering over old shingles causes more substrate irregularity, which can interfere with the bonding of the self-seal adhesive of the new shingles.

---

11 Air permeability of the roof system affects the magnitude of air pressure that is applied to the system during a wind storm.

Figure 11-23.
Blow-off of several newer shingles on a roof that had been re-covered by installing new asphalt shingles on top of old shingles (newer shingles are lighter and older shingles are darker). Hurricane Charley (Florida, 2004)

11.5.1   Asphalt Shingles

The discussion of asphalt shingles relates only to shingles with self-seal tabs. Mechanically interlocked shingles are not addressed because of their limited use.

11.5.1.1   High Winds

The key elements to the successful wind performance of asphalt shingles are the bond strength of the self-sealing adhesive; mechanical properties of the shingle; correct installation of the shingle fasteners; and enhanced attachment along the eave, hip, ridge, and rakes. In addition to the tab lifts, the number and/or location of fasteners used to attach the shingles may influence whether shingles are blown off.

Underlayment

If shingles blow off, water infiltration damage can be avoided if the underlayment remains attached and is adequately sealed at penetrations. Figures 11-24 and 11-25 show houses with underlayment that was not effective in avoiding water leakage. Reliable

> **NOTE**
>
> Neither ASCE 7, IBC, or IRC require roof assemblies to resist wind-borne debris. However, the FBC requires roof assemblies located in the High-Velocity Hurricane Zone (as defined by the FBC) to be tested for wind-borne debris or be deemed to comply with the wind-borne debris provisions as stipulated in the FBC.

> **NOTE**
>
> Storm damage investigations have revealed that gutters are often susceptible to blow-off. ANSI/SPRI GD-1, *Structural Design Standard for Gutter Systems Used with Low-Slope Roofs* (ANSI/SPRI 2010) provides information on gutter wind and water and ice loads and includes methods for testing gutter resistance to these loads. Although the standard is intended for low-slope roofs, it should be considered when designing and specifying gutters used with steep-slope roofs.
>
> ANSI/SPRI GD-1 specifies a minimum safety factor of 1.67, but a safety factor of 2 is recommended.

Figure 11-24.
Small area of sheathing
that was exposed
after loss of a few
shingles and some
underlayment. Estimated
wind speed: 140 to 160
mph. Hurricane Charley
(Florida, 2004)

Figure 11-25.
Typical underlayment
attachment;
underlayment blow-off is
common if the shingles
are blown off, as shown.
Estimated wind speed:
115 mph. Hurricane
Katrina (Louisiana, 2005)

secondary protection requires an enhanced underlayment design. Design enhancements include increased blow-off resistance of the underlayment, increased resistance to water infiltration (primarily at penetrations), and increased resistance to extended weather exposure.

If shingles are blown off, the underlayment may be exposed for only 1 or 2 weeks before a new roof covering is installed, but many roofs damaged by hurricanes are not repaired for several weeks. If a hurricane strikes a heavily populated area, roof covering damage is typically extensive. Because of the heavy workload, large numbers of roofs may not be repaired for several months. It is not uncommon for some roofs to be left for as long as a year before they are reroofed.

The longer an underlayment is exposed to weather, the more durable it must be to provide adequate water infiltration protection for the residence. Fact Sheet 7.2, *Roof Underlayment for Asphalt Shingle Roofs*, in FEMA P-499 provides three primary options for enhancing the performance of underlayment if shingles are blown off. The options in the fact sheet are listed in order of decreasing resistance to long-term weather exposure. The fact sheet provides guidance for option selection, based on the design wind speed and population of the area. The following is a summary of the enhanced underlayment options:

- **Enhanced Underlayment Option 1.** Option 1 provides the greatest reliability for long-term exposure. This option includes a layer of self-adhering modified bitumen. Option 1 has two variations. The first variation is shown in Figure 11-26. In this variation, the self-adhering sheet is applied to the sheathing, and a layer of #15 felt is tacked over the self-adhering sheet before the shingles are installed. The purpose of the felt is to facilitate future tear-off of the shingles. This variation is recommended in southern climates (e.g., south of the border between North and South Carolina). If a house is located in moderate or cold climates or has a high interior humidity (such as from an indoor swimming pool), the second variation, shown in Figure 11-27, is recommended.

> **NOTE**
>
> Some OSB has a factory-applied wax that interferes with the bonding of self-adhering modified bitumen. To facilitate bonding to waxed sheathing, a field-applied primer is needed. If self-adhering modified bitumen sheet or tape is applied to OSB, the OSB manufacturer should be contacted to determine whether a primer needs to be applied to the OSB.

In the second variation (Figure 11-27), the sheathing joints are taped with self-adhering modified bitumen. A #30 felt is then nailed to the sheathing, and a self-adhering modified bitumen sheet is applied to the felt before the shingles are installed. The second variation costs more than the first variation because the second variation requires sheathing tape, many more felt fasteners, and heavier felt. The purpose of taping the joints

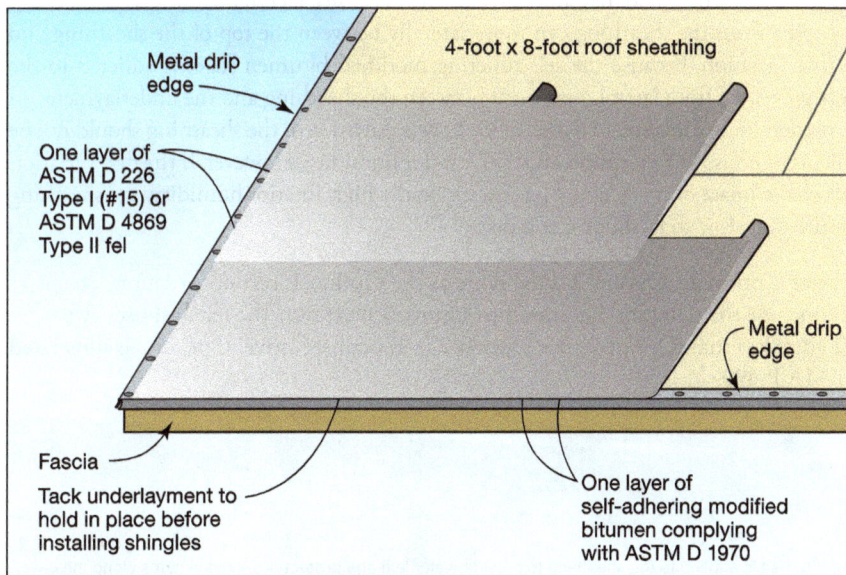

Figure 11-26. Enhanced underlayment Option 1, first variation: self-adhering modified bitumen over the sheathing

Figure 11-27.
Enhanced
underlayment Option 1,
second variation: self-
adhering modified
bitumen over the felt

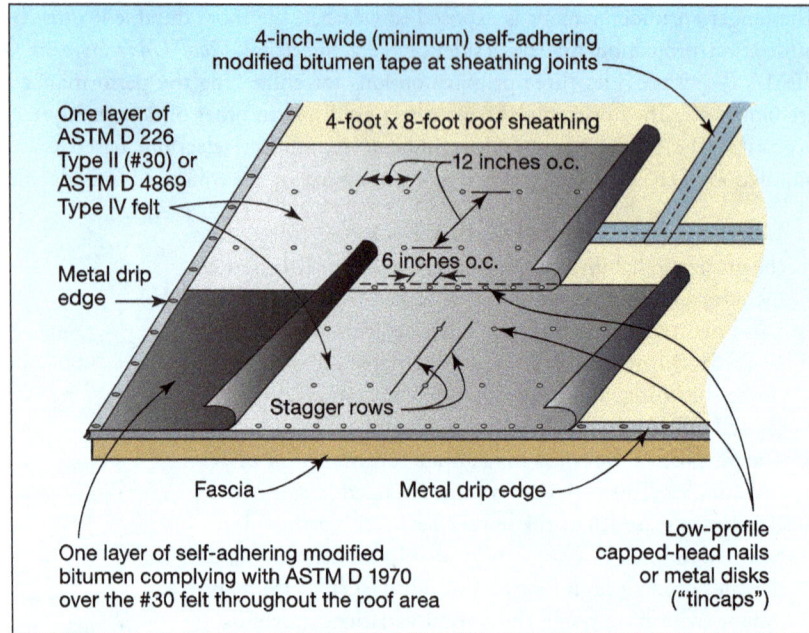

is to avoid leakage into the residence if the felt blows off or is torn by wind-borne debris. (Taping the joints is not included in the first variation, shown in Figure 11-26, because with the self-adhering modified bitumen sheet applied directly to the sheathing, sheet blow-off is unlikely, as is water leakage caused by tearing of the sheet by debris.)

The second variation is recommended in moderate and cold climates because it facilitates drying the sheathing because water vapor escaping from the sheathing can move laterally between the top of the sheathing and the nailed felt. In the first variation, because the self-adhering modified bitumen sheet is adhered to the sheathing, water vapor is prevented from lateral movement between the sheathing and the underlayment. In hot climates where the predominate direction of water vapor flow is downward, the sheathing should not be susceptible to decay unless the house has exceptionally high interior humidity. However, if the first variation is used in a moderate or cold climate or if the house has exceptionally high interior humidity, the sheathing may gain enough moisture over time to facilitate wood decay.[12]

- **Enhanced Underlayment Option 2.** Option 2 is the same as the Option 1, second variation, except that Option 2 does not include the self-adhering modified bitumen sheet over the felt and uses two layers of felt. Option 2 costs less than Option 1, but Option 2 is less conservative. Option 2 is illustrated in Fact Sheet 7.2 in FEMA P-499.

---

12 Where self-adhering modified bitumen is applied to the sheathing to provide water leakage protection from ice dams along the eave, long-term experience in the roofing industry has shown little potential for development of sheathing decay. However, sheathing decay has occurred when the self-adhering sheet is applied over all of the sheathing in cold climate areas.

■ **Enhanced Underlayment Option 3.** Option 3 is the typical underlayment scheme (i.e., a single layer of #15 felt tacked to the sheathing, as shown in Figure 11-25) with the added enhancement of self-adhering modified bitumen tape. This option provides limited protection against water infiltration if the shingles blow off. However, this option provides more protection than the typical underlayment scheme. Option 3 is illustrated in Fact Sheet 7.2 in FEMA P-499.

Figure 11-28 shows a house that used Option 3. The self-adhering modified bitumen tape at the sheathing joints was intended to be a third line of defense against water leakage (with the shingles the first line and the felt the second line). However, as shown in the inset at Figure 11-28, the tape did not provide a watertight seal. A post-storm investigation revealed application problems with the tape. Staples (arrow, inset) were used to attach the tape because bonding problems were experienced during application. Apparently, the applicator did not realize the tape was intended to prevent water from leaking through the sheathing joints. With the tape in an unbonded and wrinkled condition, it was incapable of fulfilling its intended purpose.

Self-adhering modified bitumen sheet and tape normally bond quite well to sheathing. Bonding problems are commonly attributed to dust on the sheathing, wet sheathing, or a surfacing (wax) on the sheathing that interfered with the bonding.

In addition to taping the sheathing joints in the field of the roof, the hip and ridge lines should also be taped unless there is a continuous ridge vent, and the underlayment should be lapped over the hip and ridge. By doing so, leakage will be avoided if the hip or ridge shingles blow off (see Figure 11-29). See Section 11.6 for recommendations regarding leakage avoidance at ridge vents.

Figure 11-28.
House that used enhanced underlayment Option 3 with taped sheathing joints (arrow). The self-adhering modified bitumen tape (inset) was stapled because of bonding problems. Estimated wind speed: 110 mph. Hurricane Ike (Texas, 2008)
SOURCE: IBHS, USED WITH PERMISSION

Figure 11-29.
Underlayment that was
not lapped over the hip;
water entry possible
at the sheathing joint
(arrow). Estimated
wind speed: 130 mph.
Hurricane Katrina
(Mississippi, 2005)

Shingle Products, Enhancement Details, and Application

Shingles are available with either fiberglass or organic reinforcement. Fiberglass-reinforced shingles are commonly specified because they have greater fire resistance. Fiberglass-reinforced styrene-butadiene-styrene (SBS)-modified bitumen shingles are another option. Because of the flexibility imparted by the SBS polymers, if a tab on a modified bitumen shingle lifts, it is less likely to tear or blow off compared to traditional asphalt shingles.[13] Guidance on product selection is provided in Fact Sheet 7.3, *Asphalt Shingle Roofing for High-Wind Regions*, in FEMA P-499.

The shingle product standards referenced in Fact Sheet 7.3 specify a minimum fastener (nail) pull-through resistance. However, if the basic wind speed is greater than 115 mph,[14] the Fact Sheet 7.3 recommends minimum pull-through values as a function of wind speed. If a fastener pull-through resistance is desired that is greater than the minimum value given in the product standards, the desired value needs to be specified.

ASTM D7158 addresses wind resistance of asphalt shingles.[15] ASTM D7158 has three classes: Class D, G, and H. Select shingles that have a class rating equal to or greater than the basic wind speed prescribed in the building code. Table 11-1 gives the allowable basic wind speed for each class, based on ASCE 7-05 and ASCE 7-10.

Shingle blow-off is commonly initiated at eaves (see Figure 11-30) and rakes (see Figure 11-31). Blow-off of ridge and hip shingles, as shown in Figure 11-29, is also common. For another example of blow-off of ridge

---

13 Tab lifting is undesirable. However, lifting may occur for a variety of reasons. If lifting occurs, a product that is not likely to be torn or blown off is preferable to a product that is more susceptible to tearing and blowing off.

14 The 115-mph basic wind speed is based on ASCE 7-10, Risk Category II buildings. If ASCE 7-05, or an earlier version is used, the equivalent wind speed trigger is 90 mph.

15 Fact Sheet 7.3 in FEMA P-499 references Underwriters Laboratories (UL) 2390. ASTM D7158 supersedes UL 2390.

Table 11-1. Allowable Basic Wind Speed as a Function of Class

ASTM D7158 Class[a]	Allowable Basic Wind Speed	
	Based on ASCE 7-05	Based on ASCE 7-10
D	90 mph	115 mph
G	120 mph	152 mph
H	150 mph	190 mph

(a) Classes are based on a building sited in Exposure C. They are also based on a building sited where there is no abrupt change in topography. If the residence is in Exposure D and/or where there is an abrupt change in topography (as defined in ASCE 7), the design professional should consult the shingle manufacturer.

Figure 11-30. Loss of shingles and underlayment along the eave and loss of a few hip shingles. Estimated wind speed: 115 mph. Hurricane Katrina (Louisiana, 2005)

Figure 11-31. Loss of shingles and underlayment along the rake. Estimated wind speed: 110 mph. Hurricane Ike (Texas, 2008)

and hip shingles, see Figure 11-35. Fact Sheet 7.3 in FEMA P-499 provides enhanced eave, rake, and hip/ridge information that can be used to avoid failure in these areas.

Storm damage investigations have shown that when eave damage occurs, the starter strip was typically incorrectly installed, as shown in Figure 11-32. Rather than cutting off the tabs of the starter, the starter was rotated 180 degrees (right arrow). The exposed portion of the first course of shingles (left arrow) was unbounded because the self-seal adhesive (dashed line) on the starter was not near the eave. Even when the starter is correctly installed (as shown on shingle bundle wrappers), the first course may not bond to the starter because of substrate variation. Fact Sheet 7.3 in FEMA P-499 provides information about enhanced attachment along the eave, including special recommendations regarding nailing, use of asphalt roof cement, and overhang of the shingle at the eave.

Figure 11-32. Incorrect installation of the starter course (incorrectly rotated starter, right arrow, resulted in self-seal adhesive not near the eave, dashed line). Estimated wind speed: 130 mph. Hurricane Katrina (Mississippi, 2005)

Storm damage investigations have shown that metal drip edges (edge flashings) with vertical flanges that are less than 2 inches typically do not initiate eave or rake damage. However, the longer the flange, the greater the potential for flange rotation and initiation of damage. If the vertical flange exceeds 2 inches, it is recommended that the drip edge be in compliance with ANSI/SPRI ES-1.

As with eaves, lifting and peeling failure often initiates at rakes and propagates into the field of the roof, as shown in Figure 11-33. Rakes are susceptible to failure because of the additional load exerted on the overhanging shingles and the configuration of the self-sealing adhesive. Along the long dimension of the shingle (i.e., parallel to the eave), the tab is sealed with self-sealing adhesive that is either continuous or nearly so. However, along the rake, the ends of the tab are only sealed at the self-seal lines, and the tabs are therefore typically sealed at about 5 inches on center. The result is that under high-wind loading, the adhesive at the rake end is stressed more than the adhesive farther down along the tab. With sufficient wind loading, the corner tab of the rake can begin to lift up and progressively peel, as illustrated in Figure 11-33.

Fact Sheet 7.3 in FEMA P-499 provides information about enhanced attachment along the rake, including recommendations regarding the use of asphalt roof cement along the rake. Adding dabs of cement, as shown in the Fact Sheet 7.3 in FEMA P-499 and Figure 11-33, distributes the uplift load across the ends of the rake shingles to the cement and self-seal adhesive, thus minimizing the possibility of tab uplift and progressive peeling failure.

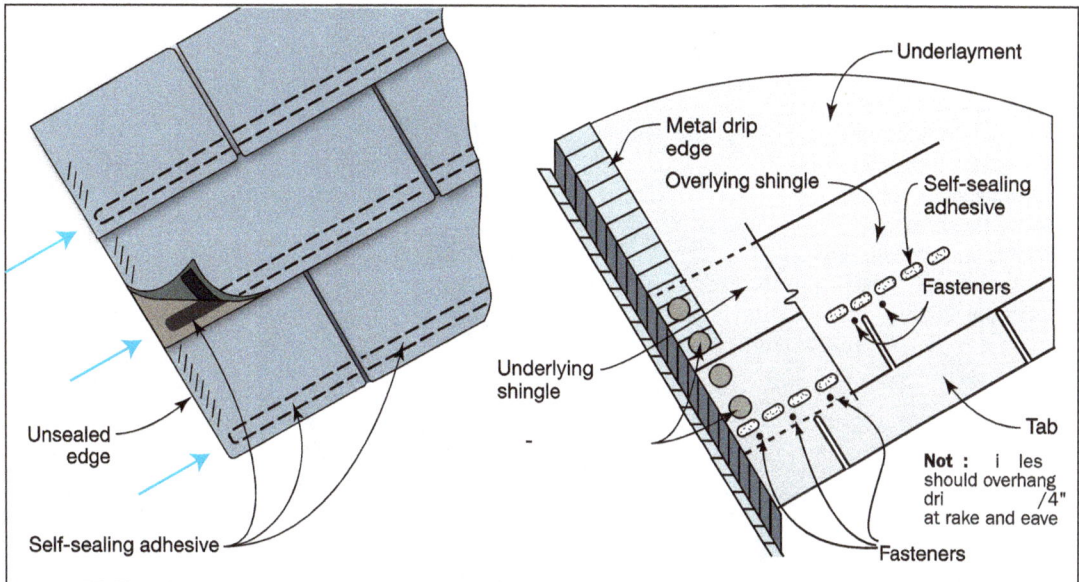

Figure 11-33.
Uplift loads along the rake that are transferred (illustrated by arrows) to the ends of the rows of self-sealing adhesive. When loads exceed resistance of the adhesive, the tabs lift and peel. The dabs of cement adhere the unsealed area shown by the hatched lines in the drawing on the left

Storm damage investigations have shown that on several damaged roofs, bleeder strips had been installed. Bleeder strips are shingles that are applied along the rake, similar to the starter course at the eave, as shown at Figure 11-34. A bleeder provides an extended straight edge that can be used as a guide for terminating the rake shingles. At first glance, it might be believed that a bleeder enhances wind resistance along the rake. However, a bleeder does not significantly enhance resistance because the concealed portion of the overlying rake shingle is the only portion that makes contact with the self-seal adhesive on the bleeder. As can be seen in Figure 11-34, the tab does not make contact with the bleeder. Hence, if the tab lifts, the shingle is placed in peel mode, which can easily break the bond with the bleeder. Also, if the tabs are not cut from the bleeder and the cut edge is placed along the rake edge, the bleeder's adhesive is too far inward to be of value.

If bleeder strips are installed for alignment purposes, the bleeder should be placed over the drip edge and attached with six nails per strip. The nails should be located 1 inch to 2 1/2 inches from the outer edge of the bleeder (1 inch is preferred if framing conditions permit). Dabs of asphalt roof cement are applied, similar to what is shown in Fact Sheet 7.3 in FEMA P-499. Dabs of asphalt roof cement are applied between the bleeder and underlying shingle, and dabs of cement are applied between the underlying and overlying shingles.

Storm damage investigations have shown that when hip and ridge shingles are blown off, there was a lack of bonding of the self-seal adhesive. Sometimes some bonding occurred, but frequently none of the adhesive had bonded. At the hip shown in Figure 11-35, the self-seal adhesive made contact only at a small area on the right side of the hip (circle). Also, at this hip, the nails were above, rather than below, the adhesive line. Lack of bonding of the hip and ridge shingles is common and is caused by substrate irregularity along the hip/ridge line. Fact Sheet 7.3 in FEMA P-499 provides recommendations regarding the use of asphalt roof cement to ensure bonding in order to enhance the attachment of hip and ridge shingles.

Figure 11-34.
A bleeder strip (double-arrow) that was used at a rake blow-off; lack of contact between the tab of the overlying shingle and the bleeder's self-seal adhesive (upper arrow). Estimated wind speed: 125 mph. Hurricane Katrina (Mississippi, 2005)

Figure 11-35.
Inadequate sealing of the self-sealing adhesive at a hip as a result of the typical hip installation procedure. Estimated wind speed: 105 mph. Hurricane Katrina (Mississippi, 2005)

Four fasteners per shingle are normally used where the basic wind speed is less than 115 mph.16 Where the basic wind speed is greater than 115 mph, six fasteners per shingle are recommended. Fact Sheet 7.3 in FEMA P-499 provides additional guidance on shingle fasteners. Storm damage investigations have shown that significant fastener mislocation is common on damaged roofs. When nails are too high above the nail line, they can miss the underlying shingle headlap or have inadequate edge distance, as illustrated

---

16 The 115-mph basic wind speed is based on ASCE 7-10, Risk Category II buildings. If ASCE 7-05 or an earlier version is used, the equivalent wind speed trigger is 90 mph.

in Figure 11-36. When laminated shingles are used, high nailing may miss the overlap of the laminated shingles; if the overlap is missed, the nail pull-through resistance is reduced (see Figure 11-37). High nailing may also influence the integrity of the self-seal adhesive bond by allowing excessive deformation (ballooning) in the vicinity of the adhesive.

The number of nails (i.e., four versus six) and their location likely play little role in wind performance as long at the shingles remain bonded. However, if they are unbounded prior to a storm, or debonded during a storm, the number and location of the nails and the shingles' nail pull-through resistance likely play an important role in the magnitude of progressive damage.

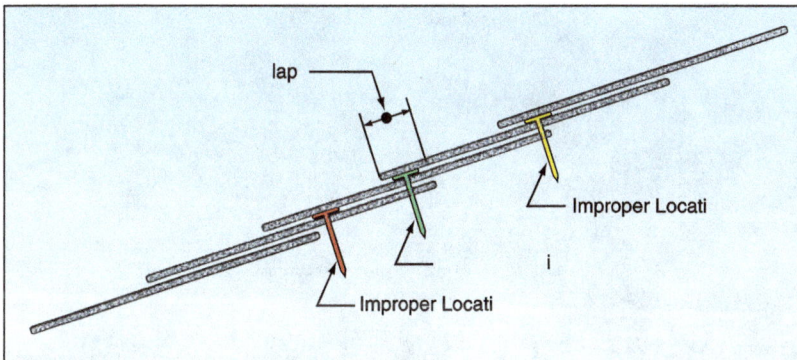

Figure 11-36.
Proper and improper location of shingle fasteners (nails). When properly located, the nail engages the underlying shingle in the headlap area (center nail). When too high, the nail misses the underlying shingle (left nail) or is too close to the edge of the underlying shingle (right nail)

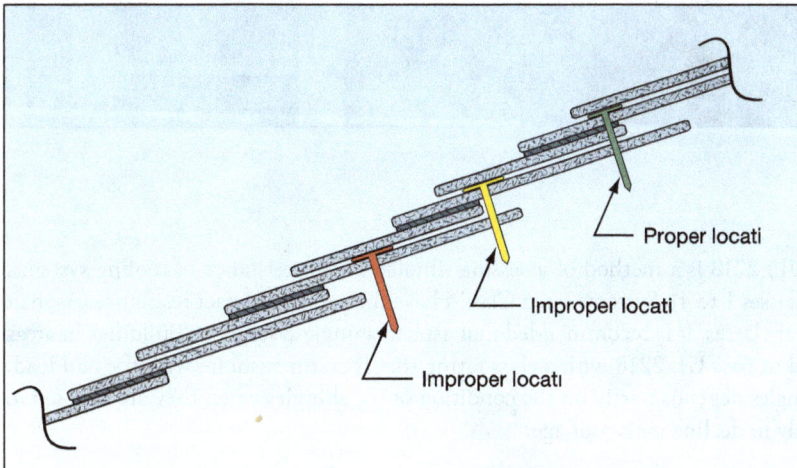

Figure 11-37.
Proper and improper location of laminated shingle fasteners (nails). With laminated shingles, properly located nails engage the underlying laminated portion of the shingle, as well as the headlap of the shingle below (right nail). When too high, the nail can miss the underlying laminated portion of the shingle but engage the headlap portion of the shingle (center nail), or the nail can miss both the underlying laminated portion of the shingle and the headlap of the underlying shingle (left nail)

Shingles manufactured with a wide nailing zone provide roofing mechanics with much greater opportunity to apply fasteners in the appropriate locations.

Shingle damage is also sometimes caused by installing shingles via the raking method. With this method, shingles are installed from eave to ridge in bands about 6 feet wide. Where the bands join one another, at every other course, a shingle from the previous row needs to be lifted up to install the end nail of the new band shingle. Sometimes installers do not install the end nail, and when that happens, the shingles are vulnerable to unzipping at the band lines, as shown in Figure 11-38. Raking is not recommended by the National Roofing Contractors Association or the Asphalt Roofing Manufacturers Association.

Figure 11-38.
Shingles that unzipped at the band lines because the raking method was used to install them. Estimated wind speed: 135 mph. Hurricane Katrina (Mississippi, 2005)

11.5.1.2   Hail

Underwriters Laboratories (UL) 2218 is a method of assessing simulated hail resistance of roofing systems. The test yields four ratings (Classes 1 to 4). Systems rated Class 4 have the greatest impact resistance. Asphalt shingles are available in all four classes. It is recommended that asphalt shingle systems on buildings in areas vulnerable to hail be specified to pass UL 2218 with a class rating that is commensurate with the hail load. Hail resistance of asphalt shingles depends partly on the condition of the shingles when they are exposed to hail. Shingle condition is likely to decline with roof age.

11.5.2   Fiber-Cement Shingles

Fiber-cement roofing products are manufactured to simulate the appearance of slate, tile, wood shingles, or wood shakes. The properties of various fiber-cement products vary because of differences in material composition and manufacturing processes.

11.5.2.1 High Winds

Because of the limited market share of fiber-cement shingles in areas where research has been conducted after high-wind events, few data are available on the wind performance of these products. Methods to calculate uplift loads and evaluate load resistance for fiber-cement products have not been incorporated into the IBC or IRC. Depending on the size and shape of the fiber-cement product, the uplift coefficient that is used for tile in the IBC may or may not be applicable to fiber-cement. If the fiber-cement manufacturer has determined that the tile coefficient is applicable to the product, Fact Sheet 7.4, *Tile Roofing for High-Wind Areas,* in FEMA P-499 is applicable for uplift loads and resistance. If the tile coefficient is not applicable, demonstrating compliance with ASCE 7 will be problematic with fiber-cement until suitable coefficient(s) have been developed.

Stainless steel straps, fasteners, and clips are recommended for roofs located within 3,000 feet of an ocean shoreline (including sounds and back bays). For underlayment recommendations, refer to the recommendation at the end of Section 11.5.4.1.

11.5.2.2 Seismic

Fiber-cement products are relatively heavy and, unless they are adequately attached, they can be dislodged during strong seismic events and fall from the roof. At press time, manufacturers had not conducted research or developed design guidance for use of these products in areas prone to large ground-motion accelerations. The guidance provided in Section 11.5.4.2 is recommended until guidance is developed for cement-fiber products.

11.5.2.3 Hail

It is recommended that fiber-cement shingle systems on buildings in areas vulnerable to hail be specified to pass UL 2218 at a class rating that is commensurate with the hail load. If products with the desired class are not available, another type of product should be considered.

11.5.3 Liquid-Applied Membranes

Liquid-applied membranes are not common on the U.S. mainland but are common in Guam, the U.S. Virgin Islands, Puerto Rico, and American Samoa.

11.5.3.1 High Winds

Investigations following hurricanes and typhoons have revealed that liquid-applied membranes installed over concrete and plywood decks have provided excellent protection from high winds if the deck remains attached to the building. This conclusion is based on performance during Hurricanes Marilyn and Georges. This type of roof covering over these deck types has high-wind-resistance reliability.

Unprotected concrete roof decks can eventually experience problems with corrosion of the slab reinforcement, based on performance observed after Hurricane Marilyn. All concrete roof decks are recommended to be covered with some type of roof covering.

11.5.3.2 Hail

It is recommended that liquid-applied membrane systems on buildings in areas vulnerable to hail be specified to pass UL 2218 or Factory Mutual Global testing with a class rating that is commensurate with the hail load.

11.5.4 Tiles

Clay and extruded concrete tiles are available in a variety of profiles and attachment methods.

11.5.4.1 High Winds

During storm damage investigations, a variety of tile profiles (e.g., S-tile and flat) of both clay and concrete tile roofs have been observed. No significant wind performance differences were attributed to tile profile or material (i.e., clay or concrete).

Figure 11-39 illustrates the type of damage that has often occurred during moderately high winds. Blow-off of hip, ridge, or eave tiles is caused by inadequate attachment. Damage to field tiles is typically caused by wind-borne debris (which is often tile debris from the eaves and hips/ridges). Many tile roofs occur over waterproof (rather than water-shedding) underlayment. Waterproof underlayments have typically been well-attached and therefore have not normally blown off after tile blow-off. Hence, many residences with tile roofs have experienced significant tile damage, but little, if any water infiltration from the roof. Figure 11-40 shows an atypical underlayment blow-off, which resulted in substantial water leakage into the house.

The four methods of attaching tile are wire-tied, mortar-set, mechanical attachment, and foam-adhesive (adhesive-set). Wire-tied systems are not commonly used in high-wind regions of the continental United States. On the roof shown in Figure 11-41, wire-tied tiles were installed over a concrete deck. Nose hooks occurred at the nose. In addition, a bead of adhesive occurred between the tiles at the headlap. Tiles at the first three perimeter rows were also attached with wind clips. The clips prevented the perimeter tiles from lifting. However, at the field of the roof, the tiles were repeatedly lifted and slammed against deck, which caused the tiles to break and blow away.

Damage investigations have revealed that mortar-set systems often provide limited wind resistance (Figure 11-42).[17] As a result of widespread poor performance of mortar-set systems during Hurricane Andrew (1992), adhesive-set systems were developed. Hurricane Charley (2004) offered the first opportunity to evaluate the field performance of this new attachment method during very high winds (see Figures 11-43 and 11-44).

Figure 11-43 shows a house with adhesive-set tile. There were significant installation problems with the foam paddies, including insufficient contact area between the patty and the tile. As can be seen in Figure 11-43, most of the foam failed to make contact with the tile. Some of the foam also debonded from the mineral surface cap sheet underlayment (see Figure 11-44).

Figure 11-45 shows tiles that were mechanically attached with screws. At the blow-off area, some of the screws remained in the deck, while others were pulled out. The ridge tiles were set in mortar.

---

17 Fact Sheet 7.4, *Tile Roofing for High-Wind Areas,* in FEMA 499 recommends that mechanical or adhesively attached methods be used in lieu of the mortar-set method.

Figure 11-39.
Blow-off of eave and hip tiles and some broken tiles in the field of the roof. Hurricane Ivan (Alabama, 2004)

Figure 11-40.
Large area of blown-off underlayment on a mortar-set tile roof. The atypical loss of waterproofing tile underlayment resulted in substantial water leakage into the house. Estimated wind speed: 140 to 160 mph. Hurricane Charley (Florida, 2004)

Figure 11-41.
Blow-off of wire-tied tiles installed over a concrete deck. Typhoon Paka (Guam, 1997)

Figure 11-42.
Extensive blow-off
of mortar-set tiles.
Hurricane Charley
(Florida, 2004)

Figure 11-43.
Blown-off adhesive-set tile. Note the very small contact area of the foam at the tile heads (left side of the tiles) and very small contact at the nose (circles). Estimated wind speed: 140 to 160 mph. Hurricane Charley (Florida, 2004)

Damage investigations have revealed that blow off of hip and ridge failures are common (see Figures 11-39, 11-45, and 11-46). Some of the failed hip/ridge tiles were attached with mortar (see Figure 11-45), while others were mortared and mechanically attached to a ridge board. At the roof shown in Figure 11-46, the hip tiles were set in mortar and attached to a ridge board with a single nail near the head of the hip tile.

Because of the brittle nature of tile, tile is often damaged by wind-borne debris, including tile from nearby buildings or tile from the same building (see Figure 11-47).

At houses on the coast, fasteners and clips that are used to attach tiles are susceptible to corrosion unless they are stainless steel. Figure 11-48 shows a 6-year-old tile roof on a house very close to the ocean that failed because the heads of the screws attaching the tile had corroded off. Stainless steel straps, fasteners, and clips are recommended for roofs within 3,000 feet of an ocean shoreline (including sounds and back bays).

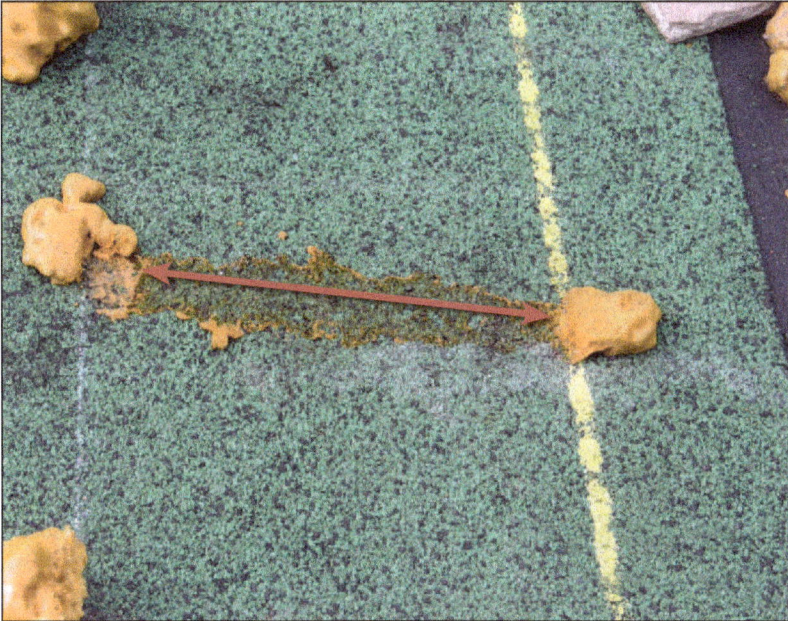

Figure 11-44.
Adhesive that debonded
from the cap sheet

Figure 11-45.
Blow-off of mechanically
attached tiles. Estimated
wind speed: 140 to 160
mph. Hurricane Charley
(Florida, 2004)

Figure 11-46.
Blow-off of hip tiles that
were nailed to a ridge
board and set in mortar.
Hurricane Ivan (Florida,
2004)

Figure 11-47.
Damage to field tiles
caused by tiles from
another area of the
roof, including a hip
tile (circle). Estimated
wind speed: 140 to 160
mph. Hurricane Charley
(Florida, 2004)

The house in Figure 11-48 had a lightning protection system (LPS), and the LPS conductors were placed under the ridge tile. Conductors are not susceptible to wind damage if they are placed under the tile and the air terminals (lightning rods) are extended through the ridge.

Figure 11-48.
The fastener heads on this mechanically attached tile roof had corroded; air terminals (lightning rods) in a lightning protection system (circle). Hurricane Ivan (Alabama, 2004)

To avoid the type of problems shown in Figures 11-39 through 11-48, see the guidance and recommendations regarding attachment and quality control in Fact Sheet 7.4, *Tile Roofing for High-Wind Areas,* in FEMA P-499. Fact Sheet 7.4 references the Third Edition of the *Concrete and Clay Roof Tile Installation Manual* (FRSA/ TRI 2001) but, as of press time, the Fourth Edition is current and therefore recommended (FRSA/TRI 2005). The Manual includes underlayment recommendations.

11.5.4.2 Seismic

Tiles are relatively heavy, and unless they are adequately attached, they can be dislodged during strong seismic events and fall away from the roof. Manufacturers have conducted laboratory research on seismic resistance of tiles, but design guidance for these products in areas prone to large ground-motion accelerations has not been developed. As shown in Figures 11-49, 11-50, and 11-51, tiles can be dislodged if they are not adequately secured.

In seismic areas where short period acceleration, Ss, exceeds 0.5g, the following are recommended:

- If tiles are laid on battens, supplemental mechanical attachment is recommended. When tiles are only loose laid on battens, they can be shaken off, as shown in Figure 11-49 where most of the tiles on the roof were nailed to batten strips. However, in one area, several tiles were not nailed. Because of the lack of nails, the tiles were shaken off the battens.

- Tiles nailed only at the head may or may not perform well. If they are attached with a smooth-shank nail into a thin plywood or OSB sheathing, pullout can occur. Figure 11-50 shows tiles that were nailed to thin wood sheathing. During the earthquake, the nose of the tiles bounced and pulled out the nails. Specifying ring-shank or screw-shank nails or screws is recommended, but even with these types of fasteners, the nose of the tile can bounce, causing enlargement of the nail hole by repeated pounding. To overcome this problem, wind clips near the nose of the tile or a bead of adhesive between the tiles at the headlap should be specified.

Figure 11-49.
Area of the roof where tiles were not nailed to batten strips. Northridge Earthquake (California, 1994)

Figure 11-50.
Tiles that were nailed to thin wood sheathing. Northridge Earthquake (California, 1994)

▪ Tiles that are attached by only one fastener experience eccentric loading. This problem can be overcome by specifying wind clips near the nose of the tile or a bead of adhesive between the tiles at the headlap.

▪ Two-piece barrel (i.e., mission) tiles attached with straw nails can slide downslope a few inches because of deformation of the long straw nail. This problem can be overcome by specifying a wire-tied system or proprietary fasteners that are not susceptible to downslope deformation.

▪ When tiles are cut to fit near hips and valleys, the portion of the tile with the nail hole(s) is often cut away. Figure 11-51 shows a tile that slipped out from under the hip tiles. The tile that slipped was trimmed to fit at the hip. The trimming eliminated the nail holes, and no other attachment was provided. The friction fit was inadequate to resist the seismic forces. Tiles must have supplemental securing to avoid displacement of these loose tiles.

Figure 11-51.
Tile that slipped out from under the hip tiles.
Northridge Earthquake (California, 1994)

- Securing rake, hip, and ridge tiles with mortar is ineffective. If mortar is specified, it should be augmented with mechanical attachment.

- Rake trim tiles fastened just near the head of the tile often slip over the fastener head because the nail hole is enlarged by repeated pounding. Additional restraint is needed for the trim pieces. Also, the design of some rake trim pieces makes them more inherently resistant to displacement than other rake trim designs.

- Stainless steel straps, fasteners, and clips are recommended for roofs within 3,000 feet of an ocean shoreline (including sounds and back bays).

11.5.4.3   Hail

Tile manufacturers assert that UL 2218 is not a good test method to assess non-ductile products such as tiles. A proprietary alternative test method is available to assess non-ductile products, but as of press time, it had not been recognized as a consensus test method.

11.5.5   Metal Panels and Metal Shingles

A variety of metal panel and shingle systems are available. Fact Sheet 7.6, *Metal Roof Systems in High-Wind Regions,* in FEMA P-499 discusses metal roofing options. Some of the products simulate the appearance of tiles or wood shakes.

11.5.5.1   High Winds

Damage investigations have revealed that some metal roofing systems have sufficient strength to resist extremely high winds, while other systems have blown off during winds that were well below the design speeds given in ASCE 7. Design and construction guidance is given in Fact Sheet 7.6 in FEMA P-499.

Figure 11-52 illustrates the importance of load path. The metal roof panels were screwed to wood nailers that were attached to the roof deck. The panels were well attached to the nailers. However, one of the nailers was inadequately attached. This nailer lifted and caused a progressive lifting and peeling of the metal panels. Note the cantilevered condenser platform (arrow), a good practice, and the broken window (circle).

Figure 11-52.
Blow-off of one of the
nailers (dashed line on
roof) caused panels
to progressively fail;
cantilevered condenser
platform (arrow);
broken window (circle).
Estimated wind speed:
130 mph. Hurricane
Katrina (Louisiana, 2005)

11.5.5.2 Hail

Several metal panel and shingle systems have passed UL 2218. Although metal systems have passed Class 4 (the class with the greatest impact resistance), they often are severely dented by the testing. Although they may still be effective in inhibiting water entry, the dents can be aesthetically objectionable. The appearance of the system is not included in the UL 2218 evaluation criteria.

11.5.6 Slate

Some fiber-cement and tile products are marketed as "slate," but slate is a natural material. Quality slate offers very long life. However, long-life fasteners and underlayment are necessary to achieve roof system longevity.

11.5.6.1 High Winds

Because of limited market share of slate in areas where research has been conducted after high-wind events, few data are available on its wind performance. However, as shown in Figure 11-53, wind damage can occur.

Methods to calculate uplift loads and evaluate load resistance for slate have not been incorporated into the IBC or IRC. Manufacturers have not conducted research to determine a suitable pressure coefficient. Demonstrating slate's compliance with ASCE 7 will be problematic until a coefficient has been developed. A consensus test method for uplift resistance has not been developed for slate.

In extreme high-wind areas, mechanical attachment near the nose of the slate should be specified in perimeter and corner zones and perhaps in the field. Because this prescriptive attachment suggestion is based on limited information, the uplift resistance that it provides is unknown.

Figure 11-53.
Damaged slate roof with nails that typically pulled out of the deck. Some of the slate broke and small portions remained nailed to the deck. Estimated wind speed: 130 mph. Hurricane Katrina (Mississippi, 2005)

Stainless steel straps, fasteners, and clips are recommended for roofs within 3,000 feet of an ocean shoreline (including sounds and back bays). For underlayment recommendations, refer to the recommendation at the end of Section 11.5.4.1.

11.5.6.2 Seismic

Slate is relatively heavy and unless adequately attached, it can be dislodged during strong seismic events and fall away from the roof. Manufacturers have not conducted research or developed design guidance for use of slate in areas prone to large ground-motion accelerations. The guidance provided for tiles in Section 11.5.4.2 is recommended until guidance has been developed for slate.

11.5.6.3 Hail

See Section 11.5.4.3.

11.5.7   Wood Shingles and Shakes

11.5.7.1 High Winds

Research conducted after high-wind events has shown that wood shingles and shakes can perform very well during high winds if they are not deteriorated and have been attached in accordance with standard attachment recommendations.

Methods to calculate uplift loads and evaluate load resistance for wood shingles and shakes have not been incorporated into the IBC or IRC. Manufacturers have not conducted research to determine suitable pressure coefficients. Demonstrating compliance with ASCE 7 will be problematic with wood shingles and shakes until such coefficients have been developed. A consensus test method for uplift resistance has not been developed for wood shingles or shakes.

For enhanced durability, preservative-treated wood is recommended for shingle or shake roofs on coastal buildings. Stainless steel fasteners are recommended for roofs within 3,000 feet of an ocean shoreline (including sounds and back bays). See Figure 11-54 for an example of shingle loss due to corrosion of the nails.

Figure 11-54.
Loss of wood shingles
due to fastener corrosion.
Hurricane Bertha (North
Carolina, 1996)

11.5.7.2   Hail

At press time, no wood-shingle assembly had passed UL 2218, but heavy shakes had passed Class 4 (the class with the greatest impact resistance) and medium shakes had passed Class 3.

The hail resistance of wood shingles and shakes depends partly on their condition when affected by hail. Resistance is likely to decline with roof age.

11.5.8   Low-Slope Roof Systems

Roof coverings on low-slope roofs need to be waterproof membranes rather than the water-shedding coverings that are used on steep-slope roofs. Although most of the low-slope membranes can be used on dead-level substrates, it is always preferable (and required by the IBC and IRC) to install them on substrates that have some slope (e.g., 1/4 inch in 12 inches [2 percent]). The most commonly used coverings on low-slope roofs are built-up, modified bitumen, and single-ply systems. Liquid-applied membranes (see Section 11.5.3), structural metal panels (see Section 11.5.5), and sprayed polyurethane foam may also be used on low-slope roofs. Information on low-slope roof systems is available in *The NRCA Roofing Manual* (NRCA 2011).

Low-slope roofing makes up a very small percentage of the residential roofing market. However, when low-slope systems are used on residences, the principles that apply to commercial roofing also apply to residential work. The natural hazards presenting the greatest challenges to low-sloped roofs in the coastal environment are high winds (see Section 11.5.8.1), earthquakes (see Section 11.5.8.2), and hail (see Section 11.5.8.3).

11.5.8.1 High Winds

Roof membrane blow-off is typically caused by lifting and peeling of metal edge flashings (gravel stops) or copings, which serve to clamp down the membrane at the roof edge. In hurricane-prone regions, roof membranes are also often punctured by wind-borne debris.

Following the criteria prescribed in the IBC will typically result in roof systems that possess adequate wind uplift resistance if properly installed. IBC references ANSI/SPRI ES-1 for edge flashings and copings. ANSI/SPRI ES-1 does not specify a minimum safety factor. Accordingly, a safety factor of 2.0 is recommended for residences.

> **NOTE**
>
> The 2009 edition of the IBC prohibits the use of aggregate roof surfacing in hurricane-prone regions.

A roof system that is compliant with IBC (and the FBC) is susceptible to interior leakage if the roof membrane is punctured by wind-borne debris. If a roof system is desired that will avoid interior leakage if struck by debris, refer to the recommendations in FEMA P-424, *Design Guide for Improving School Safety in Earthquakes, Floods and High Winds* (FEMA 2010a). Section 6.3.3.7 also provides other recommendations for enhancing wind performance.

11.5.8.2 Seismic

If a ballasted roof system is specified, its weight should be considered during seismic load analysis of the structure. Also, a parapet should extend above the top of the ballast to restrain the ballast from falling over the roof edge during a seismic event.

11.5.8.3 Hail

It is recommended that a system that has passed the Factory Mutual Research Corporation's severe hail test be specified. Enhanced hail protection can be provided by a heavyweight concrete-paver-ballasted roof system.

If the pavers are installed over a single-ply membrane, it is recommended that a layer of extruded polystyrene intended for protected membrane roof systems be specified over the membrane to provide protection if the pavers break. Alternatively, a stone protection mat intended for use with aggregate-ballasted systems can be specified.

11.6 Attic Vents

High winds can drive large amounts of water through attic ventilation openings, which can lead to collapse of ceilings. Fact Sheet 7.5, *Minimizing Water Intrusion Through Roof Vents in High-Wind Regions,* in FEMA P-499 provides design and application guidance to minimize water intrusion through new and existing attic ventilation systems. Fact Sheet 7.5 also contains a discussion of unventilated attics.

Continuous ridge vent installations, used primarily on roofs with asphalt shingles, have typically not addressed the issue of maintaining structural integrity of the roof sheathing. When the roof sheathing is used as a structural diaphragm, as it is in high-wind and seismic hazard areas, the structural integrity of the roof can be compromised by the continuous vent.

Roof sheathing is normally intended to act as a diaphragm. The purpose of the diaphragm is to resist lateral forces. To properly function, the diaphragm must have the capability of transferring the load at its boundaries from one side of the roof to the other; it normally does this through the ridge board. The continuity, or load transfer assuming a blocked roof diaphragm, is accomplished with nails. This approach is illustrated by Figure 11-55.

The problem with the continuous ridge vent installation is the need to develop openings through the diaphragm to allow air to flow from the attic space up to and through the ridge vent. For existing buildings not equipped with ridge vents, cutting slots or holes in the sheathing is required. If a saw is used to cut off 1 to 2 inches along either side of the ridge, the integrity of the diaphragm is affected. This method of providing roof ventilation should not be used without taking steps to ensure proper load transfer.

> **NOTE**
>
> When cutting a slot in a deck for a ridge vent, it is important to set the depth of the saw blade so that it only slightly projects below the bottom of the sheathing. Otherwise, as shown in Fact Sheet 7.5, the integrity of the trusses can be affected.

The two methods of providing the proper ventilation while maintaining the continuity of the blocked roof diaphragm are as follows:

1. Drill 2- to 3-inch-diameter holes in the sheathing between each truss or rafter approximately 1 1/2 inches down from the ridge. The holes should be equally spaced and should remove no more than one-half of the total amount of sheathing area between the rafters. For example, if the rafters are spaced 24 inches o.c. and 2-inch-diameter holes are drilled, they should be spaced at 6 inches o.c., which will allow about 12 square inches of vent area per linear foot when the holes are placed along either side of the ridge. This concept is illustrated in Figure 11-56.

Figure 11-55.
Method for maintaining a continuous load path at the roof ridge by nailing roof sheathing

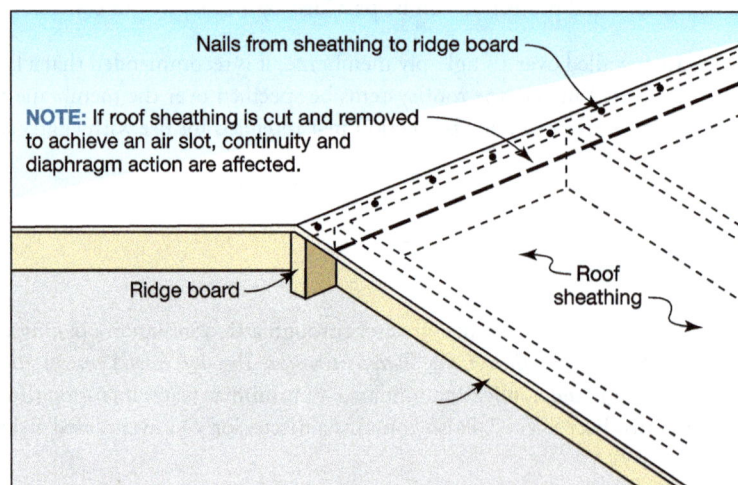

Nails from sheathing to ridge board

NOTE: If roof sheathing is cut and removed to achieve an air slot, continuity and diaphragm action are affected.

Ridge board

Roof sheathing

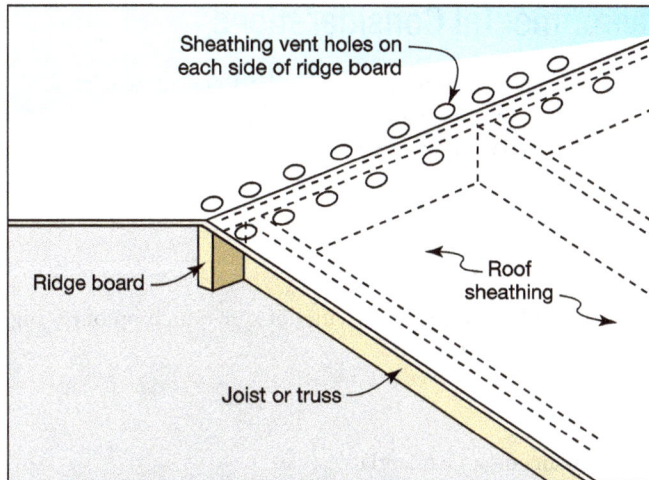

Figure 11-56.
Holes drilled in roof sheathing for ventilation and roof diaphragm action is maintained (sheathing nails not shown)

2. Install two ridge boards separated by an air space of at least 3 inches, with solid blocking between the ridge boards at each rafter or truss. Stop the sheathing at the ridge board and fully nail the sheathing as required. The ridge vent must be wide enough to cover the 3-inch gap between the ridge boards. The ridge board and blocking must be nailed to resist the calculated shear force.

For new construction, the designer should detail the ridge vent installation with the proper consideration for the load transfer requirement. Where high-diaphragm loads may occur, a design professional should be consulted regarding the amount of sheathing that can be removed or other methods of providing ventilation while still transferring lateral loads. The need to meet these requirements may become a significant problem in large or complex residential buildings where numerous ventilation openings are required. In these instances, ridge vents may need to be augmented with other ventilating devices (e.g., off-ridge vents or gable end vents).

Many ridge vent products are not very wide. When these products are used, it may be difficult to provide sufficiently large openings through the sheathing and maintain diaphragm integrity if holes are drilled through the sheathing. Manufacturers' literature often illustrates large openings at the ridge with little or no consideration for the transfer of lateral loads.

NOTE

When continuous ridge vents are used, it is not possible to continue the underlayment across the ridge. Hence, if wind-driven rain is able to drive through the vent or if the ridge vent blows off, water will leak into the house. It is likely that the ridge vent test standard referenced in Fact Sheet 7.5 in *FEMA P-499* is inadequate. One option is to avoid vent water infiltration issues by designing an unventilated attic (where appropriate, as discussed in Fact Sheet 7.5). The other option is to specify a vent that has passed the referenced test method and attach the vent with closely spaced screws (with spacing a function of the design wind speed).

11.7 Additional Environmental Considerations

In addition to water intrusion and possible resulting decay, sun (heat and ultraviolet [UV] radiation) and wind-driven rain must also be considered in selecting materials to be used in coastal buildings. The coastal environment is extremely harsh, and materials should be selected that not only provide protection from the harsh elements but also require minimal maintenance.

11.7.1 Sun

Buildings at or near the coast are typically exposed to extremes of sun, which produces high heat and UV radiation. This exposure has the following effects:

- The sun bleaches out many colors
- Heat and UV shorten the life of many organic materials
- Heat dries out lubricants such as those contained in door and window operating mechanisms

To overcome these problems:

- Use materials that are heat/UV-resistant
- Shield heat/UV susceptible materials with other materials
- Perform periodic maintenance and repair (refer to Chapter 14)

11.7.2 Wind-Driven Rain

Wind-driven rain is primarily a problem for the building envelope. High winds can carry water droplets into the smallest openings and up, into, and behind flashings, vents, and drip edges. When buildings are constructed to provide what is considered to be complete protection from the effects of natural hazards, any small "hole" in the building envelope becomes an area of weakness into which sufficiently high wind can drive a large amount of rain.

11.8 References

AAMA (American Architectural Manufacturers Association). 2008. *Standard/Specification for Windows, Doors, and Unit Skylights*. AAMA/WDMA/CSA 101/I.S.2/A440-08.

AAMA. 2008. *Voluntary Specification for Field Testing of Newly Installed Fenestration Products*. AAMA 502-08.

AAMA. 2009. *Voluntary Specification for Rating the Severe Wind-Driven Rain Resistance of Windows, Doors, and Unit Skylights*. AAMA 520-09.

AAMA. *Fenestration Anchorage Guidelines*. AAMA TIR-A14.

AAMA. *Voluntary Guideline for Engineering Analysis of Window and Sliding Glass Door Anchorage Systems.* AAMA 2501.

ANSI/SPRI (American National Standards Institute / Single-Ply Roofing Industry). 2003. *Wind Design Standard for Edge Systems Used with Low Slope Roofing Systems.* ANSI/SPRI ES-1.

ANSI/SPRI. 2010. *Structural Design Standard for Gutter Systems Used with Low-Slope Roofs.* ANSI/SPRI GD-1.

ASCE (American Society of Civil Engineers). 2005. *Minimum Design Loads for Buildings and Other Structures.* ASCE Standard ASCE 7-05.

ASCE. 2010. *Minimum Design Loads for Buildings and Other Structures.* ASCE Standard ASCE 7-10.

ASTM. *Standard Practice for Determining Resistance of Solar Collector Covers to Hail by Impact with Propelled Ice Balls.* ASTM E822.

ASTM. *Standard Specification for Performance of Exterior Windows, Curtain Walls, Doors, and Impact Protective Systems Impacted by Windborne Debris in Hurricanes.* ASTM E1996.

ASTM. *Standard Test Method for Field Determination of Water Penetration of Installed Exterior Windows, Skylights, Doors, and Curtain Walls, by Uniform or Cyclic Static Air Pressure Difference.* ASTM E1105.

ASTM. *Standard Test Method for Performance of Exterior Windows, Curtain Walls, Doors, and Impact Protective Systems Impacted by Missile(s) and Exposed to Cyclic Pressure Differentials.* ASTM E1886.

ASTM. *Standard Test Method for Structural Performance of Exterior Windows, Doors, Skylights, and Curtain Walls by Uniform Static Air Pressure Difference.* ASTM E330.

ASTM. *Standard Test Method for Structural Performance of Exterior Windows, Doors, Skylights, and Curtain Walls by Cyclic Air Pressure Differential.* ASTM E1233.

ASTM. *Standard Test Method for Wind Resistance of Asphalt Shingles (Uplift Force/Uplift Resistance Method).* ASTM D7158.

DASMA (Door & Access Systems Manufacturers Association International). 2005. *Standard Method for Testing Sectional Garage Doors and Rolling Doors: Determination of Structural Performance Under Missile Impact and Cyclic Wind Pressure.* ANSI/DASMA 115. Available at http://www.dasma.com. Accessed January 2011.

DASMA. 2005. *Standard Method For Testing Sectional Garage Doors and Rolling Doors: Determination of Structural Performance Under Uniform Static Air Pressure Difference.* ANSI/DASMA 108. Available at http://www.dasma.com. Accessed January 2011.

DASMA. 2010. *Connecting Garage Door Jambs to Building Framing.* Technical Data Sheet #161. Available at http://www.dasma.com/PubTechData.asp. Accessed January 2011.

FEMA (Federal Emergency Management Agency). 1992. *Building Performance: Hurricane Andrew in Florida – Observations, Recommendations, and Technical Guidance.* FEMA FIA 22.

FEMA. 1993. *Building Performance: Hurricane Iniki in Hawaii – Observations, Recommendations, and Technical Guidance.* FEMA FIA 23.

FEMA. 1998. *Typhoon Paka: Observations and Recommendations on Building Performance and Electrical Power Distribution System, Guam, U.S.A.* FEMA-1193-DR-GU.

FEMA. 1999. *Building Performance Assessment Team (BPAT) Report – Hurricane Georges in Puerto Rico, Observations, Recommendations, and Technical Guidance.* FEMA 339.

FEMA. 2005a. *Mitigation Assessment Team Report: Hurricane Charley in Florida.* FEMA 488.

FEMA. 2005b. *Hurricane Ivan in Alabama and Florida: Observations, Recommendations and Technical Guidance.* FEMA 489.

FEMA. 2006. *Hurricane Katrina in the Gulf Coast.* FEMA 549.

FEMA. 2008. *Home Builder's Guide to Construction in Wildfire Zones.* FEMA P-737.

FEMA. 2009. *Hurricane Ike in Texas and Louisiana.* FEMA P-757.

FEMA. 2010a. *Design Guide for Improving School Safety in Earthquakes, Floods and High Winds.* FEMA P-424.

FEMA. 2010b. *Home Builder's Guide to Coastal Construction Technical Fact Sheet Series.* FEMA P-499.

Fernandez, G., F. Masters, and K. Gurley. 2010. "Performance of Hurricane Shutters under Impact by Roof Tiles," *Engineering Structures* Vol. 32, Issue 10, pp. 3384–3393.

FMA/AAMA (Fenestration Manufacturers Association/American Architectural Manufacturers Association). 2007. *Standard Practice for the Installation of Windows with Flanges or Mounting Fins in Wood Frame Construction.* FMA/AAMA 100-07.

FMA/AAMA. 2009. *Standard Practice for the Installation of Windows with Frontal Flanges for Surface Barrier Masonry Construction for Extreme Wind/Water Conditions.* FMA/AAMA 200-09.

FRSA/TRI (Florida Roofing, Sheet Metal and Air Conditioning Contractors Association, Inc./The Roofing Institute). 2001. *Concrete and Clay Roof Tile Installation Manual. Third Edition.*

FRSA/TRI. 2005. *Concrete and Clay Roof Tile Installation Manual. Fourth Edition.*

ICC (International Code Council). 2008. *2007 Florida Building Code: Building.*

ICC. 2009a. *International Building Code* (2009 IBC). Country Club Hills, IL: ICC.

ICC. 2009b. *International Residential Code* (2009 IRC). Country Club Hills, IL: ICC.

ICBO (International Council of Building Officials). *Uniform Building Code.*

Lopez, C., F.J. Masters, and S. Bolton. 2011. "Water Penetration Resistance of Residential Window and Wall Systems Subjected to Steady and Unsteady Wind Loading," *Building and Environment* 46, Issue 7, pp. 1329–1342.

McDonald, J.R. and T.L. Smith. 1990. *Performance of Roofing Systems in Hurricane Hugo.* Institute for Disaster Research, Texas Tech University.

NRCA (National Roofing Contractors Association). 2011. *The NRCA Roofing Manual.*

Salzano, C.T., F.J. Masters, and J.D. Katsaros. 2010. "Water Penetration Resistance of Residential Window Installation Options for Hurricane-prone Areas," *Building and Environment* 45, Issue 6, pp. 1373–1388.

Smith, T.L. 1994. "Causes of Roof Covering Damage and Failure Modes: Insights Provided by Hurricane Andrew," *Proceedings of the Hurricanes of 1992*, ASCE.

Installing Mechanical Equipment and Utilities

This chapter provides guidance on design considerations for elevators, exterior-mounted and interior mechanical equipment, and utilities (electric, telephone, and cable TV systems and water and wastewater systems). Protecting mechanical equipment and utilities is a key component of successful building performance during and after a disaster event.

CROSS REFERENCE

For resources that augment the guidance and other information in this Manual, see the Residential Coastal Construction Web site (http://www.fema.gov/rebuild/mat/fema55.shtm).

12.1 Elevators

Elevators are being installed with increasing frequency in elevated, single-family homes in coastal areas. The elevators are generally smaller than elevators in non-residential buildings but are large enough to provide handicap accessibility and accommodate small household furniture and equipment.

Small (low-rise) residential elevators that are added as part of a post-construction retrofit are usually installed in a shaft independent of an outside wall. Residential elevators designed as part of new construction can be installed in a shaft in the interior of the structure. In either case, the elevator shaft must have a landing, which is usually at the ground level, and a cab platform near the top. The bottom or pit of an elevator with a landing at the lower level is almost always below the BFE.

Appendix H in NFIP Technical Bulletin 4, *Elevator Installation for Buildings Located in Special Flood Hazard Areas in Accordance with the National Flood Insurance Program* (FEMA 2010a), discusses the installation of elevator systems and equipment in the floodplain. As explained in the bulletin, elevator shafts and enclosures that extend below the BFE in coastal areas must be designed to resist hydrostatic, hydrodynamic, and wave forces as well as erosion and scour, but are not required to include hydrostatic openings or breakaway walls. In addition, elevator accessory equipment should be installed above the BFE, replaced with flood damage-resistant elements, or treated with flood damage-resistant paint or coatings to minimize flood damage.

For safety reasons, commercial and large (high-rise) elevators are designed with "fire recall" circuitry that sends the elevator to a designated floor during a fire so emergency services personnel can use the elevators. However, during flooding, this feature may expose the cab directly to floodwaters. Therefore, for elevators in coastal buildings, the elevator must be equipped with a float switch that sends the elevator cab to a level above the BFE. In addition, the design professional must ensure that the elevator stops at a level above the BFE when the power is lost. This can be accomplished by installing an emergency generator or a battery descent feature that is integrated into the float switch, as described in NFIP Technical Bulletin 4.

Finally, although elevators and elevator equipment are permitted for building access and may be covered by flood insurance, their presence, location, and size can affect flood insurance premiums. For buildings in Zone V, the NFIP considers an elevator enclosure a building enclosure or an obstruction, which may be subject to an insurance rate loading depending on:

- Square footage of the enclosure
- Value of the elevator equipment
- Location of the elevator equipment in relation to the BFE

12.2   Exterior-Mounted Mechanical Equipment

Exterior-mounted mechanical equipment can include exhaust fans, vent hoods, air conditioning units, duct work, pool motors, and well pumps. High winds, flooding, and seismic events are the natural hazards that can cause the greatest damage to exterior-mounted mechanical equipment.

12.2.1   High Winds

Equipment is typically damaged because it is not anchored or the anchorage is inadequate. Damage may also be caused by inadequate equipment strength or corrosion. Relatively light exhaust fans and vent hoods are commonly blown away during high winds. Air conditioning condensers, which are heavier than fans and vent hoods, can also be blown off of buildings.

Considering the small size of most exhaust fans, vent hoods, and air-conditioning units used on residential buildings, the following prescriptive attachment recommendations should be sufficient for most residences:

- For curb-mounted units, #14 screws with gasketed washers
- For curbs with sides smaller than 12 inches, one screw at each side of the curb

- For curbs between 12 and 24 inches, two screws per side

- For curbs between 24 and 36 inches, three screws per side

- For buildings within 3,000 feet of the ocean, stainless steel screws

- For units that have flanges attached directly to the roof, #14 pan-head screws, a minimum of two screws per side, and a maximum spacing of 12 inches o.c.

- Air conditioning condenser units, 1/2-inch bolts at the four corners of base of each unit

If the equipment is more than 30 inches above the curb, the attachment design should be based on calculated wind loads. ASCE 7-10 contains provisions for determining the horizontal and lateral force and the vertical uplift force on rooftop equipment for buildings with a mean roof height less than or equal to 60 feet. The lateral force is based on the vertical area of the equipment as projected on a vertical plane normal to the direction of the wind. The uplift force is based on the horizontal area of the equipment as projected on a horizontal plane above the equipment and parallel to the direction of the wind.

Until equipment manufacturers produce more wind-resistant equipment, job-site strengthening of vent hoods is recommended. One approach is to use 1/8-inch-diameter stainless steel cables. Two or four cables are recommended, depending on design wind conditions. Alternatively, additional heavy straps can be screwed to the hood and curb.

To avoid corrosion problems in equipment within 3,000 feet of the ocean shoreline (including sounds and backbays), nonferrous metal, such as aluminum, stainless steel, or steel with minimum G-90 hot-dip galvanized coating, is recommended for the equipment, equipment stands, and equipment anchors. Stainless steel fasteners are also recommended. See Section 11.6 for guidance regarding attic vents.

12.2.2 Flooding

Flood damage to mechanical equipment is typically caused by the failure to elevate equipment sufficiently, as shown in Figure 12-1. Figure 12-2 shows proper elevation of an air-conditioning condenser in a flood-prone area.

Exterior-mounted mechanical equipment in one- to four-family buildings is normally limited to the following:

- Air-conditioning condensers

- Ductwork (air supply and return)

- Exhaust fans

- Pool filter motors

- Submersible well pumps

Floodwaters can separate mechanical equipment from the supports and sever the connection to mechanical or electric

CROSS REFERENCE

For additional information, see FEMA 348, *Protecting Building Utilities From Flood Damage – Principles and Practices for the Design and Construction of Flood-Resistant Building Utility Systems* (FEMA 1999), and Fact Sheet 8.3, *Homebuilder's Guide to Coastal Construction*, in FEMA P-499 (FEMA 2010b).

Figure 12-1.
Condenser damaged as
a result of insufficient
elevation, Hurricane
Georges (U.S. Gulf Coast,
1998)

Figure 12-2.
Proper elevation of
an air-conditioning
condenser in a
floodprone area;
additional anchorage is
recommended

systems. Mechanical equipment can also be damaged or destroyed when inundated by floodwaters, especially saltwater. Although a short period of inundation may not destroy some types of mechanical equipment, any inundation of electric equipment causes, at a minimum, significant damage to wiring and other elements.

Minimizing flood damage to mechanical equipment requires elevating it above the DFE. Because of the uncertainty of wave heights and the probability of wave run-up, the designer should consider additional elevation above the DFE for this equipment.

NOTE

Although the 2012 IBC and 2012 IRC specify that flood damage-resistant materials be used below the BFE, in this Manual, flood damage-resistant materials are recommended below the DFE.

In Zone V, mechanical equipment must be installed either on a cantilevered platform supported by the first floor framing system or on an open foundation. A cantilevered platform is recommended. However, if the platform is not cantilevered, it is strongly recommended that the size of the elements, depth, and structural integrity of the open foundation that is used to support mechanical equipment be the same as the primary building foundation. Although smaller diameter piles could be used because the platform load is minimal, the smaller piles are more susceptible to being broken by floodborne debris, as shown in Figure 12-3.

In Zone A, mechanical equipment must be elevated to the DFE on open or closed foundations or otherwise protected from floodwaters entering or accumulating in the equipment elements. For buildings constructed over crawlspaces, the ductwork of some heating, ventilation, and air-conditioning systems are routed through the crawlspace. The ductwork must be installed above the DFE or be made watertight in order to minimize flood damage. Many ductwork systems today are constructed with insulated board, which is destroyed by flood inundation.

Figure 12-3.
Small piles supporting a platform broken by floodborne debris

12.2.3 Seismic Events

Residential mechanical equipment is normally fairly light. Therefore, with some care in the design of the attachment of the equipment for resistance to shear and overturning forces, these units should perform well during seismic events. Because air-conditioning units that are mounted on elevated platforms experience higher accelerations than ground-mounted units, extra attention should be given to attaching these units in areas that are prone to large ground accelerations.

12.3 Interior Mechanical Equipment

Interior mechanical equipment includes but is not limited to furnaces, boilers, water heaters, and distribution ductwork. High winds normally do not affect interior mechanical equipment. Floodwaters, however, can cause significant damage to furnaces, boilers, water heaters, and distribution ductwork. Floodwaters can extinguish gas-powered flames, short circuit the equipment's electric system, and inundate equipment and ductwork with sediment.

The following methods of reducing flood damage to interior equipment are recommended:

- Elevate the equipment and the ductwork above the DFE by hanging the equipment from the existing first floor or placing it in the attic or another location above the DFE.

- In areas other than Zone V (where enclosure of utilities below the BFE is not recommended), build a waterproof enclosure around the equipment, allowing access for maintenance and replacement of equipment parts.

12.4 Electric Utility, Telephone, and Cable TV Systems

Electric utilities serving residential buildings in coastal areas are frequently placed in harsh and corrosive environments. Such environments increase maintenance and shorten the lifespan of the equipment. Common electric elements of utilities in residential buildings that might be exposed to severe wind or flood events, which increase maintenance and shorten the lifespan further, are electric meters, electric service laterals and service drops from the utility company, electric panelboards, electric feeders, branch circuit wiring, receptacles, lights, security system wiring and equipment, and telephone and cable television wiring and equipment.

The primary method of protecting elements from flooding is to elevate them above the DFE, but elevation is not always possible. Floodplain management requirements and other code requirements sometimes conflict. One conflict that is difficult to fully resolve is the location of the electric meter. Figure 12-4 shows a bank of meters and electric feeds that failed during Hurricane Opal.

Utility companies typically require electric meters to be mounted where they can be easily read for billing purposes; meters are usually centered approximately 5 feet above grade. They are normally required by utility regulations to be no higher than eye level. However, this height is often below the DFE for coastal homes, and the placement therefore conflicts with floodplain management requirements that meters be installed above the DFE. Since meter sockets typically extend 12 inches below the center of the meter, design floods

Figure 12-4.
Electric service meters
and feeders that were
destroyed by floodwaters
during Hurricane Opal
(1995)

that produce 4 feet of flooding can cause water to enter the meter socket and disrupt the electric service. When a meter is below the flood level, electric service can be exposed to floodborne debris, wave action, and flood forces. Figure 12-5 shows an electric meter that is easily accessible by the utility company but is above the DFE.

Since many utility companies no longer manually read meters, there may be flexibility in meter socket mounting, preferably above the design flood. The use of automatic meter reading ("smart meters") by electric utility companies is increasing. The designer should consult the utility to determine whether smart meters can be placed higher than meters that must be read manually.

Similar situations often exist with other electrical devices. For example, switches for controlling access and egress lighting and security sensors occasionally need to be placed below the DFE. The following methods are recommended when necessary to reduce the potential for damage to electric wiring and equipment and to facilitate recovery from a flood event:

- **Wiring methods.** Use conduit instead of cable. Placing insulated conductors in conduits allows flood-damaged wiring to be removed and replaced. The conduit, after being cleaned and dried, can typically be reused. In saltwater environments, non-metallic conduits should be used.

- **Routing and installation.** Install main electric feeders on piles or other vertical structural elements to help protect them from floating debris forces. Since flood damage is often more extensive on the seaward side of a building, routing feeders on the landward side of the structural elements of the building can further reduce the potential for damage. Do not install wiring or devices on breakaway walls. Figure 12-5 is an illustration of recommended installation techniques for electric lines, plumbing, and other utility elements.

- **Design approach.** Install the minimum number of electric devices below the DFE that will provide compliance with the electric code. Feed the branch circuit devices from wiring above the DFE to minimize the risk of flood inundation.

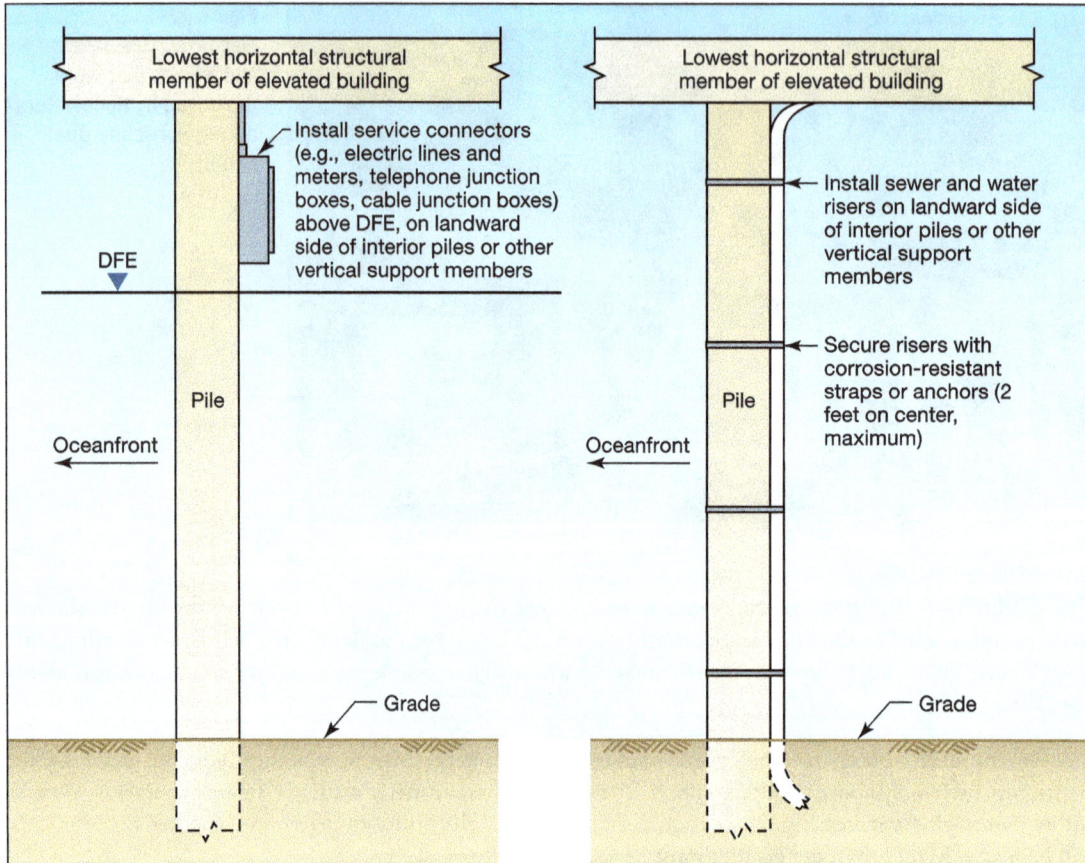

Figure 12-5.
Recommended installation techniques for electric and plumbing lines and utility elements

- **Service style.** Feed the building from underground service laterals instead of overhead electric service drops. When overhead services are needed, avoid penetrating the roof with the service mast to reduce the potential for roof damage and resulting water infiltration. Figure 12-6 illustrates the vulnerability of roof damage and resulting water infiltration when an electric service mast penetrates a roof.

- **Panel location.** Install branch circuit and service panelboards above the DFE. If required to meet utility National Electrical Code requirements, provide a separate service disconnect remote to the panel.

Fact Sheet 8.3, *Protecting Utilities,* in FEMA P-499 contains other recommendations for reducing the vulnerability of utilities that supply buildings.

Direct wind damage to exterior-mounted electric utility equipment (see Figure 12-6) is infrequent in part because of the small size of most equipment (e.g., disconnect switches, conduit). Exceptions are satellite dishes, photovoltaic panels, and electric service penetrations through the roof. Satellite dish and photovoltaic panel failures are typically caused by the design professional's failure to perform wind load calculations and provide for adequate anchorage.

**Figure 12-6.
Damage caused by
dropped overhead
service, Hurricane
Marilyn (U.S. Virgin
Islands, 1995)**

12.4.1 Emergency Power

Because a severe wind event often interrupts electric service, designers and homeowners need to make a decision about the need for backup power.

Emergency power can be provided by permanently installed onsite generators or by temporary generators brought to the site after the event. For permanently installed units, the following is recommended:

CROSS REFERENCE

For guidance on determining the proper size of an emergency generator, see Section VI-D of FEMA 259, *Engineering Principles and Practices for Retrofitting Flood Prone Residential Buildings* (FEMA 2001).

- Locate the generator above the DFE.

- If located on the exterior of the building, place the unit to prevent engine exhaust fumes from being drawn into doors, windows, or any air intake louvers into the building. If located on the inside of the building, provide ventilation for combustion air and cooling air and provision for adequately discharging exhaust fumes.

- Locate the fuel source above the DFE and store an amount of fuel adequate for the length of time the generator is expected to operate.

- Install the generator where its noise and vibration will cause the least disruption.

- Determine the expected load (e.g., heat, refrigeration, lights, sump pumps, sewer ejector pumps). Non-fuel-fired heating systems and most cooling systems require large generators. Capacity considerations may limit the generator to providing only freeze protection and localized cooling.

- Install manual or automatic transfer switches that prevent backfeeding power from the generator into the utility's distribution system. Backfeeding power from generators into the utility's distribution system can kill or injure workers attempting to repair damaged electrical lines.

- Provide an "emergency load" subpanel to supply critical circuits. Do not rely on extension cords. Supply the emergency panel from the load side of a manual or automatic transfer switch.

WARNING

Do not "backfeed" emergency power through the service panel. Utility workers can be killed!

- Determine whether operation of the generator will be manual or automatic. Manual operation is simpler and less expensive. However, a manual transfer switch requires human intervention. Owners should not avoid or delay evacuation to tend to an emergency power source.

- Size the generator, transfer switches, and interconnecting wiring for the expected load. The generator should be large enough to operate all continuous loads and have ample reserve capacity to start the largest motor load while maintaining adequate frequency and voltage control and maintaining power quality.

12.5   Water and Wastewater Systems

Water and wastewater systems include wells, septic systems, sanitary systems, municipal water connections, and fire sprinkler systems.

12.5.1   Wells

For protection of well systems from a severe event (primarily a flood), the design must include a consideration of the following, at a minimum:

- Floodwaters that enter aquifers or saturate the soil can contaminate the water supply. FEMA P-348, *Principles and Practices for Flood-Resistant Building Utilities* (FEMA 1999), recommends installing a watertight encasement that extends from at least 25 feet below grade to at least 1 foot above grade.

- Non-submersible well pumps must be above the DFE.

- If water is to be available following a disaster, an alternative power source must be provided.

- The water supply line riser must be protected from hydrodynamic and floodborne debris impact damage; the supply line must be on the landward side of a pile or other vertical structural member or inside an enclosure designed to withstand the forces from the event (see Figure 12-5).

- Backflow valves must be installed to prevent floodwaters from flowing into the water supply when water pressure in the supply system is lost.

12.5.2   Septic Systems

Leach fields and septic tanks, and the pipes that connect them, are highly susceptible to erosion and scour, particularly in Coastal A Zone and Zone V with velocity flow risks. The best way to protect leach fields and other onsite sewage management elements is to locate them outside the floodplain.

WARNING

In some areas, high groundwater levels may preclude the installation of septic tanks below the level of expected erosion and scour.

If septic systems cannot be located outside the floodplain, the design of septic systems for protection from severe events must include a consideration of the following, at a minimum:

- If the septic tank is dislodged from its position in the ground, the piping will be disconnected, releasing sewage into floodwaters. Also, the tank could damage the nearest structure. Therefore, bury the system below the expected depth of erosion and scour, if possible, and ensure the tank is anchored to prevent a buoyancy failure.

- The sewage riser lines and septic tank risers must be protected from water and debris flow damage; risers should be on the landward side of a pile or other vertical structural member or inside an enclosure designed to withstand the forces from the event (see Figure 12-5).

If leach fields, pipes, and tanks cannot be located outside the floodplain, one possible way to protect them is to bury them below the expected scour depth. However, many local health codes or ordinances restrict or even prohibit the placement of septic elements in the floodplain. In these cases, alternate sewage management systems must be used.

Because leach fields rely on soil to absorb moisture, saturated soil conditions can render leach fields inoperable. This problem and its potential mitigating measures depend on complex geotechnical considerations. Therefore, a geotechnical engineer and/or a qualified sewer designer should be consulted for the design and installation of leach fields.

12.5.3   Sanitary Systems

To protect sanitary systems from a severe event, the design must include a consideration of the following, at a minimum:

- Sanitary riser lines must be protected from water and debris flow damage; risers should be on the landward side of a pile or other vertical structural member or inside an enclosure designed to withstand the forces from the event (see Figure 12-5).

- When the line breaks at the connection of the building line and main sewer line, raw sewage can flow back out of the line, contaminating the soil near the building. A check valve in the line may help prevent this problem.

12.5.4 Municipal Water Connections

If water risers are severed during a coastal event, damage to the water supply system can include waste from flooded sewer or septic systems intruding into the water system, sediment filling some portion of the pipes, and breaks in the pipes at multiple locations. Protecting municipal water connections is accomplished primarily by protecting the water riser that enters the building from damage by debris. See Section 12.5.1 for more information.

12.5.5 Fire Sprinkler Systems

Protecting the fire sprinkler system is similar to protecting the other systems discussed in Section 12.5. The primary issue is to locate the sprinkler riser such that the location provides shielding from damage. In addition, there must be consideration to the location of shutoff valves and other elements so that if an unprotected portion of the fire water supply line is damaged, the damage is not unnecessarily added to the damage caused by the natural hazard event.

12.6 References

ASCE (American Society of Civil Engineers). 2010. *Minimum Design Loads for Buildings and Other Structures*. ASCE Standard ASCE 7-10.

FEMA (Federal Emergency Management Agency). 1999. *Principles and Practices for Flood-Resistant Building Utilities*. FEMA P-348.

FEMA. 2001. *Engineering Principles and Practices for Retrofitting Flood Prone Residential Buildings*. FEMA 259.

FEMA. 2010a. *Elevator Installation for Buildings Located in Special Flood Hazard Areas in Accordance with the National Flood Insurance Program*. FEMA NFIP Technical Bulletin 4.

FEMA. 2010b. *Homebuilder's Guide to Coastal Construction*. FEMA P-499.

FEMA. 2010c. *National Flood Insurance Program Flood Insurance Manual*.

ICC (International Code Council). 2011a. *International Building Code*. 2012 IBC. Country Club Hills, IL: ICC.

ICC. 2011b. *International Residential Code for One-and Two-Family Dwellings*. 2012 IRC. Country Club Hills, IL: ICC.

13

Constructing the Building

This chapter provides guidance on constructing residential buildings in coastal areas, which presents challenges that are usually not present in more inland locations (risk of high winds and coastal flooding and a corrosive environment) and other challenges such as the need to elevate the building.

Considerations related to these challenges include the need to:

- Perform more detailed inspections of connection details than those performed in noncoastal areas to ensure the details can withstand the additional hazards found in coastal areas

- Include with the survey staking the building within property line setbacks and at or above the design flood elevation (DFE) (see Section 4.5 for additional coastal survey regulatory requirements)

- Ensure that all elements of the building will be able to withstand the forces associated with high winds, coastal flooding, or other hazards required of the design

- Ensure that the building envelope is constructed to minimize and withstand the intrusion of air and moisture during high-wind events (see Section 11.3.1.4)

- Provide durable exterior construction that can withstand a moist and sometimes salt-laden environment

- Protect utilities, which may include placing them at or above the DFE

CROSS REFERENCE

For resources that augment the guidance and other information in this Manual, see the Residential Coastal Construction Web site (http://www.fema.gov/rebuild/mat/fema55.shtm).

Constructing coastal residential buildings on elevated pile foundations present the following additional challenges:

- The difficulty of constructing a driven pile foundation to accepted construction plan tolerances

- The difficulty of constructing a building on an elevated post-and-beam foundation, which is more difficult than building on a continuous wall foundation

This chapter discusses the construction aspects of the above challenges and other aspects of the coastal construction process, including the construction items that are likely to require the most attention from the builder in order for the design intent to be achieved.

Although much of the discussion in this chapter is related to constructing the building to meet the architect's and engineer's design intent for existing and future conditions (such as erosion and sea-level rise), durability of the building elements is also important. Wood decay, termite infestation, metal corrosion, and concrete and masonry deterioration can weaken a building significantly, making it hazardous to occupy under any conditions and more likely to fail in a severe natural hazard event.

Builders may find that the permitting and inspection procedures in coastal areas are more involved than those in inland areas. Not only must all Federal, State, and local Coastal Zone Management and other regulatory requirements be met, the design plans and specifications may need to be sealed by a design professional. Building permit submittals must often include detailed drawings and other types of information for all elements of the wind-resisting load path, including sheathing material, sheathing nailing, strap and tiedown descriptions, bolted connections, and pile description and placement. The placement of utilities at or above the DFE, breakaway walls, and flood equalization openings must be clearly shown. Site inspections are likely to focus on the approved plans, and building officials may be less tolerant of deviations from these approved construction documents than those in noncoastal areas. Inspection points are also discussed.

13.1 Foundation Construction

Constructing a foundation in a coastal environment includes designing the layout, selecting the foundation type, selecting the foundation material with consideration for durability, and installing the foundation. Although pile foundations are the most common foundation type in Zone V and should be used in Coastal A Zones, shallow foundations, both masonry and concrete, may be acceptable elsewhere. Whether masonry, concrete, wood, or steel, all coastal foundation materials must be designed and installed to withstand the likelihood of high winds, moisture, and salt-laden air. See Chapter 10 for guidance on the design of coastal foundations.

13.1.1 Layout

Surveying and staking must be done accurately in order to establish the building setback locations, the DFE, and the house plan and support locations. Figure 13-1 is a site layout with pile locations, batter boards, and setbacks and is intended to show the constraints a builder may face when laying out a pile-supported structure on a narrow coastal lot. There may be conflicts between what the contractor would like to do to prepare the site and what the environmental controls dictate can be done on the site. For example, leveling the site, especially altering dunes, and removing existing vegetation may be restricted. Furthermore, these

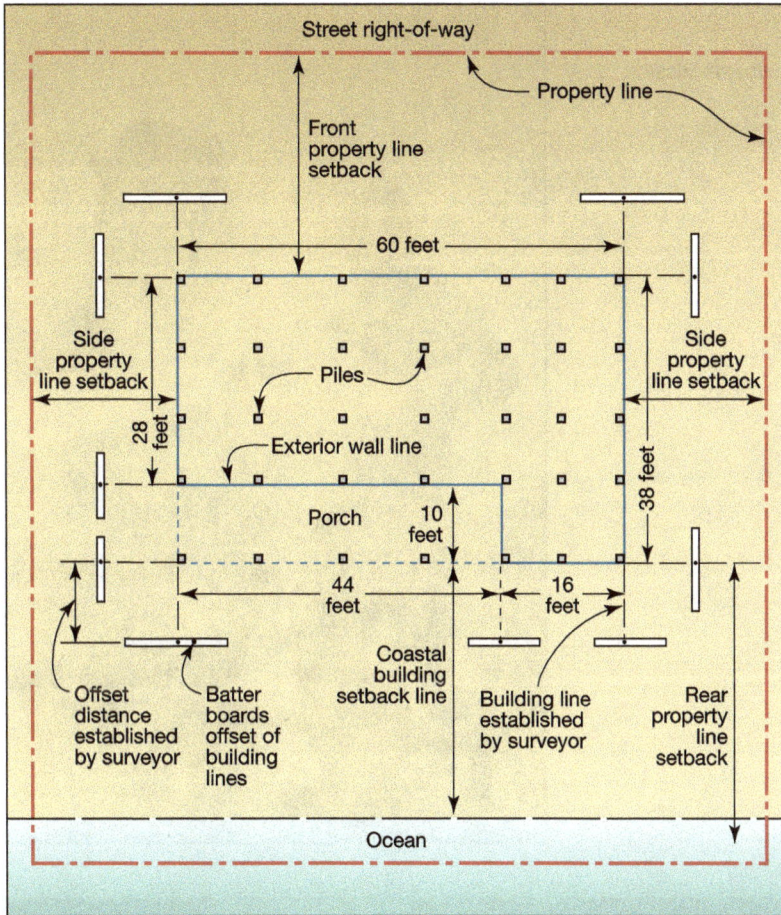

**Figure 13-1.
Site layout**

restrictions may limit access by pile drivers and other heavy equipment. Similarly, masonry and concrete foundations may require concrete pumping because of limited access to the traditional concrete mix truck and chute.

In an elevated building with a pile foundation, the layout of the horizontal girders and beams should anticipate the fact that the final plan locations of the tops of the piles will likely not be precise. Irregularities in the piles and soil often prevent the piles from being driven perfectly plumb. The use of thick shims or overnotching for alignment at bolted pile-girder connections may have a significant adverse effect on the connection capacity and should be avoided.

Figure 13-2 shows the typical process of pile notching; the use of a chain saw for this process can lead to inaccuracies at this early stage of construction. Figure 13-3 shows a wood pile that is overnotched. Figure 13-4 shows a pile that has been properly notched to support the floor girder and cut so plenty of wood remains at the top of the pile.

Figure 13-2.
Typical pile notching process
SOURCE: PATTY MCDANIEL, USED WITH PERMISSION

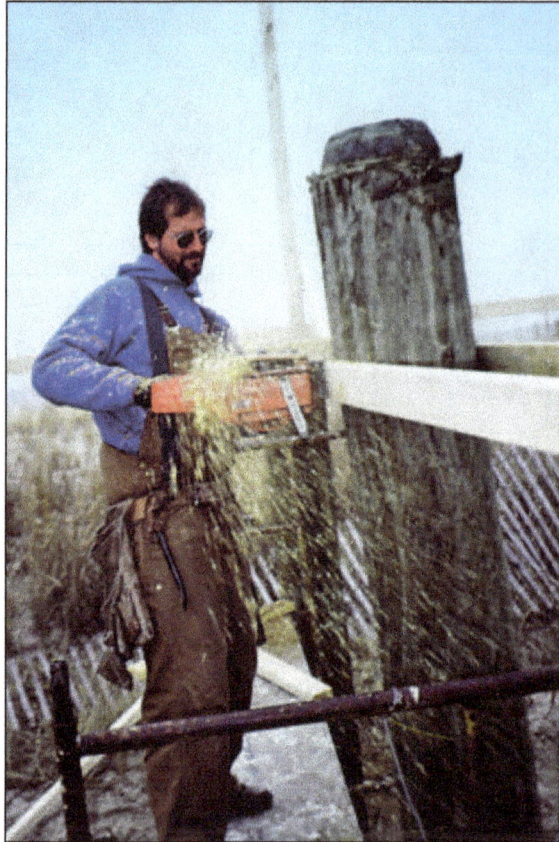

Figure 13-3.
Improper overnotched
wood pile
SOURCE: PATTY MCDANIEL,
USED WITH PERMISSION

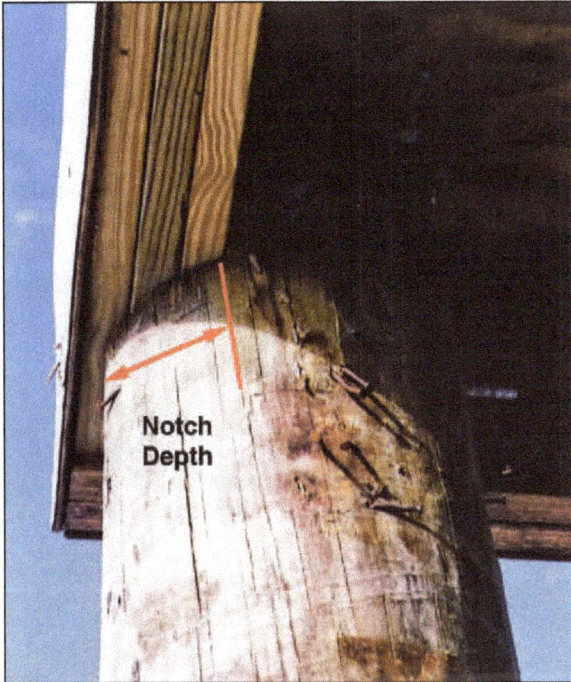

Figure 13-4.
Properly notched pile; outer member of this three-
member beam supported by the through-bolt
rather than the beam seat

A rule of thumb regarding notching is to notch no more than 50 percent of the pile cross section, but in no case should notching be in excess of that specified by the design professional. Section 13.2 presents information concerning the reinforcement of overnotched and misaligned piles.

The primary floor girders spanning between pile or foundation supports should be oriented parallel to the primary flow of potential floodwater and wave action if possible. This orientation (normally at right angles to the shoreline) allows the lowest horizontal structural member perpendicular to flow to be the floor joists. Thus, in an extreme flood, the girders are not likely be subjected to the full force of the floodwater and debris along their more exposed surfaces.

The entire structure is built on the first floor, and it is therefore imperative that the first floor be level and square. The "squaring" process normally involves taking diagonal measurements across the outer corners and shifting either or both sides until the diagonal measurements are the same, at which point the building is square. An alternative is to take the measurements of a "3-4-5" triangle and shift the floor framing until the "3-4-5" triangular measurement is achieved.

13.1.2   Pile Foundations

Pile foundations are the most common foundation type in Zone V coastal buildings and should also be used in Coastal A Zones where scour and erosion conditions along with potentially destructive wave forces make it inadvisable to construct buildings on shallow foundations.

In many coastal areas, the most common type of pile foundation is the elevated wood pile foundation in which the tops of the piles extend above grade to about the level of the DFE (see Figure 13-5).

Figure 13-5.
Typical wood pile
foundation

Horizontal framing girders connected to the tops of the piles form a platform on which the house is built. Appendix B of ICC 600-2008 contains some girder designs for use with foundations discussed in FEMA P-550, *Recommended Residential Construction for the Gulf Coast* (FEMA 2006). In addition, the 2012 IRC contains prescriptive designs of girder and header spans. Furthermore, Fact Sheet 3.2, *Pile Installation*, in FEMA P-499 (FEMA 2011) presents basic information about pile design and installation, including pile types, sizes and lengths, layout, installation methods, bracing, and capacities. For more information on pile-to-beam connections, see Fact Sheet 3.3, *Wood Pile-to-Beam Connections*, in FEMA P-499, which presents basic construction guidance for various construction methods. The discussion in this section is focused on the construction of an elevated wood pile foundation.

Precautions should be taken in handling and storing pressure-preservative-treated round or square wood piles. They should not be dragged along the ground or dropped. They should be stored well-supported on skids so that there is air space beneath the piles and the piles are not in standing water. Additional direction and precautions for pile handling, storage, and construction are found in Section 10.5 of this Manual and AWPA Standard M4-91.

The effectiveness of pile foundations and the pile load capacity is related directly to the method of installation. The best method is to use a pile driver, which uses leads to hold the pile in position while a single- or double-acting diesel- or air-powered hammer drives the pile into the ground. Pile driving is often used with auguring to increase pile embedment. Augurs are used to drill the first several feet into the soil, and the piles are then driven to refusal. Auguring has the added benefit of improving pile alignment.

WARNING

The amount of long-term and storm-induced erosion expected to occur at the site (see Section 3.5 in Volume I of this Manual) must be determined before any assumptions about soils are made or analyses of the soils are conducted. Only the soils that will remain after erosion can be relied on to support the foundation members.

The pile driver method is cost-effective in a development when a number of houses are constructed at one time but may be expensive for a single building. The drop hammer method is a lower cost alternative and is considered a type of pile driving, as discussed in Section 10.5.4. A drop hammer consists of a heavy weight that is raised by a cable attached to a power-driven winch and then dropped onto the end of the pile.

A less desirable but frequently used method of inserting piles into sandy soil is "jetting." Jetting involves forcing a high-pressure stream of water through a pipe advanced along the side of the pile. The water blows a hole in the sand into which the pile is continuously pushed or dropped until the required depth is reached. Unfortunately, jetting loosens the soil around the pile and the soil below the tip, resulting in a lower load capacity.

CROSS REFERENCE

See Section 10.5.4 for a discussion of pile capacities for various installation methods.

Holes for piles may be excavated by an auger if the soil is sufficiently clayey or silty. In addition, some sands may contain enough clay or silt to permit augering. This method can be used by itself or in conjunction with pile driving. If the hole is full-sized, the pile is dropped in and the void backfilled. Alternatively, an undersized hole can be excavated and a pile driven into it. When the soil conditions are appropriate, the hole stays open long enough to drop or drive in a pile. In general, piles dropped or driven into augered holes may not have as much capacity as those driven without augering.

If precast concrete piles or steel piles are used, only a regular pile driver with leads and a single- or double-acting hammer should be used. For any pile driving, the building jurisdiction or the engineer-of-record will probably require that a driving log be kept for each pile. The log will show the number of inches per blow as the driving progresses—a factor used in determining the pile capacity, as shown in Equation 13-1. As noted in Section 10.3, the two primary determinants of pile capacity are the depth of embedment in the soil and the soil properties.

Piles must be able to resist vertical loads (both uplift and gravity) and lateral loads. Sections 8.5 and 8.10 contain guidance on determining pile loads. It is common practice to estimate the ultimate vertical load bearing capacity of a single pile on the basis of the driving resistance. Several equations are available for making such estimates. However, the results are not always reliable and may over-predict or under-predict the capacity, so the equations should be used with caution. One method of testing the recommended capacity based on an equation is to load test at least one pile at each location of known soil variation.

The designer should also keep in mind that constructing a pile foundation appropriately for the loads it must resist in the coastal environment may drastically reduce future costs by helping to avoid premature failure. Many factors in addition to vertical and lateral loads must be taken into account in the coastal environment. For example, erosion and scour can add stress on the foundation members and change the capacity to which the piles should be designed. The complex and costly repairs to the home shown in Figure 10-2 could have been avoided if all forces and the reduced pile capacity resulting from erosion and scour had been considered in the pile foundation design.

Equation 13.1 can be used to determine pile capacity for drop hammer pile drivers. Equations for other pile driver configurations are provided in U.S. Department of the Navy Design Manual 7.2, *Foundation and Earth Structures Design* (USDN 1982).

Σ

EQUATION 13.1. PILE DRIVING RESISTANCE FOR DROP HAMMER PILE DRIVERS

$$Q_{all} = \frac{2WH}{(S+1)}$$

where:

Q_{all} = allowable pile capacity (in lb)

W = weight of the striking parts of the hammer (in lb)

H = effective height of the fall (in ft)

S = average net penetration, given as in. per blow for the last 6 in. of driving

Lateral and uplift load capacity of piles varies greatly with the soils present at the site. Pile foundation designs should be based on actual soil borings at the site (see Section 10.3.3.2). Variation in the final locations of the pile tops can complicate subsequent construction of floor beams and bracing. The problem is worsened by piles with considerable warp, non-uniform soil conditions, and material buried below the surface of the ground such as logs, gravel bars, and abandoned foundations. Builders should inquire about subsurface conditions at the site of a proposed building before committing to the type of pile or the installation method (see Section 10.3.3). A thorough investigation of site conditions can help prevent costly installation errors.

The soils investigation should determine the following:

- Type of foundations that have been installed in the area in the past

- Type of soil that might be expected (based on past soil borings and soil surveys)

- Whether the proposed site has been used for any other purpose and if so, the likelihood of buried materials present on the site

Scour and erosion both reduce pile capacities and erosion can increase flood loads on a pile. Scour and erosion must be considered in a properly designed pile foundation. Additional guidance on the effects of scour and erosion on piles is provided in Section 8.5.11 and Section 10.5.5.

13.1.3 Masonry Foundation Construction

The combination of high winds and moist and sometimes salt-laden air can have a damaging effect on masonry construction by forcing moisture into the smallest cracks or openings in the masonry joints. The entry of moisture into reinforced masonry construction can lead to corrosion of the reinforcement and subsequent cracking and spalling if proper protection of the reinforcement is not provided, as required by TMS 402/ACI 530/ASCE 5 and TMS 602/ACI 530.1/ASCE 6. Moisture resistance

WARNING

Open masonry foundations in earthquake hazard areas require special reinforcement detailing and pier proportions to meet the requirement for increased ductility.

is highly influenced by the quality of the materials and the quality of the masonry construction at the site. Masonry material selection is discussed in Section 9.4 of this Manual.

The quality of masonry construction depends on many considerations. Masonry units and packaged mortar and grout materials should be stored off the ground and covered. Mortar and grouts must be carefully batched and mixed. As the masonry units are placed, head and bed joints must be well mortared and tooled. The 2012 IRC provides grouting requirements. Masonry work in progress must be well protected.

Moisture penetration or retention must be carefully controlled where masonry construction adjoins other materials. As in any construction of the building envelope in the coastal environment, flashing at masonry must be continuous, durable, and of sufficient height and extent to impede the penetration of expected wind-driven precipitation. For more information on moisture barrier systems, see Fact Sheet 1.9, *Moisture Barrier Systems*, in FEMA P-499. Because most residential buildings with masonry foundations have other materials (e.g., wood, concrete, steel, vinyl) attached to the foundation, allowance must be made for shrinkage of materials as they dry out and for differential movement between the materials. Expansion and contraction joints must be placed so that the materials can move easily against each other.

NOTE

Tooled concave joints and V-joints provide the best moisture resistance.

Masonry is used for piers, columns, and foundation walls. As explained in Section 10.2.1, the National Flood Insurance Program (NFIP) regulations require open foundations (e.g., piles, piers, posts, columns) for buildings constructed in Zone V. Buildings in Zone A may be constructed on any foundation system. However, because of the history of observed damage in Coastal A Zone and the magnitude of the flood and wind forces that can occur in these areas, this Manual recommends that only open foundation systems be constructed in Coastal A Zones. Figure 13-6 shows an open masonry foundation with only two rows of piers. It is unlikely that this foundation system could resist the overturning caused by the forces described in Chapter 8 and shown in Example 8-10. Fact Sheet 3.4, *Reinforced Masonry Pier Construction*, in FEMA P-499 provides recommendations on pier construction best practices. Fact Sheet 4.2, *Masonry Details*, in FEMA P-499 provides details on masonry wall-to-foundation connections.

Reinforced masonry has much more strength and ductility than unreinforced masonry for resisting large wind, water, and earthquake forces. This Manual recommends that permanent masonry foundation construction in and near coastal flood hazard areas (both Zone A and Zone V) be fully or partially reinforced and grouted solid regardless of the purpose of the construction and the design loads. Grout should be in conformance with the requirements of the 2012 IBC. Knockouts should be placed at the bottom of fully grouted cells to ensure that the grout completely fills the cells from top to bottom. Knockouts are required only for walls (or piers) exceeding 5 feet in height.

NOTE

In areas not subject to earthquake hazards, breakaway walls below elevated buildings may be constructed using unreinforced and ungrouted masonry.

For CMUs, shrinkage cracking can be minimized by using Type I moisture-controlled units and keeping them dry in transit and on the job. Usually, for optimum crack control, Type S mortar should be used for below-grade applications and Type N mortar for above-grade applications. The 2012 IBC specifies grout proportions by volume for masonry construction.

Figure 13-6.
Open masonry foundation

13.1.4   Concrete Foundation Construction

Concrete foundation or superstructure elements in coastal construction almost always require steel reinforcement. Figure 13-7 shows a concrete foundation, and Figure 13-8 shows a house being constructed with concrete. Completed cast-in-place exterior concrete elements should generally provide 1-1/2 inches or more of concrete cover over the reinforcing bars. Minimum cover values vary according to bar size and exposure to earth or weather per ACI 318-08. This thickness of concrete cover serves to protect the reinforcing bars from corrosion, as does an epoxy coating. The bars are also protected by the natural alkalinity of the concrete. However, if saltwater penetrates the concrete cover and reaches the reinforcing steel, the concrete alkalinity will be reduced by the salt chloride and the steel can corrode if it is not otherwise protected. As the corrosion forms, it expands and cracks the concrete, allowing the additional entry of water and further corrosion. Eventually, the corrosion of the reinforcement and the cracking of the concrete weaken the concrete structural element, making it less able to resist loads caused by natural hazards.

During placement, concrete normally requires vibration to eliminate air pockets and voids in the finished surface. The vibration must be sufficient to eliminate the air without separating the concrete or water from the mix.

Figure 13-7.
Concrete foundation

Figure 13-8.
Concrete house

To ensure durability and long life in coastal, saltwater-affected locations, it is especially important to carefully carry out concrete construction in a fashion that promotes durability. "Material Durability in Coastal Environments," available on the Residential Coastal Construction Web site (http://www.fema.gov/rebuild/mat/fema55.html) describes the 2012 IBC requirements for more durable concrete mixes with lower water-cement ratios and higher compressive strengths (5,000 pounds/square inch) to be used in a saltwater environment. The 2012 IBC also requires that additional cover thickness be provided. Proper placement, consolidation, and curing are also essential for durable concrete. The concrete mix water-cement ratio required by 2012 IBC or by the design should not be exceeded by the addition of water at the site.

It is likely that concrete will have to be pumped at many sites because of access limitations or elevation differences between the top of the forms and the concrete mix truck chute. Pumping concrete requires some minor changes in the mix so that the concrete flows smoothly through the pump and hoses. Plasticizers should be used to make the mix pumpable; water should not be used to improve the flow of the mix. Concrete suitable for pumping must generally have a slump of at least 2 inches and a maximum aggregate size of 33 to 40 percent of the pump pipeline diameter. Pumping also increases the temperature of the concrete, thus changing the curing time and characteristics of the concrete depending on the outdoor temperature.

> **NOTE**
>
> ACI 318-08 specifies minimum amounts of concrete cover for various construction applications. Per the Exception to 1904.3 in the 2012 IBC, concrete mixtures for any R occupancies need only comply with the freeze/thaw requirements (as traditionally tabulated in the 2012 IBC and 2012 IRC), not the permeability and corrosion requirements of ACI 318-08.

Freeze protection may be needed, particularly for columns and slabs, if pouring is done in cold temperatures. Concrete placed in cold weather takes longer to cure, and the uncured concrete may freeze, which adversely affects its final strength. Methods of preventing concrete from freezing during curing include:

- Heating adjacent soil before pouring on-grade concrete

- Warming the mix ingredients before batching

- Warming the concrete with heaters after pouring (avoid overheating)

- Placing insulating blankets over and around the forms after pouring

- Selecting a cement mix that will shorten curing time

Like masonry, concrete is used for piers, columns, and walls; the recommendation in Section 13.1.3 regarding open foundations in Coastal A Zones also applies to concrete foundations. In addition, because the environmental impact of salt-laden air and moisture make the damage potential significant for concrete, this Manual recommends that all concrete construction in and near coastal flood hazard areas (both Zone V and Zone A) be constructed with the more durable 5,000-pounds/square inch minimum compressive strength concrete regardless of the purpose of the construction and the design loads.

13.1.5   Wood Foundation Construction

All of the wood used in the foundation piles, girders, beams, and braces must be preservative-treated wood or, when allowed, naturally decay-resistant wood. Section 9.4 discusses materials selection for these wood elements. Piles must be treated with waterborne arsenicals, creosote, or both. Girders and braces may be treated with waterborne arsenicals, pentachlorophenol, or creosote. Certain precautions apply to working with any of these treated wood products, and additional precautions apply for pentachlorophenol- and creosote-treated wood (see Section 13.1.5.1). Additional information is available in Consumer Information Sheets where the products are sold.

Wood foundations are being constructed in some parts of the country as part of a basement or crawlspace. These foundation elements have walls constructed with pressure-preservative-treated plywood and footings constructed with wide pressure- preservative-treated wood boards such as 2x10s or 2x12s. Because the NFIP regulations allow continuous foundation walls (with the required openings) in Coastal A Zones, continuous wood foundations might seem to be acceptable in these areas. However, because of the potential forces from waves less than 2 feet high (as discussed in Section 10.8), a wood foundation supported on a wood footing is not recommended in Coastal A Zones.

When working with treated wood, the following health and safety precautions should be taken:

- Avoid frequent or prolonged inhalation of the sawdust.

- When sawing and boring, wear goggles and a dust mask.

- Use only treated wood that is visibly clean and free of surface residue should be used for patios, decks, and walkways.

- Before eating or drinking, wash all exposed skin areas thoroughly.

- If preservatives or sawdust accumulate on clothes, wash the clothes (separately from other household clothing) before wearing them again.

- Dispose of the cuttings by ordinary trash collection or burial. The cuttings should not be burned in open fires or in stoves, fireplaces, or residential boilers because toxic chemicals may be produced as part of the smoke and ashes. The cuttings may be burned only in commercial or industrial incinerators or boilers in accordance with Federal and State regulations.

- Avoid frequent or prolonged skin contact with pentachlorophenol or creosote-treated wood; when handling it, wear long-sleeved shirts and long pants and use gloves impervious to the chemicals (e.g., vinyl-coated gloves).

- Do not use pentachlorophenol-pressure-treated wood in residential interiors except for laminated beams or for building elements that are in ground contact and are subject to decay or insect infestation and where two coats of an appropriate sealer are applied. Sealers may be applied at the installation site. Urethane, shellac, latex epoxy enamel, and varnish are acceptable sealers.

- Do not use creosote-treated wood in residential interiors. Coal tar pitch and coal tar pitch emulsion are effective sealers for outdoor creosote-treated wood-block flooring. Urethane, epoxy, and shellac are acceptable sealers for all creosote-treated wood.

13.1.6   Foundation Material Durability

Ideally, all of the pile-and-beam foundation framing of a coastal building is protected from rain by the overhead structure, even though all of the exposed materials should be resistant to decay and corrosion. In practice, the overhead structure includes both enclosed spaces (such as the main house) and outside decks. The spaces between the floor boards on an outside deck allow water to pass through and fall on the framing below. A worst case for potential rain and moisture penetration exists when less permeable decks collect water and channel it to fall as a stream onto the framing below. In addition, wind-driven rain and ocean spray penetrates into many small spaces, and protection of the wood in these spaces is therefore important to long-term durability of the structure.

The durability of the exposed wood frame can be improved by detailing it to shed water during wetting and to dry readily afterward. Decay occurs in wetted locations where the moisture content of the exposed, untreated interior core of treated wood elements remains above the fiber saturation point—about 30 percent. The moisture content of seasoned, surface-dry 2x lumber (S-DRY) is less than or equal to 19 percent content when it arrives at the job site, but the moisture content is quickly reduced as the wood dries in the finished building. The moisture content of the large members (i.e., greater than 3 times) is much higher than 19 percent when they arrive at the job site, and the moisture content takes months to drop below 19 percent.

The potential for deterioration is greatest at end grain surfaces. Water is most easily absorbed along the grain, allowing it to penetrate deep into the member where it does not readily dry. Figure 13-9 illustrates deterioration in the end of a post installed on a concrete base. This is a typical place for wood deterioration to occur. Even when the end grain is more exposed to drying, the absorptive nature of the end grain creates an exaggerated shrink/swell cycling, resulting in checks and splits, which in turn allow increased water penetration.

Exposed pile tops present the vulnerable horizontal end grain cut to the weather. Cutting the exposed top of a pile at a slant does not prevent decay and may even channel water into checks. Water enters checks and splits in the top and side surfaces of beams and girders. It can then penetrate into the untreated core and cause decay. These checks and splits occur naturally in large sawn timbers as the wood dries and shrinks over time. They are less common in glue-laminated timbers and built-up sections. It is generally, but not universally, agreed that caulking the checks and splits is unwise because caulking is likely to promote water retention more than keep water out. The best deterrent is to try to keep the water from reaching the checks and splits.

Framing construction that readily collects and retains moisture, such as pile tops, pile-beam connections, and horizontal girder and beam top surfaces, can be covered with flashing or plywood. However, there should always be an air gap between the protected wood and the flashing so that water vapor passing out of the wood is not condensed at the wood surface. For example, a close-fitting cap of sheet metal on a pile top can cause water vapor coming out of the pile top to condense and cause decay. The cap can also funnel water into the end grain penetrations of the vertical fasteners.

Figure 13-9.
Wood decay at the base of a post supported by concrete

When two flat wood surfaces are in contact in a connection, the contact surface tends to retain any water directed to it. The wider the connection's least dimension, the longer the water is retained and the higher the likelihood of decay. Treated wood in this contact surface is more resistant to decay but only at an uncut surface. The least dimension of the contact surface should be as small as possible. When the contact surfaces are for structural bearing, only as much bearing surface as needed should be provided, considering both perpendicular-to-grain and parallel-to-grain bearing design stresses. For example, deck boards on 2x joists have a smaller contact surface least dimension than deck boards on 4x joists. A beam bolted alongside an unnotched round wood pile has a small least dimension of the contact surface. Figure 13-10 illustrates the least-dimension concept.

Poor durability performance has been observed in exposed sistered members. When sistered members must be used in exposed conditions, they should be of ground-contact-rated treated wood, and the top surface should be covered with a self-adhering modified bitumen ("peel and stick") flashing membrane. This material is available in rolls as narrow as 3 inches. The membranes seal around nail penetrations to keep water out. In contrast, sheet-metal flashings over sistered members, when penetrated by nails, can channel water into the space between the members.

Other methods of improving exposed structural frame durability include:

- Using drip cuts to avoid horizontal water movement along the bottom surface of a member. Figure 13-11 shows this type of cut.

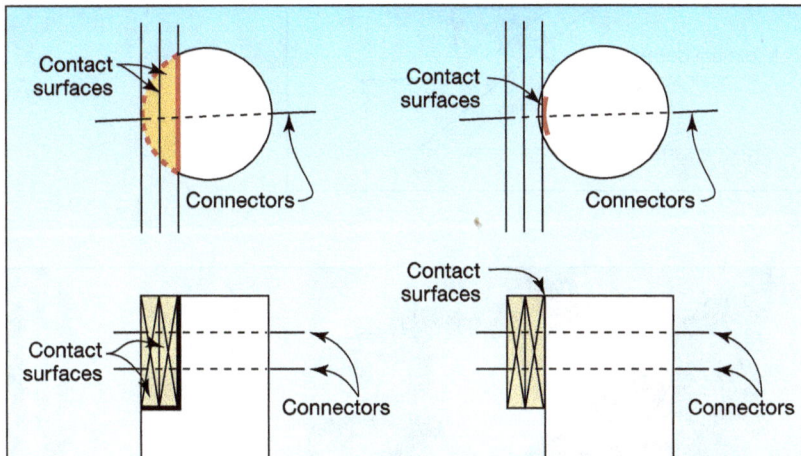

Figure 13-10.
Examples of minimizing the least dimension of wood contact surfaces

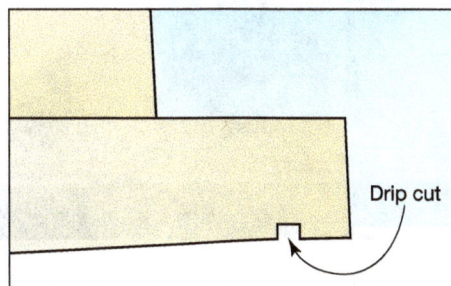

Figure 13-11.
Drip cut to minimize horizontal water movement along the bottom surface of a wood member

▪ Avoiding assemblies that form "buckets" and retain water adjacent to wood.

▪ Avoiding designs that result in ledges below a vertical or sloped surface. Ledges collect water quite readily, and the resulting ponding from rain or condensation alternating with solar radiation causes shrink-swell cycling, resulting in checks that allow increased water penetration.

▪ To the extent possible, minimizing the number of vertical holes in exposed horizontal surfaces from nails, lags, and bolts.

▪ When possible, avoiding the use of stair stringers that are notched for each stair. Notching exposes the end grain, which is then covered by the stair. As a result, the stair tends to retain moisture at the notch where the bending stress is greatest at the minimum depth section. Figure 13-12 illustrates stair stringer exposure, and Figure 13-13 shows the type of deterioration that can result.

Figure 13-12.
Exposure of end grain in stair stringer cuts

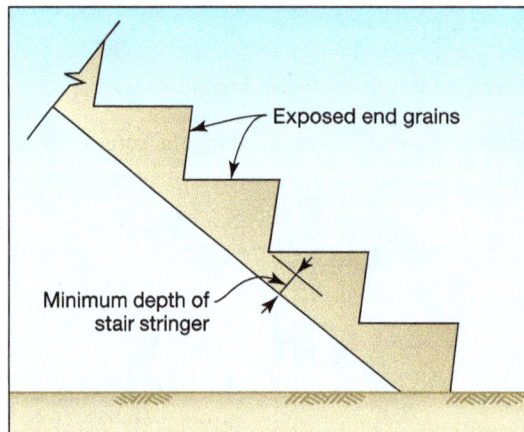

Figure 13-13.
Deterioration in a notched stair stringer

■ Using the alternative stair stringer installation shown in Figure 13-14 when the stair treads are either nailed onto a cleat or the stringer is routed out so the tread fits into the routed-out area. Even these alternatives allow water retention at end grain surfaces, and these surfaces should therefore be field-treated with wood preservative.

■ Caulking joints at wood connections to keep water out. Caulk only the top joints in the connection. Recaulk after the wood has shrunk, which can take up to a year for larger members.

■ When structurally possible, considering using spacers or shims to separate contact surfaces. A space of about 1/16-inch discourages water retention by capillary action but can easily fill with dirt and debris. A 1/4- to 1/2-inch space is sufficient to allow water and debris to clear from the interface. This spacing has structural limitations; a bolted connection with an unsupported shim has much less shear capacity than an unspaced connection because of increased bolt bending and unfavorable bearing stress distribution in the wood.

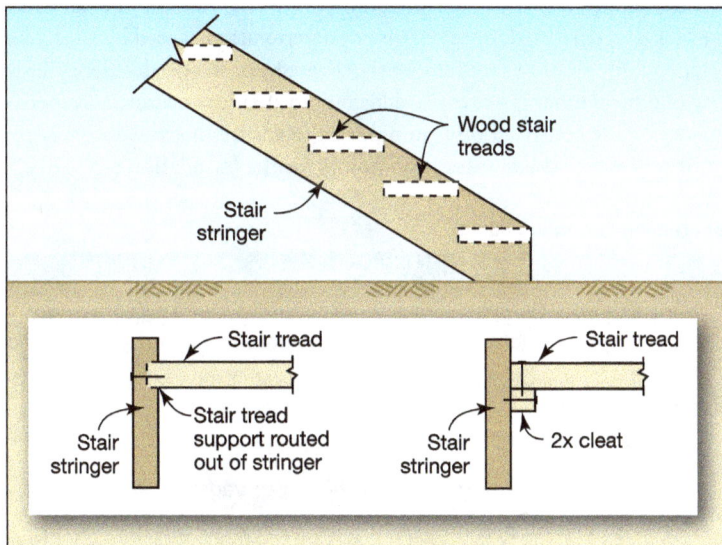

Figure 13-14.
Alternative method of installing stair treads

13.1.7   Field Preservative Treatment

Field cuts and bores of pressure-preservative-treated piles, timbers, and lumber are inevitable in coastal construction. Unfortunately, these cuts expose the inner untreated part of the wood member to possible decay and infestation. Although field preservative treatments are much less effective than pressure-preservative treatment, the decay and infestation potential can be minimized by treating the cuts and bolt holes with field-applied preservative.

13.1.8 Substitutions

During construction, a builder may find that materials called for in the construction plans or specifications are not available or that the delivery time for those materials is too long and will delay the completion of the building. These conflicts require decisions about substituting one type of construction material for another. Because of the high natural hazard forces imposed on buildings near the coast and the effects of the severe year-round environment in coastal areas, substitutions should be made only after approval by a design professional and, if necessary, the local building official.

> **WARNING**
>
> When substitutions are proposed, the design professional's approval should be obtained before the substitution is made. The ramifications of the change must be evaluated, including the effects on the building elements, constructability, and long-term durability. Code and regulatory ramifications should also be considered.

13.1.9 Foundation Inspection Points

If the foundation is not constructed properly, many construction details in the foundation can cause failure during a severe natural hazard event or premature failure because of deterioration caused by the harsh coastal environment. Improperly constructed foundations are frequently covered up, so any deficiency in the load-carrying or distributing capacity of one member is not easily detected until failure occurs. It is therefore very important to inspect the foundation while construction is in progress to ensure that the design is completed as intended. Table 13-1 is a list of suggested critical inspection points for the foundation.

Table 13-1. Foundation and Floor Framing Inspection Points

Inspection Point	Reason
1. Pile-to-girder connection	Ensures that pile is not overnotched, that it is field-treated, and that bolts are properly installed with washers and proper end and edge distance
2. Joist-to-girder connection	Verifies presence of positive connection with properly nailed, corrosion-resistant connector
3. Joist blocking	Ensures that the bottom of the joist is prevented from bending/buckling
4. Sheathing nailing – number, spacing, depth	Ensures that sheathing acts as a shear diaphragm
5. Material storage – protection from elements prior to installation	Ensures that the wood does not absorb too much moisture prior to installation—exposure promotes checks and splits in wood, warp, and separation in plywood
6. Joist and beam material – excessive crown or lateral warping, large splits	Ensures that new floors are installed level and eliminates need to repair large splits in new material

13.1.10 Top Foundation Issues for Builders

The top foundation-related issues for builders are as follows:

- Piles, piers, or columns must be properly aligned.

- Piles, piers, or columns must be driven or placed at the proper elevation to resist failure and must extend below the expected depth of scour and erosion.

- Foundation materials must be damage-resistant to flooding (pressure-treated wood, masonry, or concrete).

- The support at the top of the foundation element must be adequate to properly attach the floor framing system. Notching of a wood foundation element should not exceed the specifications in the construction documents.

- Breakaway walls should not be overnailed to the foundation. They are intended to fail. Utilities and other obstructions should not be installed behind these walls, and the interior faces should not be finished.

- For masonry or concrete foundation elements (except slabs-on-grade), the proper size of reinforcing, proper number of steel bars, and proper concrete cover over the steel should be used.

- Concrete must have the proper mix to meet the specialized demands of the coastal environment.

- Exposed steel in the foundation corrodes; corrosion should be planned for by installing hot-dipped galvanized or stainless steel.

- Areas of pressure-treated wood that have been cut or drilled retain water and decay; these cut areas should be treated in the field.

13.2   Structural Frame

Structural framing includes framing the floors, walls, and roof and installing critical connections between each element.

13.2.1   Structural Connections

One of the most critical aspects of building in a coastal area is the method that is used to connect the structural members. A substantial difference usually exists between connections acceptable in inland construction and those required to withstand the natural hazard forces and environmental conditions in coastal areas. Construction in noncoastal, nonseismic areas must normally support only vertical dead and live loads and modest wind loads. In most coastal areas, large forces are applied by wind, velocity flooding, wave impact, and floating debris. The

> **WARNING**
>
> The connections described in this Manual are designed to hold the building together in a design event. Builders should never underestimate the importance of installing connectors according to manufacturers' recommendations. Installing connectors properly is extremely important.

calculated forces along the complete load path usually require that the builder provide considerable lateral and uplift capacity in and between the roof, walls, floors, girders, and piles. Consequently, builders should be sure to use the specified connectors or approved substitutes. Connectors that look alike may not have the same capacity, and a connector designed for gravity loads may have little uplift resistance. Fact Sheet 4.1, *Load Path*, in FEMA P-499 describes load paths and highlights important connections in a typical wind load path.

The nails required for the connection hardware may not be regularly found on the job site. For example, full-diameter 8d to 20d short nails are commonly specified for specific hurricane/seismic connection hardware.

Figure 13-15.
Connector failure caused
by insufficient nailing

WARNING

Proper nail selection and
installation are critical. Builders
should not substitute different
nails or nailing patterns without
approval from the designer.

For full strength, these connections require that all of the holes in the hardware be nailed with the proper nails. In the aftermath of investigated hurricanes, failed connector straps and other hardware have often been found to have been attached with too few nails, nails of insufficient diameter, or the wrong type of nail. Figure 13-15 shows a connector that failed because of insufficient nailing.

As mentioned previously, connection hardware must be corrosion-resistant. If galvanized connectors are used, additional care must be taken during nailing. When a hammer strikes the connector and nail during installation, some of the galvanizing protection is knocked off. One way to avoid this problem is to use corrosion-resistant connectors that do not depend on a galvanized coating, such as stainless steel or wood (see Section 9.2.3). Only stainless steel nails should be used with stainless steel connectors. An alternative to hand-nailing is to use a pneumatic hammer that "shoots" nails into connector holes.

NOTE

Additional information about pneumatic nail guns can be obtained from the International Staple, Nail and Tool Association, 512 West Burlington Ave., Suite 203, LaGrange, IL 60525-2245. A report prepared by National Evaluation Service, Inc., NER-272, *Power-Driven Staples and Nails for Use in All Types of Building Construction* (NES 1997), presents information about the performance of pneumatic nail guns and includes prescriptive nailing schedules.

All connections between members in a wood-frame building are made with nails, bolts, screws, or a similar fastener. Each fastener is installed by hand. The predominant method of installing nails is by pneumatic nail gun. Many nail guns use nails commonly referred to as "sinkers." Sinkers are slightly smaller in diameter and thus have lower withdrawal and shear capacities than common nails of the same size. Nail penetration is governed by air pressure for pneumatic nailers, and nail penetration is an important quality control issue for builders. Many prescriptive codes have nailing schedules for various building elements such as shearwalls and diaphragms.

Another critical connection is the connection of the floor to the piles. Pile alignment and notching are critical not only to successful floor construction but also to the structural adequacy during a natural hazard event (see Section 13.1.1). Construction problems related to these issues are also inevitable, so solutions to pile misalignment and overnotching must be developed. Figure 13-16 illustrates a method of reinforcing an overnotched pile, including one that is placed on a corner. The most appropriate solution to pile misalignment is to re-drive a pile in the correct location. An alternative is illustrated in Figure 13-17, which shows a method of supporting a beam at a pile that has been driven "outside the layout" of the pile foundation. Figure 13-18

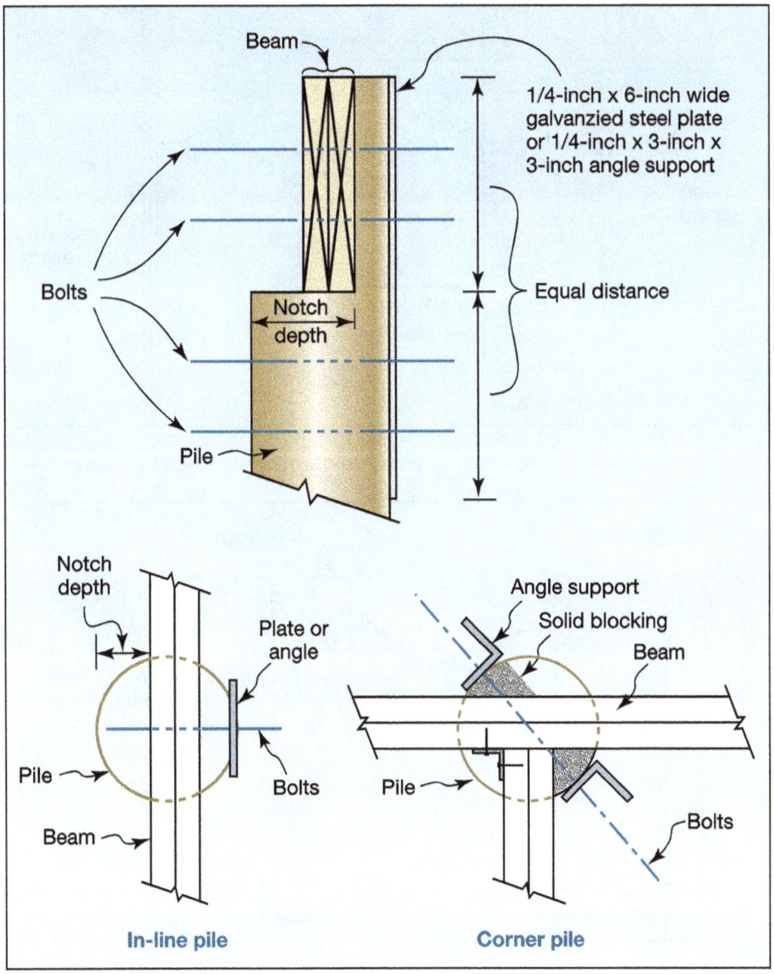

Figure 13-16.
Reinforcement of overnotched piles

Figure 13-17.
Beam support at
misaligned piles

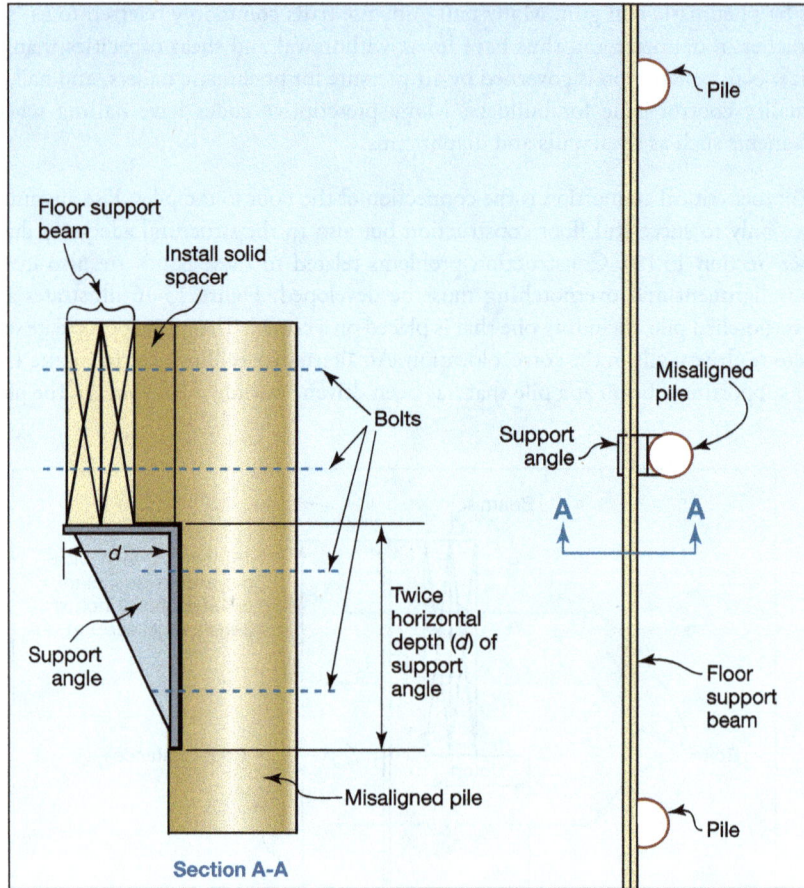

Section A-A

Floor support
beam

Install solid
spacer

Bolts

Twice
horizontal
depth (*d*) of
support
angle

Support
angle

Misaligned pile

Pile

Misaligned
pile

Support
angle

A   A

Floor
support
beam

Pile

Figure 13-18.
Proper pile notching for
two-member and four-
member beams

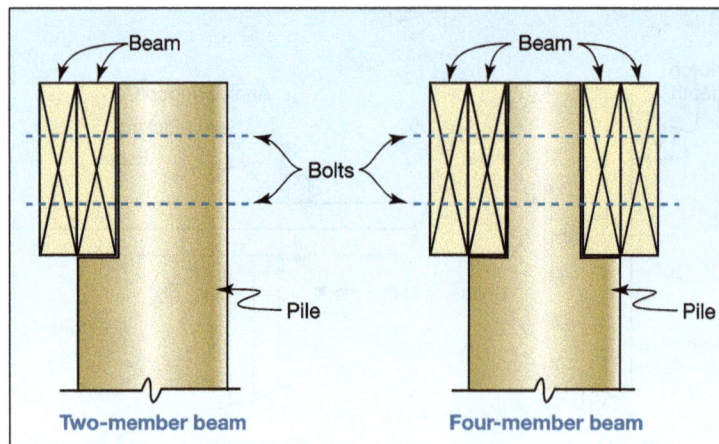

Beam

Beam

Bolts

Pile

Pile

Two-member beam   **Four-member beam**

illustrates the proper pile notching for both two-member and four-member beams. See Section 13.1.1 for more information on pile notching.

After the "square" foundation has been built, the primary layout concerns about how the building will perform under loads are confined to other building elements being properly located so that load transfer paths are complete.

13.2.2   Floor Framing

The connection between wood floor joists and the supporting beams and girders is usually a bearing connection for gravity forces with a twist strap tie for uplift forces. Figure 13-19 shows a twist tie connection. This connection is subjected to large uplift forces from high winds. In addition, the undersides of elevated structures, where these connectors are located, are particularly vulnerable to salt spray; the exposed surfaces are not washed by rain, and they stay damp longer because of their sheltered

CROSS REFERENCE

See NFIP Technical Bulletin 8-96, *Corrosion Protection for Metal Connectors in Coastal Areas* (FEMA 1996).

location. Consequently, the twist straps and the nails used to secure them must be hot-dipped galvanized or stainless steel. One way to reduce the corrosion potential for metal connectors located under the building is to cover the connectors with a plywood bottom attached to the undersides of the floor joists. (The bottom half of the joist-to-girder twist straps will still be exposed, however.) This covering will help keep insulation in the floor joist space as well as protect the metal connectors.

Because the undersides of Zone V buildings are exposed, the first floor is more vulnerable to uplift wind and wave forces, as well as to the lateral forces of moving water, wave impact, and floating debris. These loads cause compressive and lateral forces in the normally unbraced lower flange of the joist. Solid blocking or 1x3 cross-bridging at 8-foot centers is recommended for at least the first floor joists unless substantial sheathing (at least 1/2-inch thick) has been nailed well to the bottom of these joists. Figure 13-19 also shows solid blocking between floor joists.

Figure 13-19.
Proper use of metal twist strap ties (circled); solid blocking between floor joists

Floor framing materials other than 2x sawn lumber are becoming popular in many parts of the country. These materials include wood floor trusses and wood I-joists. Depending on the shape of the joist and the manufacturer, the proper installation of these materials may require some additional steps. For instance, some wood I-joists require solid blocking at the end of the joist where it is supported so that the plywood web is not crushed or does not buckle. Figure 13-20 illustrates the use of plywood web I-joists. As shown in the figure, the bottom flanges of the joists are braced with a small metal strip that helps keep the flange from twisting. Solid wood blocking is a corrosion-resistant alternative to the metal braces.

Floor surfaces in high-wind, flood, or seismic hazard areas are required to act as a diaphragm. For the builder, this means that the floor joists and sheathing are an important structural element. Therefore, the following installation features may require added attention:

- Joints in the sheathing should fully bear on top of a joist, not a scabbed-on board used as floor support

- Nailing must be done in accordance with a shear diaphragm plan

- Construction adhesive is important for preventing "squeaky" floors, but the adhesive must not be relied on for shear resistance in the floor

Joints in the sheathing across the joists must be fully blocked with a full-joist-height block. Horizontal floor diaphragms with lower shear capacities can be unblocked if tongue-and-groove sheathing is used.

13.2.2.1 Horizontal Beams and Girders

Girders and beams can be solid sawn timbers, glue-laminated timbers (see Figure 13-20), or built-up sections (see "Material Durability in Coastal Environments" on the Residential Coastal Construction Web site at http://www.fema.gov/rebuild/mat/fema55.html). The girders span between the piles and support the beams and joists. The piles are usually notched to receive the girders. To meet the design intent, girders, beams, and joists must be square and level, girders must be secured to the piles, and beams and joists must be secured to the girders.

Figure 13-20.
Engineered joists used as floor joists with proper metal brace to keep the bottoms of the joists from twisting; engineered wood beam

The layout process involves careful surveying, notching, sawing, and boring. The bottom of the notch provides the bearing surface for downward vertical loads. The bolted connection between the girder and the vertical notch surface provides capacity for uplift loads and stability. Girder splices are made as required at these connections. Splices in multiple-member girders may be made away from the pile but should be engineered so that the splices occur at points of zero bending moment. This concept is illustrated in Figure 13-21.

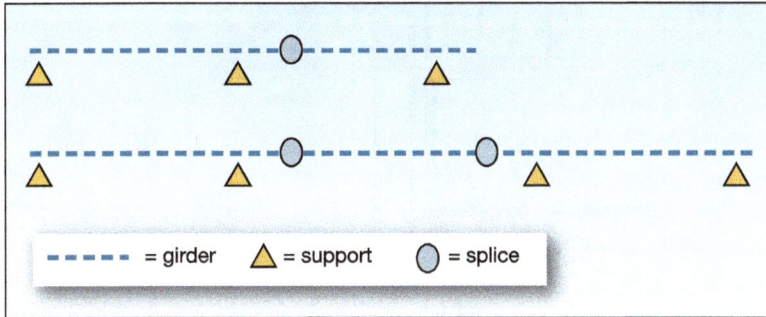

Figure 13-21.
Acceptable locations for splices in multiple-member girders

13.2.2.2 Substitution of Floor Framing Materials

The considerations discussed in Section 13.1.8 for substitution of foundation materials also apply to substitutions of floor framing materials.

13.2.2.3 Floor Framing Inspection Points

Proper connections between elements of the floor framing help to guarantee that the load path is continuous and the diaphragm action of the floor is intact. If floor framing is not constructed properly, many construction details in the floor framing can become structural inadequacies during a severe natural hazard event or cause premature failure because of deterioration caused by the harsh coastal environment. Table 13-1 is a list of suggested critical inspection points in foundations and a guide for floor framing inspections.

13.2.3   Wall Framing

Exterior walls and designated interior shear walls are an important part of the building's vertical and lateral force-resisting system. All exterior walls must be able to withstand in-plane (i.e., parallel to the wall surface), gravity, and wind uplift tensile forces, and out-of-plane (i.e., normal or perpendicular to the wall surface) wind forces. Exterior and designated interior shear walls must be able to withstand shear and overturning forces transferred through the walls to and from the adjacent roof and floor diaphragms and framing.

The framing of the walls should be of the specified material and fastened in accordance with the design drawings and standard code practice. Exterior wall and designated shear wall sheathing panels must be of the specified material and fastened with accurately placed nails whose size, spacing, and durability are in accordance with the design. Horizontal sheathing joints in shear walls must be solidly blocked in accordance with shear wall capacity tables. Shear transfer can be better accomplished if the sheathing extends the full height from the bottom of the floor joist to the wall top plate (see Figure 13-22), but sheathing this long is often unavailable.

Figure 13-22.
Full-height sheathing to improve transfer of shear

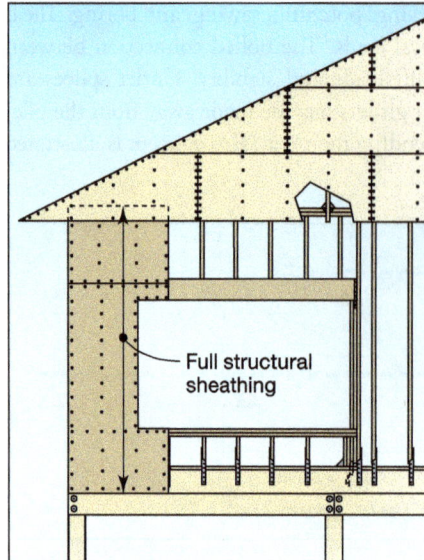

Full structural sheathing

The design drawings may show tiedown connections between large shearwall vertical posts and main girders. Especially in larger, taller buildings, these connections must resist thousands of pounds of overturning forces during high winds. See Section 8.7 for information regarding the magnitude of these forces. The connections must be accomplished with careful layout, boring, and assembly. Shear transfer nailing at the top plates and sills must be in accordance with the design. Proper nailing and attachment of the framing material around openings is very important; see Section 9.2.1 for a discussion of the difficulty of transferring large shear loads when there are large openings in the shearwall.

It is very important that shearwall sheathing (e.g., plywood, oriented strand board [OSB]) with an exterior exposure be finished appropriately with pigmented finishes such as paint, which last longer than unpigmented finishes, or semitransparent penetrating stains. It is also important that these finishes be properly maintained. Salt crystal buildup in surface checks in siding can damage the siding. Damage is typically worse in siding that is sheltered from precipitation because the salt crystals are never washed off with fresh rainwater.

To meet the design intent, walls must:

- Be plumb and square to each other and to the floor

- Be lined up over solid support such as a beam, floor joists, or a perimeter band joist

- Not have any more openings than designated by the plans

- Not have openings located in places other than designated on the plans

- Consist of material expected to resist corrosion and deterioration

- Be properly attached to the floors above and below the wall, including the holddown brackets required to transfer overturning forces

In addition, all portions of walls designed as shearwalls must be covered with sheathing nailed in accordance with either the plans or a specified prescriptive standard.

13.2.3.1 Substitution of Wall Framing Materials

The considerations discussed in Section 13.1.8 for substitution of foundation materials also apply to substitutions of wall framing materials.

13.2.3.2 Wall Framing Inspection Points

Proper connections between elements of the wall framing help to guarantee a continuous load path and the diaphragm action of the walls is intact. If not completed properly, there are many construction details in the floor framing that can become structural inadequacies and fail during a severe natural hazard event or cause premature failure because of deterioration caused by the harsh coastal environment. Table 13-2 is a list of suggested critical inspection points that can be used as a guide for wall framing inspections.

Table 13-2. Wall Inspection Points

Inspection Point	Reason
1. Wall framing attachment to floors	Ensures that nails are of sufficient size, type, and number
2. Size and location of openings	Ensures performance of shear wall
3. Wall stud blocking	Ensures that there is support for edges of sheathing material
4. Sheathing nailing – number, spacing, depth of nails	Ensures that sheathing acts as a shear diaphragm
5. Material storage – protection from elements prior to installation	Ensures that the wood does not absorb too much moisture prior to installation—exposure promotes checks and splits in wood, warp, and separation in plywood
6. Stud material – excessive crown (crook) or lateral warping (bow)	Maintains plumb walls and eliminates eccentricities in vertical loading
7. Header support over openings	Ensures that vertical and lateral loads will be transferred along the continuous load path

13.2.4 Roof Framing

Proper roof construction is very important in high-wind and earthquake hazard areas. Reviews of wind damage to coastal buildings reveal that most damage starts with the failure of roof elements. The structural integrity of the roof depends on a complete load path, including the resistance to uplift of porch and roof overhangs, gable end overhangs, roof sheathing nailing, roof framing nailing and strapping, roof member-to-wall strapping, and gable end-wall bracing.

WARNING

The most common roof structure failure is the uplift failure of porch, eave, and gable end overhangs. The next most common is roof sheathing peeling away from the framing. Nailing the sheathing at the leading edge of the roof, the gable edge, and the joints at the hip rafter or ridge is very important, as is securing the roof framing to prevent uplift. This failure point (leading edge of sheathing at gable edge, ridge, and hip) is also the most likely place for progressive failure of the entire structure to begin.

All of this construction must use the specified wood materials, straps, and nails. The appropriate nails must be used in all of the holes in the straps so that the straps develop their full strength. Sheathing nails must be of the specified length, diameter, and head, and the sheathing must be nailed at the correct spacing. In addition, sheathing nails must penetrate the underlying roof framing members and must not be overdriven, which frequently occurs when pneumatic nail guns are used. When prefabricated roof trusses are used, handling precautions must be observed, and the trusses must be laterally braced as specified by the design professional or manufacturer.

Fact Sheets 7.1 through 7.4 in FEMA P-499 discuss roof construction, including sheathing installation, asphalt shingle roofing, and tile roofing.

To meet the design intent, roofs must meet the following requirements:

- Roof trusses and rafters must be properly attached to the walls

- Roof sheathing must be nailed according to either the construction plans or a specified prescriptive standard

- Roofs must consist of materials expected to resist corrosion and deterioration, particularly the connectors

13.2.4.1   Substitution of Roof Framing Materials

The considerations discussed in Section 13.1.8 for substitution of foundation materials also apply to substitutions of roof framing materials.

13.2.4.2   Roof Frame Inspection Points

Proper connections between elements of the roof frame help to guarantee a continuous load path and the diaphragm action of the walls is intact. If not completed properly, there are many construction details in the roof framing that can become structural inadequacies and fail during a severe natural hazard event or cause premature failure because of deterioration caused by the harsh coastal environment. Table 13-3 contains suggestions of critical inspection points as a guide for roof framing inspections.

> **WARNING**
>
> Do not substitute nails, fasteners, or connectors without approval of the designer.

13.2.5   Top Structural Frame Issues for Builders

The top structural frame issues for builders are as follows:

- Connections between structural elements (e.g., roofs to walls) must be made so that the full natural hazard forces are transferred along a continuous load path.

- Care must be taken to nail elements so that the nails are fully embedded.

- Compliance with manufacturers' recommendations on hardware use and load ratings is critically important.

Table 13-3. Roof Frame Inspection Points

Inspection Point	Reason
1. **Roof framing attachment to walls**	Ensures that the sufficient number, size, and type of nails are used in the proper connector
2. **Size and location of openings**	Ensures performance of roof as a diaphragm
3. **"H" clips or roof frame blocking**	Ensures that there is support for edges of the sheathing material
4. **Sheathing nailing** – number, spacing, depth of nails	Ensures that sheathing acts as a shear diaphragm and is able to resist uplift
5. **Material storage** – protection from elements prior to installation	Ensures that the wood does not absorb too much moisture prior to installation—exposure promotes checks and splits in wood, warp, and separation in plywood
6. **Rafter or ceiling joist material** – excessive crown or lateral warping	Maintains level ceilings
7. **Gable-end bracing**	Ensures that bracing conforms to design requirements and specifications

- Only material that is rated and specified for the expected use and environmental conditions should be used.

- Builders should understand that the weakest connections fail first and that it is therefore critical to pay attention to every connection. The concept of continuous load path must be considered for every connection in the structure.

- Exposed steel in the structural frame corrodes even in places such as the attic space. The builder should plan for it by installing hot-dipped galvanized or stainless steel hardware and nails.

- Compliance with suggested nailing schedules for roof, wall, and floor sheathing is very important.

13.3   Building Envelope

The building envelope comprises the exterior doors, windows, skylights, non-load-bearing walls, wall coverings, soffits, roof systems, and attic vents. The floor is also considered a part of the envelope in buildings elevated on open foundations. Building envelope design is discussed in detail in Chapter 11. The key to successful building envelope construction is having a detailed plan that is followed carefully by the builder, as described below.

A suitable design must be provided that is sufficiently specified and detailed to allow the builder to understand the design intent and to give the contractor adequate and clear guidance. Lack of sufficient and clear design guidance regarding the building envelope is common. If necessary, the contractor should seek additional guidance from the design professional or be responsible for providing design services in addition to constructing the building.

The building must be constructed as intended by the design professional (i.e., the builder must follow the drawings and specifications). Examples are:

- Installing flashings, building paper, or air infiltration barriers so that water is shed at laps

- Using the specified type and size of fasteners and spacing them as specified

- Eliminating dissimilar metal contact

- Using materials that are compatible with one another

- Installing elements in a manner that accommodates thermal movements so that buckling or jacking out of fasteners is avoided

- Applying finishes to adequately cleaned, dried, and prepared substrates

- Installing backer rods or bond breaker tape at sealant joints

- Tooling sealant joints

For products or systems specified by performance criteria, the contractor must exercise care in selecting those products or systems and in integrating them into the building envelope. For example, if the design professional specifies a window by requiring that it be capable of resisting a specified wind pressure, the contractor should ensure that the type of window that is being considered can resist the pressure when tested in accordance with the specified test (or a suitable test if a test method is not specified). Furthermore, the contractor needs to ensure that the manufacturer, design professional, or other qualified entity provides guidance on how to attach the window frame to the wall so that the frame can resist the design pressures.

When the selection of accessory items is left to the discretion of the contractor, without prescriptive or performance guidance, the contractor must be aware of and consider special conditions at the site (e.g., termites, unusually severe corrosion, and high earthquake or wind loads) that should influence the selection of the accessory items. For example, instead of using screws in plastic sleeves to anchor elements to a concrete or masonry wall, a contractor can use metal expansion sleeves or steel spikes intended for anchoring to concrete, which should provide a stronger and more reliable connection, or the use of plastic shims at metal doors may be appropriate to avoid termite attack.

Adequate quality control (i.e., inspection by the contractor's personnel) and adequate quality assurance (i.e., inspection by third parties such as the building official, the design professional, or a test lab) must be provided. The amount of quality control/quality assurance depends on the magnitude of the natural hazards being designed for, complexities of the building design, and the type of products or systems being used. For example, installing windows that are very tall and wide and make up the majority of a wall should receive more inspection than isolated, relatively small windows. Inspecting roof coverings and windows is generally more critical than inspecting most wall coverings because of the general susceptibility of roofing and glazing to wind and the resulting damage from water infiltration that commonly occurs when these elements fail.

13.3.1 Substitution of Building Envelope Materials

The considerations discussed in Section 13.1.8 for substitution of foundation materials also apply to substitutions of envelope materials. Proposed substitutions of materials must be thoroughly evaluated and must be approved by the design professional (see Section 13.1.8). The building envelope must be installed in a manner that will not compromise the building's structural integrity. For example, during construction, if a window larger than originally intended is to be installed because of delivery problems or other reasons, the contractor should obtain the design professional's approval prior to installation. The larger window may unacceptably reduce the shear capacity of the wall, or different header or framing connection details may be necessary. Likewise, if a door is to be located in a different position, the design professional should evaluate the change to determine whether it adversely affects the structure.

13.3.2   Building Envelope Inspection Points

Table 13-4 is a list of suggested critical inspection points that can be used as a guide for building envelope inspections. Fact Sheet 6.1, *Window and Door Installation,* in FEMA P-499 discusses proper window and door installation and inspection points.

Table 13-4. Building Envelope Inspection Points

Inspection Point	Reason
1. **Siding attachment** to wall framing	Ensures there are sufficient number, type, and spacing of nails
2. **Attachment of windows and doors** to the wall framing	Ensures there are sufficient number, type, and spacing of either nails or screws
3. **Flashings** around wall and roof openings, roof perimeters, and at changes in building shape	Prevents water penetration into building envelope
4. **Roof covering attachment to sheathing,** including special connection details	Minimizes potential for wind blowoff. In high-seismic-load areas, attention to attachment of heavy roof coverings, such as tile, is needed to avoid displacement of the covering.
5. **Attachments of vents and fans** at roofs and walls	Reduces chance that vents or fans will blow off and allow wind-driven rain into the building

13.3.3   Top Building Envelope Issues for Builders

The top building envelope issues for builders are as follows:

- Many manufacturers do not rate their products in a way that it is easy to determine whether the product will really be adequate for the coastal environment and the expected loads. Suppliers should be required to provide information about product reliability in the coastal environment.

- Wind-driven rain finds a way into a building if there is an open path. Sealing openings and shedding water play significant parts in building a successful coastal home.

- Window and door products are particularly vulnerable to wind-driven rain leakage and air infiltration. These products should be tested and rated for the expected coastal conditions.

- The current high-wind techniques of adding extra roof surface sealing or attachments at the eaves and gable end edges should be used.

- Coastal buildings require more maintenance than inland structures. The maintenance requirement needs to be considered in the selection of materials and the care with which they are installed.

13.4   References

ACI (American Concrete Institute). 2008. *Building Code Requirements for Structural Concrete*. ACI 318-08.

AWPA (American Wood Protection Association). 1991. *Care of Pressure-Treated Wood Products*. AWPA Standard M4-91. Woodstock, MD.

AWPA. 1994. *Standards*. Woodstock, MD.

FEMA (Federal Emergency Management Agency). 1996. *Corrosion Protection for Metal Connectors in Coastal Areas*. NFIP Technical Bulletin 8-96.

FEMA. 2006. *Recommended Residential Construction for the Gulf Coast*. FEMA P-550.

FEMA. 2011. *Home Builder's Guide to Coastal Construction Technical Fact Sheets*. FEMA P-499.

ICC (International Code Council). 2008. *Standard for Residential Construction in High-Wind Regions*, ICC 600-2008. ICC: Country Club Hills, IL.

ICC. 2011a. *International Building Code*. 2012 IBC. Country Club Hills, IL: ICC.

ICC. 2011b. *International Residential Code for One-and Two-Family Dwellings*. 2012 IRC. Country Club Hills, IL: ICC.

NES (National Evaluation Service, Inc.). 1997. *Power-Driven Staples and Nails for Use in All Types of Building Construction*. National Evaluation Report NER-272.

TMS (The Masonry Society). 2008. *Building Code Requirements and Specification for Masonry Structures and Commentaries*. TMS 402-08/ACI 530-08/ASCE 5-08 and TMS 602-08/ACI 530.1-08/ASCE 6-08.

USDN (U.S. Department of the Navy). 1982. *Foundation and Earth Structures Design*. Design Manual 7.2.

Maintaining the Building

This chapter provides guidance on maintaining the building structure and envelope.

For maximum performance of a building in a coastal area, the building structure and envelope (i.e., exterior doors, windows, skylights, exterior wall coverings, soffits, roof systems, and attic vents) must not be allowed to deteriorate. Significant degradation by corrosion, wood decay, termite attack, or weathering increases the building's vulnerability to damage from natural hazards. Figure 14-1 shows a post that appears on the exterior to be in

CROSS REFERENCE

or resources that augment the guidance and other information in this Manual, see the Residential Coastal Construction Web site (http://www.fema.gov/rebuild/mat/fema55.shtm).

Figure 14-1.
Pile that appears acceptable from the exterior but has interior decay

acceptable condition but is weakened by interior decay, which can be determined only through a detailed inspection. This post failed under the loads imposed by a natural hazard event.

Long-term maintenance and repair demands are influenced directly by decisions about design, materials, and construction methods during building design and construction. Using less durable materials will increase the frequency and cost of required maintenance and repair. The design and detailing of various building systems (e.g., exposed structural, window, or roof systems) also significantly influence maintenance and repair demands.

COST CONSIDERATION

Maintenance and repair costs are related directly to original design decisions, materials selection, and construction methods.

14.1   Effects of Coastal Environment

The coastal environment can cause severe damage to the building structure and envelope. The damage arises primarily from salt-laden moisture, termites, and weathering.

14.1.1   Corrosion

The corrosive effect of salt-laden, wind-driven moisture in coastal areas cannot be overstated. Salt-laden, moist air can corrode exposed metal surfaces and penetrate any opening in the building. The need to protect metal surfaces through effective design and maintenance (see Section 14.2.6 for maintenance of metal connectors) is very important for the long-term life of building elements and the entire building. Stainless steel is recommended because many galvanized (non-heavy-gauge) products and unprotected steel products do not last in the harsh coastal environment.

Corrosion is most likely to attack metal connectors (see Section 14.2.6) that are used to attach the parts of the structure to one another, such as floor joists to beams and connectors used in cross-bracing below the finished lowest floor. Galvanized connectors coated with zinc at the rate of 0.9 ounce per square foot of surface area (designated G-90) can corrode in coastal environments at a rate of 0.1 to 0.3 millimeter/year. At this rate, the zinc protection will be gone in 7 years. A G-185 coated connector, which provides twice as much protection as G-90, can corrode in less than 20 years. More galvanized protection (more ounces of zinc per square foot of surface area to be protected) increases service life.

CROSS REFERENCE

For additional information on corrosion, see Section 9.4.5 in this Manual and FEMA Technical Bulletin 8-96, *Corrosion Protection of Metal Connectors in Coastal Areas for Structures Located in Special Flood Hazard Areas* (FEMA 1996).

Corrosion can also affect fasteners for siding and connectors for attaching exterior-mounted heating, ventilation, and air-conditioning units, electrical boxes, lighting fixtures, and any other item mounted on the exterior of the building. These connectors (nails, bolts, and screws) should be stainless steel or when they must be replaced, replaced with stainless steel. These connectors are small items, and the increased cost of stainless steel is small.

14.1.2 Moisture

There are many sources of exterior moisture from outside the home in the coastal environment. Whenever an object absorbs and retains moisture, the object may decay, mildew, or deteriorate in other ways. Figure 14-2 shows decay behind the connection plate on a beam.

Significant sources of interior moisture, such as kitchens, baths, and clothes dryers, should be vented to the outside in such a way that condensation does not occur on interior or exterior surfaces.

**Figure 14-2.
Wood decay behind a
metal beam connector**

Connectors should be designed to shed water to prevent water from accumulating between the connector and the material the connector is attached to. Trapped moisture increases the moisture content of the material and potentially leads to decay. Moisture is most likely to enter at intersections of materials where there is a hole in the building envelope (e.g., window, door) of where two surfaces are joined (e.g., roof to wall intersection). If properly installed, the flashings for the openings and intersections should not require maintenance for many years. However, flashings are frequently not properly installed or installed at all, creating an ongoing moisture intrusion problem.

The potential for wood framing in crawlspaces in low-lying coastal areas to decay is high. Moisture migration into the floor system can be reduced if the floor of the crawlspace is covered with a vapor barrier of at least 6-millimeter polyethylene. Where required by the local building code, wood framing in the crawlspace should be preservative-treated or naturally decay-resistant. The building code may have ventilation requirements.

Many existing crawlspaces are being converted to "conditioned crawlspaces." A moisture barrier is placed on the floor and walls of the crawlspace interior, insulation is added to the floor system (commonly sprayed-on polyurethane foam), and conditioned air is introduced into the space. In order for a conditioned crawlspace to be successful in low-lying coastal areas, moisture control must be nearly perfect so that the moisture content of the floor system does not exceed 20 percent (the minimum water content in wood that promotes mold growth). Conditioned crawlspaces are typically not practical in a floodplain where flood vents are required.

Sprinkler systems used for landscaping and other exterior water distribution systems (e.g., fountains) must be carefully tested so they do not create or increase water collection where metal connectors are fastened. Water collection can be prevented easily during installation of the exterior water distribution system by making sure the water distribution pattern does not increase the moisture that is present in the building materials.

14.1.3   Weathering

The combined effects of sun and water on many building materials, particularly several types of roof and wall coverings, cause weathering damage, including:

- Fading of finishes
- Accelerated checking and splitting of wood
- Gradual loss of thickness of wood
- Degradation of physical properties (e.g., embrittlement of asphalt shingles)

In combination, the effects of weathering reduce the life of building materials unless they are naturally resistant to weathering or are protected from it, either naturally or by maintenance. Even finishes intended to protect exterior materials fade in the sun, sometimes in only a few years.

14.1.4   Termites

The likelihood of termite infestation in coastal buildings can be reduced by maintenance that makes the building site drier and otherwise less hospitable to termites, specifically:

- Storing firewood and other wood items that are stored on the ground, including wood mulch, well away from the building
- Keeping gutters and downspouts free of debris and positioned to direct water away from the building
- Keeping water pipes, water fixtures, and drainpipes in good repair
- Avoiding dampness in crawlspaces by providing adequate ventilation or installing impervious ground cover membranes
- Avoiding frequent plant watering adjacent to the house and trimming plants away from the walls

If any wood must be replaced under the house in or near contact with the ground, the new wood should be treated. Removing moisture and treating the cellulose in wood, which is the termite's food source, are the most frequently used remedies to combat termites.

14.2   Building Elements That Require Frequent Maintenance

To help ensure that a coastal building is properly maintained, this Manual recommends that buildings be inspected annually by professionals with the appropriate expertise. The following building elements should be inspected annually:

- Building envelope – wall coverings, doors, windows, shutters, skylights, roof coverings, soffits, and attic vents

- Foundation, attic, and the exposed structural frame

- Exterior-mounted mechanical and electrical equipment

Table 14-1 provides a maintenance inspection checklist. Items requiring repair or replacement should be documented and the required work scheduled.

Table 14-1. Maintenance Inspection Checklist

Item	Element	Condition			Repair/Replace	
		Good	Fair	Poor	Yes	No
Foundation	**Wood pile** – decay, termite infestation, severe splits, connection to framing					
	Sill plates – deterioration, splits, lack of attachment to foundation					
	Masonry – deteriorated mortar joints, cracked block, step cracks indicating foundation settlement					
	Concrete – spalling, exposed or corroding reinforcing steel, $\geq$ 1/4-inch vertical cracks or horizontal cracks with lateral shift in the concrete across the crack					
Exterior Walls	**Siding** – deterioration, nail withdrawal, discoloration, buckling, attachment to studs (nails missing, withdrawn, or not attached to studs), sealant cracked/dried out					
	Trim – deterioration, discoloration, separation at joints, sealant cracked/ dried out					
Porches/ Columns	**Top and bottom connections to framing** – corrosion in connectors					
	Base of wood columns – deterioration					
Floors	**Joists or beams** – decay, termite infestation, corrosion at tiedown connectors, splits, excessive holes or notching, excessive sagging					
	Sheathing – deterioration, "squeaky" floors, excessive sagging, attachment to framing (nails missing, withdrawn, or not attached to framing)					

Table 14-1. Maintenance Inspection Checklist (concluded)

Item	Element	Condition			Repair/Replace	
		Good	Fair	Poor	Yes	No
Floors	**Sheathing under floors** – attachment to framing, nail corrosion fastening sheathing to floor joists, buckling/warping caused by excessive moisture					
Windows/ Doors	**Glazing** – cracked panes, condensation between panes of insulated glass, nicks in glass surface, sealant cracked/dried out					
	Trim – deterioration, discoloration, separation at joints, caulking dried out or separated					
	Shutters – permanent shutters should be operated at least twice/year and temporary panels should be checked once/year for condition					
Roof	**Asphalt shingles** – granule loss, shingles curled, nails withdrawing from sheathing, de-bonding of tabs along eaves and corners					
	Wood shakes – splits, discoloration, deterioration, moss growth, attachment to framing (nails missing, withdrawn, or not attached to framing)					
	Metal – corrosion, discoloration, connection of fasteners or fastening system adequacy					
	Flashings – corrosion, joints separated, nails withdrawing					
Attic	**Framing** – condition of truss plates sagging or bowed rafters or truss chords, deterioration of underside of roof sheathing, evidence of water leaks, adequate ventilation					

Other items that should be inspected include cavities through which air can freely circulate (e.g., above soffits and behind brick or masonry veneers) and, depending on structural system characteristics and access, the structural system. For example, painted, light-gauge, cold-formed steel framing is vulnerable to corrosion, and the untreated cores of treated timber framing are vulnerable to decay and termite damage. Depending on visual findings, it may be prudent to determine the condition of concealed items through nondestructive or destructive tests (e.g., test cuts).

The following sections provide information on the building elements that require frequent maintenance in coastal environments: glazing, siding, roofs, outdoor mechanical and electrical equipment, decks and exterior wood, and metal connectors.

14.2.1   Glazing

Glazing includes glass or a transparent or translucent plastic sheet in windows, doors, skylights, and shutters. Glazing is particularly vulnerable to damage in hurricane-susceptible coastal areas because high winds create wind-borne debris that can strike the glazing. Maintenance suggestions for glazing include the following:

- Checking glazing gaskets/sealants for deterioration and repairing or replacing as needed. Broken seals in insulated glass are not uncommon in coastal areas.

- Checking wood frames for decay and termite attack, and checking metal frames for corrosion. Frames should be repainted periodically (where appropriate), and damaged wood should be replaced. Maintaining the putty in older wood windows minimizes sash decay. Metal frames should be cleaned of corrosion or pitting and the operation of the windows tested on some frequency.

- Checking vinyl frames for cracks especially in the corners and sealing any cracks with a sealant to prevent water entry into the window frame. Vinyl may become discolored from the ultraviolet (UV) rays of the sun.

- Checking for signs of water damage (e.g., water stains, rust streaks from joints) and checking sealants for substrate bond and general condition. Repair or replace as needed.

- Checking glazing for stress cracks in corners. Stress cracks might be an indication of either settlement of the house or of lateral movement that is causing excessive stress in the lateral load system.

- Checking shutters for general integrity and attachment and repainting periodically where appropriate.

- Replacing or strengthening the attachment of the shutter system to the building as appropriate.

- Checking the shutters for ease of operation. Sand can easily get into the hinges and operators and render shutters inoperable.

- Checking locks and latches frequently for corrosion and proper operation. Lock mechanisms are vulnerable to attack by salt-laden air. Applying a lubricant or rust inhibitor improves the operation of these mechanisms over the short term.

- Installing double hung and awning windows, which generally perform better than sliding or jalousie windows in the coastal environment, primarily because the sliding and jalousie windows allow more water, sand, and air infiltration because of the way the windows open and close.

- Replacing sliding and jalousie windows to reduce infiltration.

14.2.2   Siding

Solar UV degradation occurs at a rate of about 1/16 inch over 10 years on exposed wood. This rate of degradation is not significant for dimension lumber, but it is significant for plywood with 1/8-inch veneers. If the exterior plywood is the shearwall sheathing, the loss will be significant over time. Maintenance suggestions for siding materials include the following:

- Protecting plywood from UV degradation with pigmented finishes rather than clear finishes. Pigmented finishes are also especially recommended for exposed shearwall sheathing.

- Protecting wood siding with a protective sealant—usually a semi-transparent stain or paint. The coating should be re-applied regularly because the degradation will occur nearly linearly if re-application is done but will progress faster if allowed to weather with no regular sealing.

- Keeping siding surfaces free of salt and mildew and washing salt from siding surfaces not washed by rain, taking care to direct the water stream downward. Mildew should be washed as needed from siding using commercially available products or the homemade solution of bleach and detergent described in Finishes for Exterior Wood: Section, Application and Maintenance (Williams et al. 1996). Power washing is another technique to keep the siding clean as long as the siding sealant is not removed. Mildew grows on almost any surface facing north, no matter how small the surface.

- Caulking seams, joints, and building material discontinuities with a sealant intended for severe exterior exposures and renew the sealant every 5 years at a minimum or when staining or painting the siding and trim. Sealant applied at large wood members should be renewed about 1 year after the wood has shrunk away from the caulked joint.

- Caulking carefully to avoid closing off weep or water drainage holes below windows or in veneers that are intended to drain will prevent sealing the moisture inside the wall cavity, which can lead to significant, long-term deterioration.

- Renailing siding when nails withdraw (pop out) and renailing at a new location so the new nail does not go into the old nail hole.

- Ensuring vinyl siding has the ability to expand and contract with temperature changes. Buckling in the siding is an indication that the siding was installed too tightly to the wall sheathing with an insufficient amount of room under the siding nails to allow for the normal horizontal movement of the siding.

14.2.3 Roofs

Roof coverings are typically the building envelope material most susceptible to deterioration from weathering. Also, depending on roof system design, minor punctures or tears in the roof covering can allow water infiltration, which can lead to serious damage to the roof system and other building elements. Maintenance suggestions for roof materials include the following:

- Checking the general condition of the roof covering. Granule loss from asphalt shingles is always a sign of some deterioration, as is curling and clawing (reverse curling) although some minor loss is expected even from new shingles.

- Dabbing roofing cement under the tabs of the first layer of shingles, including the base course, to help ensure that this layer stays down in high winds.

- Dabbing roofing cement under any shingle tabs that have lifted up from the existing tack strip.

- Checking the nails that attach the shingles to the roof for corrosion or pullout.

- Checking metal flashings and replacing or repairing as necessary.

- Cleaning dirt, moss, leaves, vegetative matter, and mildew from wood shakes.

■ Cleaning corroded surfaces of ferrous metal roofs and applying an appropriate paint or sealer.

■ Checking the attachment of the roof surface to the deck. Screws and nails can become loose and may require tightening. Gasketed screws should be added to tighten the metal deck to the underlayment. Some roofing systems are attached to the underlayment with clips that can corrode—these clips should be inspected and any corroded clips replaced, but in many cases, the clips will be concealed and will require some destructive inspection to discover the corroded clips.

■ Removing debris from the roof and ensuring that drains, scuppers, gutters, and downspouts are not clogged.

■ Removing old asphalt shingles before recovering. This is recommended because installing an additional layer of shingles requires longer nails, and it is difficult to install the new asphalt shingles so that they lay flat over the old. New layers installed over old layers can therefore be susceptible to wind uplift and damage, even in relatively low wind speeds. New layers installed over old shingles could void the warranty for the new shingles.

■ Checking attachments of eave and fascia boards. Deterioration in these boards will likely allow flashings attached to them to fail at lower than design wind speeds.

14.2.4   Exterior-Mounted Mechanical and Electrical Equipment

Most outdoor mechanical and electrical equipment includes metal parts that corrode in the coastal environment. Life expectancy improves if the salt is washed off the outside of the equipment frequently. This occurs naturally if the equipment is fully exposed to rainwater, but partially protected equipment is subject to greater corrosion because of the lack of the natural rinsing action.

Using alternative materials that do not include metal parts can also help reduce the problems caused by corrosion. In all cases, electrical switches should be the totally enclosed type to help prevent moisture intrusion into the switch, even if the switch is located on a screened porch away from the direct effects of the weather. Building owners should expect the following problems in the coastal environment:

■ Electrical contacts can malfunction and either short out or cause intermittent operation

■ Housings for electrical equipment; heating, ventilation, and air-conditioning condensers; ductwork; and other elements deteriorate more rapidly

■ Fan coils for outside condensers can deteriorate more rapidly unless a coastal environment is specified

■ Typical metal fasteners and clips used to secure equipment can deteriorate more rapidly in a coastal environment than a non-coastal environment

14.2.5   Decks and Exterior Wood

The approach to maintaining exterior wood 2x members is different from the approach for thicker members. The formation of small checks and splits in 2x wood members from cyclical wetting and drying can be reduced by using water-repellent finishes. The formation of larger checks and splits in thicker wood members is caused more by long-term drying and shrinking and is not as significantly reduced by the use of water-repellent finishes. Installation of horizontal 2x members with the cup (concave surface) down minimizes water retention and wood deterioration.

- Cyclical wetting and drying, such as from dew or precipitation, causes the exterior of a wood member to swell and shrink more quickly than the interior. This causes stress in the surface, which leads to the formation of checks and splits. This shrink-swell cycling is worst on southern and western exposures. Checks and splits, especially on horizontal surfaces, provide paths for water to reach the interior of a wood member and remain, where they eventually cause decay. Maintaining a water-repellent finish, such as a pigmented paint, semi-transparent stain, or clear finish, on the wood surface can reduce the formation of checks and splits. These finishes are not completely water- or vapor- repellent, but they significantly slow cyclical wetting and drying. Of the available finishes, pigmented paints and semi-transparent stains have the longest lifetime; clear finishes must be reapplied frequently to remain effective. Matte clear finishes are available that are almost unnoticeable on bare wood. These finishes are therefore attractive for decking and other "natural" wood, but they must be renewed when water no longer beads on the finished surface.

- Wood deck surfaces can be replaced with synthetic materials, which are sold under a variety of trade names. Many of these products should be attached with stainless screws or hidden clips to preservative-treated framing.

- Moisture-retaining debris can collect between deck boards and in the gaps in connections. Periodic cleaning of this debris from between wood members, especially at end grains, allows drying to proceed and inhibits decay.

- Larger timbers can also be vulnerable to checks, splits, and other weather-related problems. The best way to maintain larger timbers is to keep water away from joints, end grain surfaces, checks, and splits. Much can be learned by standing under the house (given sufficient headroom) during a rain with the prevailing wind blowing to see where the water goes. Measures, such as preservative treatments, can then be taken or renewed to minimize the effect of this water on the larger timbers.

- Connections of deck band boards to the structure should be inspected periodically for moisture intrusion. These connections frequently leak from wind-driven rain and moisture accumulation. Leakage can occur at the flashing to structure interface or at the bolts connecting the band board to the structure.

14.2.6 Metal Connectors

Most sheet-metal connectors, such as tiedown straps, joist hangers, and truss plates used in structural applications in the building, should be specified to last the lifetime of the building without the need for maintenance. However, the use of corrosion-prone connectors is a common problem in existing coastal houses.

- Galvanized connectors may have corrosion issues. If galvanized connectors remain gray, the original strength is generally unaffected by corrosion. When most of the surface of the connector turns rust red, the sacrificial galvanizing has

CROSS REFERENCE

The selection of metal connectors for use within the building envelope and in exposed locations is addressed in Section 9.4 of this Manual.

been consumed and the corrosion rate of the unprotected steel can be expected to accelerate by up to a factor of 50 times. Figure 14-3 illustrates severe corrosion under an exterior deck.

- Sheet-metal connectors can be susceptible to rapid corrosion and are frequently without reserve strength. During routine inspections, any sheet-metal connectors found to have turned rust red or to show severe, localized rusting sufficient to compromise their structural capacity should be replaced immediately. However, the replacement of sheet metal connectors is usually difficult for a number of reasons: the connection may be under load, the nails or bolts used to secure connectors are usually hard to remove, and the location of a connector often makes removal awkward.

- Salt-laden air can increase corrosion rates in building materials. Covering exposed connectors with a sheathing material reduces their exposure and therefore increases their life expectancy.

> **WARNING**
>
> Using corrosion-prone sheet metal connectors increases maintenance requirements and potentially compromises structural integrity.

Figure 14-3.
Severely corroded deck connectors

14.3 Hazard-Specific Maintenance Techniques

The maintenance practices described above for minimizing corrosion, wood decay, termite infestation, and UV degradation will improve the resistance of a coastal building to flood, wind, and seismic damage by maintaining the strength of the structural elements. The additional measures described in the following sections will further maintain the building's resistance to natural hazards.

14.3.1 Flooding

When designing for the lateral force capacity of an unbraced or braced pile foundation, the designer should allow for a certain amount of scour. Scour in excess of the amount allowed for reduces the embedment of the piles and causes them to be overstressed in bending during the maximum design flood, wind, or earthquake. As allowed by local regulations and practicality, the grade level should be maintained at the original design elevation.

Scour and long-term beach erosion may affect pile maintenance requirements. If tidal wetting was not anticipated in the original design, the piles may have received the level of preservative treatment required only for ground contact and not the much higher marine treatment level that provides borer resistance. If the pile foundation is wetted by high tides or runup, borer infestation is possible. Wrapping treatments that minimize borer infestation are available for the portions of the piles above grade that are subject to wetting.

14.3.2 Seismic and Wind

Many seismic and wind tiedowns at shearwall vertical chords use a vertical threaded rod as the tension member. Each end of the threaded rod engages the tiedown hardware or a structural member. Over time, cross-grain shrinkage in the horizontal wood members between the threaded rod connections loosens the threaded rod, allowing more rocking movement and possible damage to the structure. Whenever there is an opportunity to access the tiedowns, the nuts on the rods should be tightened firmly. New proprietary tiedown systems that do this automatically are available.

Shearwall sill plates bearing directly on continuous footings or concrete slabs-on-grade, if used in coastal construction, are particularly susceptible to decay in moist conditions. Figure 14-4 shows a deteriorated sill plate. Even if the decay of the preservative-treated sill plate is retarded, the attached untreated plywood can easily decay and the shearwall will lose strength. Conditions that promote sill and plywood decay include an outside soil grade above the sill, stucco without a weep screed at the sill plate, and sources of excessive interior water vapor. Correcting these conditions helps maintain the strength of the shearwalls.

Figure 14-4.
Deteriorated wood sill
plate

14.4   References

FEMA (Federal Emergency Management Agency). 1996. *Corrosion Protection of Metal Connectors in Coastal Areas for Structures Located in Special Flood Hazard Areas*. NFIP Technical Bulletin 8-96.

Shifler, D.A. 2000. *Corrosion and Corrosion Control in Saltwater Environments*. Pennington, NJ: Electrochemical Society.

Williams, R., M. Knaebe, and W. Feist. 1996. *Finishes for Exterior Wood: Section, Application and Maintenance*. Forest Products Society.

Retrofitting Buildings for Natural Hazards

This chapter provides guidance on retrofitting existing residential structures to resist or mitigate the consequences of natural hazards in the coastal environment. The natural hazards that are addressed are wildfires, seismic events, floods, and high winds. Specific retrofitting methods and implementation are discussed briefly, and resources with more in-depth information are provided. Some retrofitting methods are presented together with broader, non-retrofitting mitigation methods when retrofitting and non-retrofitting methods are presented together in the referenced guidance. For retrofitting to mitigate high winds, the new three-tiered wind retrofit program that is provided in FEMA P-804, *Wind Retrofit Guide for Residential Buildings* (FEMA 2010c), is discussed. The program includes systematic and programmatic guidance.

Retrofitting opportunities present themselves every time maintenance is performed on a major element of a building. Retrofitting that increases resistance to natural hazards should focus on improvements that provide the largest benefit to the owner.

CROSS REFERENCE

For resources that augment the guidance and other information in this Manual, see the Residential Coastal Construction Web site (http://www.fema.gov/rebuild/mat/fema55.shtm).

NOTE

FEMA's Hazard Mitigation Assistance (HMA) grant programs provide funding for eligible mitigation activities that reduce disaster losses and protect life and property from future disaster damage. Currently, FEMA administers the following HMA grant programs: Hazard Mitigation Grant Program, Pre-Disaster Mitigation, Flood Mitigation Assistance, Repetitive Flood Claims, and Severe Repetitive Loss.

If an existing building is inadequate to resist natural hazard loads, retrofitting should be considered.

15.1   Wildfire Mitigation

Thousands of residential and non-residential buildings are damaged or destroyed every year by wildfires, resulting in more than $200 million in property damage annually. More than $100 million is spent every year on fire suppression and even more on recovering from catastrophic natural and manmade hazards. Studies cited by IBHS in *Mega Fires* (IBHS 2008) have shown that financial losses can be prevented if simple measures are implemented to protect existing buildings.

FEMA offers funding through the HMGP and the PDM Program for wildfire mitigation projects. Projects funded through these programs involve retrofits to buildings that help minimize the loss of life and damage to the buildings from wildfire. Eligible activities for wildfire mitigation per FEMA's *Hazard Mitigation Assistance Unified Guidance* (FEMA 2010a) may include:

> **TERMINOLOGY: RETROFITTING**
>
> Retrofitting is a combination of adjustments or additions to existing building features that are intended to eliminate or reduce the potential for damage from natural hazards. Retrofitting is a specific type of hazard mitigation.

- Provision of defensible space through the creation of perimeters around residential and non-residential buildings and structures by removing or reducing flammable vegetation. The three concentric zones of defensible space are shown in Figure 15-1.

Zone 2: Prune and remove dead and dying branches from individual and well-spaced clumps of trees and shrubs

Zone 2: Place woodpiles at least 30 feet from the building and store the wood in a vegetation-free zone such as a graveled area

Zone 1: Remove combustible litter on roofs and gutters and trim tree branches that overhang the roof and chimney

Zone 1: Eliminate all combustible materials within 30 feet of the home

Zone 3: Reduce fuels by thinning and pruning vegetation horizontally and vertically

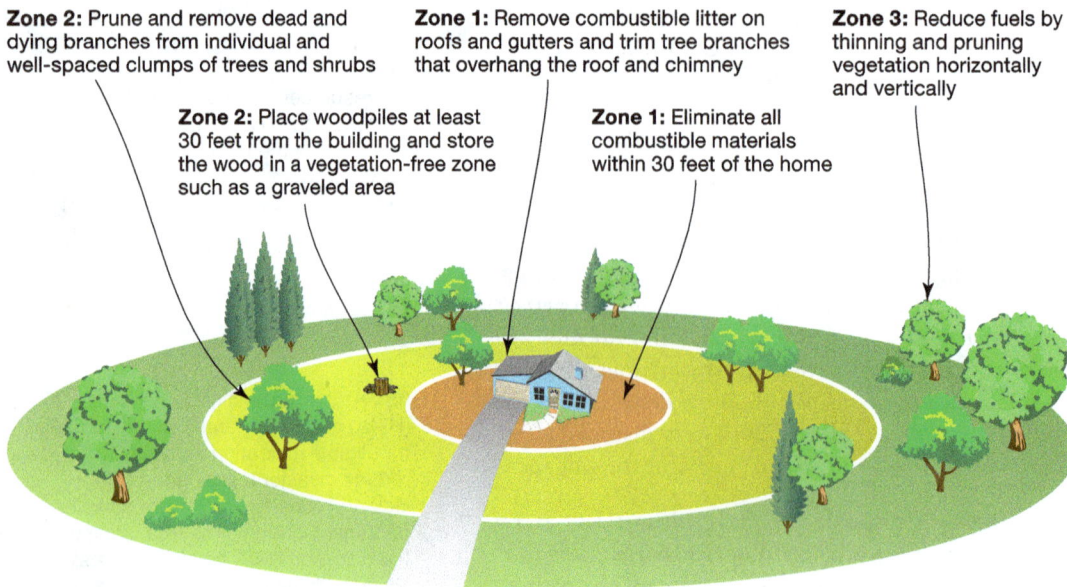

Figure 15-1.
The three concentric zones of defensible space
SOURCE: ADAPTED FROM FEMA P-737

▪ Application of non-combustible building envelope assemblies that can minimize the impact of wildfires through the use of ignition-resistant materials and proper retrofitting techniques. The components of the building envelope are shown in Figure 15-2.

▪ Reduction of hazardous fuels through vegetation management, vegetation thinning, or reduction of flammable materials. These actions protect life and property that are outside the defensible space perimeter but close to at-risk structures. Figure 15-3 shows a fire that is spreading vertically through vegetation.

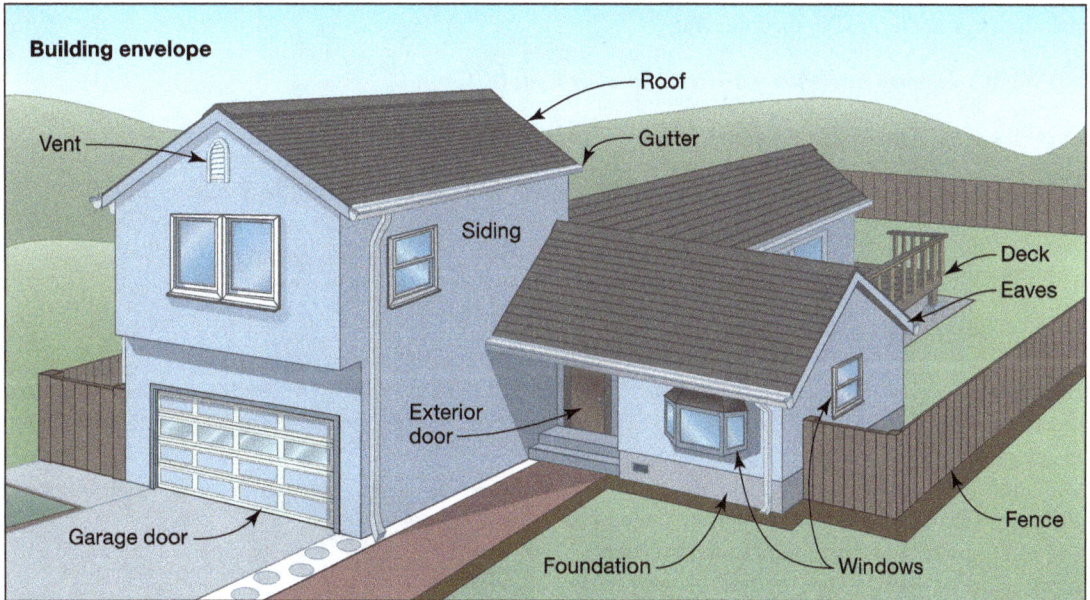

Figure 15-2.
The building envelope
SOURCE: ADAPTED FROM FEMA P-737

Figure 15-3.
Fire spreads vertically through vegetation

FEMA may fund above-code projects in communities with applicable fire-related codes. For homes and structures constructed or activities completed prior to the adoption of local building codes, FEMA may fund mitigation that meets or exceeds the codes currently in effect. For communities without fire codes, FEMA may fund mitigation when the materials and technologies are in accordance with the ICC, FEMA, U.S. Fire Administration, and the National Fire Protection Association (NFPA). Firewise recommendations, as appropriate. The Firewise program provides resources for communities and property owners to use in the creation of defensible space. Additional fire-related information and tools can be found at http://www.firewise.org and http://www.nfpa.org.

Wildfire mitigation is required to be in accordance with the applicable fire-related codes and standards, including but not limited to the following:

- IWUIC, *International Wildland-Urban Interface Code* (ICC)

- NFPA 1144, *Standard for Reducing Structure Ignition Hazards from Wildland Fire*

- NFPA 1141, *Standard for Fire Protection Infrastructure for Land Development in Suburban and Rural Areas*

- NFPA 703, *Standard for Fire-Retardant Treated Wood and Fire-Retardant Coatings for Building Materials*

- *Code for Fire Protection of Historical Structures* (NFPA)

FEMA P-737, *Home Builder's Guide to Construction in Wildfire Zones* (FEMA 2008a), is a Technical Fact Sheet Series (see Figure 15-4) that provides information about wildfire behavior and recommendations for building design and construction methods in the wildland/urban interface. The fact sheets cover mitigation topics for existing buildings including defensible space, roof assemblies, eaves, overhangs, soffits, exterior walls, vents, gutters, downspouts, windows, skylights, exterior doors, foundations, decks and other attached structures, landscape fencing and walls, fire sprinklers, and utilities and exterior equipment. Implementation of the recommended design and construction methods in FEMA P-737 can greatly increase the probability that a building will survive a wildfire.

Home Builder's Guide to Construction in Wildfire Zones

Technical Fact Sheet Series

FEMA P-737 / September 2008

FEMA

Figure 15-4.
FEMA P-737, *Home Builder's Guide to Construction in Wildlife Zones: Technical Fact Sheet Series*

Since it may not be financially possible for the homeowner to implement all of the measures that are recommended in FEMA P-737, homeowners should consult with local fire and building code officials or their fire management specialists to perform a vulnerability assessment and develop a customized, prioritized list of recommendations for remedial work on defensible space and the building envelope. Helpful information about the vulnerabilities of the building envelope is available at http://firecenter.berkeley.edu/building_in_ wildfire_prone_areas. The homeowner can use the Homeowner's Wildfire Assessment survey on this Web site to learn about the risks a particular building has and the measures that can be taken to address them.

15.2   Seismic Mitigation

Seismic hazard, which is well documented and defined in the United States, is mitigated in existing residential buildings primarily through retrofitting. Although modifications to existing residential structures have the potential to reduce earthquake resistance, it is possible to take advantage of these modifications to increase resistance through earthquake retrofits (upgrades). FEMA has produced documents, including those referenced below, that address the evaluation and retrofit of buildings to improve performance during seismic events. For nationally applicable provisions governing seismic evaluation and rehabilitation, the design professional should reference ASCE 31 and ASCE 41.

In addition, FEMA offers funding for seismic retrofits through the HMGP and the PDM Program to reduce the risk of loss of life, injury, and damage to buildings. Seismic retrofits, which are classified as structural and non-structural, are subject to the same HMGP and PDM funding processes as wind retrofits (see Section 15.4.3).

FEMA 232, *Homebuilders' Guide to Earthquake Resistant Design and Construction* (FEMA 2006) (see Figure 15-5), contains descriptions of eight earthquake upgrades that address common seismic weaknesses in existing residential construction. The upgrades are foundation bolting, cripple wall bracing, weak- and soft-story bracing, open-front bracing, hillside house bracing, split-level floor interconnection, anchorage of masonry chimneys, and anchorage of concrete and masonry walls. The upgrades are summarized below. For in-depth information on these upgrades, see FEMA 232.

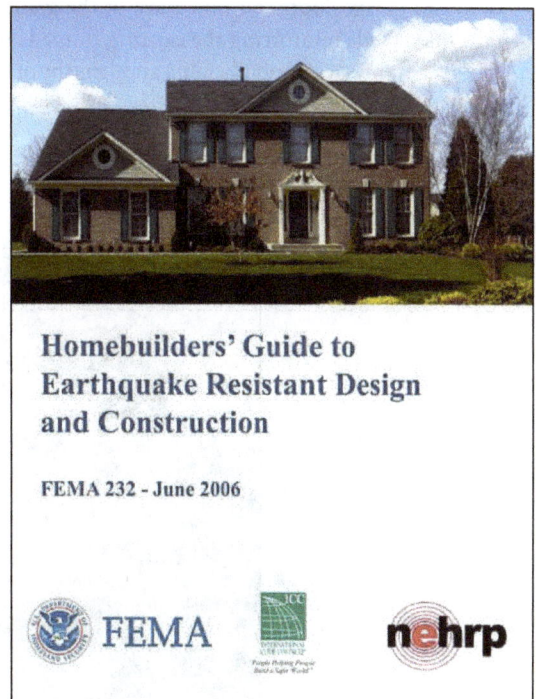

Figure 15-5.
FEMA 232, *Homebuilders Guide to Earthquake Resistant Design and Construction*

■ **Foundation bolting.** Inadequate attachment of the sill plate to the foundation can allow the framed structure to separate and shift off the foundation. Sill plate anchor bolts (either adhesive or expansion type depending on the foundation material) can be added provided there is sufficient access to the top surface of the sill plate. Alternately, proprietary anchoring hardware is available that is typically attached to the face of the foundation wall for greater ease of installation when access is limited. Reinforcing sill plate anchorage offers a generally high benefit in return for low cost.

■ **Cripple wall bracing.** Another relatively inexpensive foundation-level retrofit is bracing the cripple walls. Cripple walls are framed walls occasionally installed between the top of the foundation and first-floor framing in the above-grade wall sections of basements and crawl spaces. Because of their location, cripple walls are particularly vulnerable to seismic loading, as shown in Figure 15-6. These walls can be braced through the prescribed installation of wood structural panel sheathing to the interior and/or exterior wall surface.

■ **Weak- and soft-story bracing.** Although first-story framed walls must bear greater seismic loads than the roof and walls above, they frequently have more openings and therefore less bracing. As a result, first-story framed walls, and any other level with underbraced wall sections, may be referred to as weak or soft stories. These walls can be retrofit by removing the interior finishes at wall corners and installing hold-down anchors between the corner studs and continuous reinforced foundation below. If renovations or repairs require removing larger areas of interior wall sheathing, additional hold-down anchors can be installed to tie in the floor or roof framing above. Additional wall bracing can be achieved by adding blocking for additional nailing and wood structural panel sheathing.

■ **Open-front bracing.** An open-front configuration is one in which braced exterior walls are absent or grossly inadequate. Frequently, open-front configurations are found in garage entry walls where overhead garage doors consume most of the available wall area, as shown in Figure 15-7. Possible retrofits include reinforcing the existing framed end walls and replacing the framed wall ends with steel moment frames; common heights and lengths of steel moment frames are available commercially.

Figure 15-6.
A house with severe damage due to cripple wall failure

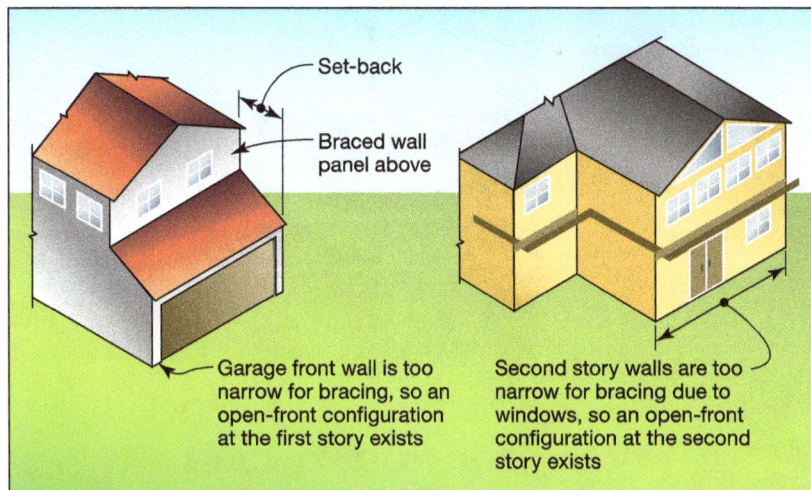

Figure 15-7.
Common open-front configurations in one- and two- family detached houses

■ **Hillside house bracing.** Houses built on steep hillsides are vulnerable to damage when the floor system separates from the uphill foundation or foundation wall. Retrofitting to mitigate this type of seismic damage requires an engineered design that should include anchoring each floor system to the uphill foundation and the supplemental anchorage, strapping, and bracing of cripple walls.

■ **Split-level floor interconnection.** Houses with vertical offsets between floor elevations on a common wall or support are exposed to seismic damage that is similar to hillside houses. The potential for separation of the floor system from the common wall may be reduced by adequately anchoring floor framing on either side of the common wall. Prescriptive solutions may apply where a direct tension tie can be provided between both floors, but an engineered design may be necessary where greater floor offsets exist.

■ **Anchorage of masonry chimneys.** Unreinforced masonry chimneys can be anchored to the roof and adjacent or surrounding floor systems with metal straps but will still be subject to brittle failure. The benefit of this type of anchoring may be limited to collapse prevention. A chimney collapse reportedly caused one fatality in the 1992 earthquake in Landers, CA, but other mitigation measures may be more cost-effective. These measures include the practical approaches provided on the Association of Bay Area Governments Web site (http://quake.abag.ca.gov/residents/chimney).

■ **Anchorage of concrete and masonry walls.** Floor systems in houses with full-height concrete or masonry walls may be supported by a weight-bearing ledger strip only. With this type of existing construction, a tension connection can be installed between the walls and floor system to provide the necessary direct anchorage. An engineering evaluation and design are recommended for this type of seismic retrofit.

FEMA 530, *Earthquake Safety Guide for Homeowners* (FEMA 2005) (Figure 15-8) includes guidance similar to FEMA 232 on seismic structural retrofits along with tips on strengthening a variety of existing foundation types. One non-structural retrofit in FEMA 530 is to brace water heaters, which can cause gas leaks, fires, or flooding if toppled during an earthquake. Written for the homeowner, FEMA 530 provides information on the relative cost of prevention versus the cost of post-disaster repair or replacement and on plans, permitting, and selecting contractors.

Figure 15-8.
FEMA 530, *Earthquake Safety Guide for Homeowners*

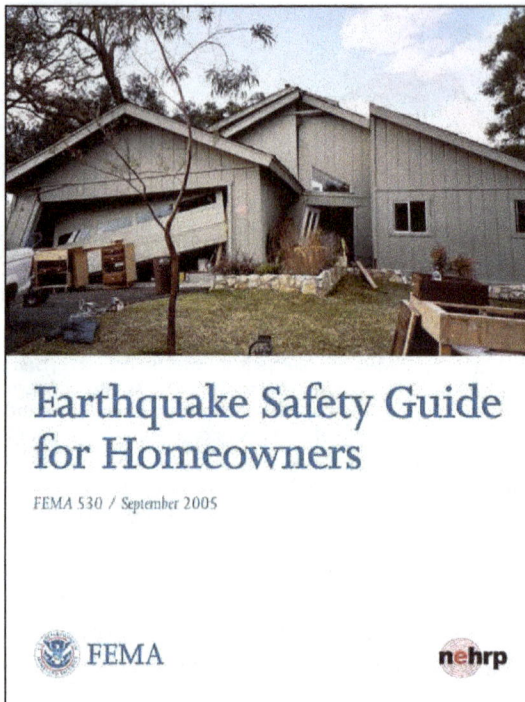

Earthquake Safety Guide for Homeowners

FEMA 530 / September 2005

FEMA   nehrp

15.3 Flood Mitigation

FEMA 259, *Engineering Principles and Practices of Retrofitting Floodprone Structures* (FEMA 2011), addresses retrofitting flood-prone residential structures. The objective of the document is to provide engineering design and economic guidance to engineers, architects, and local code officials about what constitutes technically feasible and cost-effective retrofitting measures for flood-prone residential structures.

The focus in this chapter in regard to retrofitting for the flood hazard is retrofitting one- to four-family residences that are subject to flooding without wave action. The retrofitting measures that are described in this section include both active and passive efforts and wet and dry floodproofing. Active efforts require human intervention preceding the flood event and may include activities such as engaging protective shields at openings. Passive efforts do not require human intervention. The flood retrofitting measures are elevating the building in place, relocating the building, constructing barriers (levees and floodwalls), dry floodproofing (sealants, closures, sump pumps, and backflow valves), and wet floodproofing (using flood damage-resistant materials and protecting utilities and contents).

Flood retrofitting projects may be eligible for funding through the following FEMA Hazard Mitigation Programs: HMGP, PDM, Flood Mitigation Assistance, Repetitive Flood Claims, and Severe Repetitive Loss. More information on obtaining funding for flood retrofitting is available in *Hazard Mitigation Assistance Unified Guidance* (FEMA 2010a).

15.3.1 Elevation

Elevating a building to prevent floodwaters from reaching damageable portions of the building is an effective retrofitting technique. The building is raised so that the lowest floor is at or above the DFE to avoid damage from the design flood. Heavy-duty jacks are used to lift the building. Cribbing is used to support the building while a new or extended foundation is constructed. In lieu of constructing new support walls, open

> **CROSS REFERENCE**
>
> For definitions of DFE and BFE, see Section 8.5.1 of this Manual.

Figure 15-9.
Home elevated on piles

foundations such as piers, columns, posts, and piles are often used (see Figure 15-9). Elevating the building on fill may be an option. Closed foundations are not permitted in Zone V and are not recommended in Coastal A Zones. See Table 10-1 for the types of foundations that are acceptable in each flood zone.

The advantages and disadvantages of elevation are listed in Table 15-1.

Table 15-1. Advantages and Disadvantages of Elevation

Advantages	Disadvantages
• Brings a substantially damaged or improved building into compliance with the NFIP if the lowest horizontal member is elevated to the BFE • Reduces flood risk to the structure and its contents • Eliminates the need to relocate vulnerable items above the flood level during flooding • Often reduces flood insurance premiums • Uses established techniques • Requires qualified contractors who are often readily available • Reduces the physical, financial, and emotional strain that accompanies flood events • Does not require the additional land that may be needed for floodwalls or levees	• May be cost-prohibitive • May adversely affect the building's appearance • Prohibits the building from being occupying during a flood • May adversely affect access to the building • Cannot be used in areas with high-velocity water flow, fast-moving ice or debris flow, or erosion unless special measures are taken • May require additional costs to bring the building up to current building codes for plumbing, electrical, and energy systems • Requires a consideration of forces from wind and seismic hazards

SOURCE: FEMA 259

BFE = base flood elevation
NFIP = National Flood Insurance Program

15.3.2 Relocation

Relocation involves moving a structure to a location that is less prone to flooding or flood-related hazards such as erosion. The structure may be relocated to another portion of the current site or to a different site. The surest way to eliminate the risk of flood damage is to relocate the structure out of the floodplain. Relocation normally involves preparing the structure for the move (see Figure 15-10), placing it on a wheeled vehicle, transporting it to the new location, and setting it on a new foundation.

Relocation is an appropriate measure in high hazard areas where continued occupancy is unsafe and/or owners want to be free of the risk of flooding. Relocation is also a viable option in communities that are considering using the resulting open space for more appropriate floodplain activities. Relocation may offer an alternative to elevation for substantially damaged structures that are required under local regulations to meet NFIP requirements. Table 15-2 lists the advantages and disadvantages of relocation.

Figure 15-10.
Preparing a building for
relocation

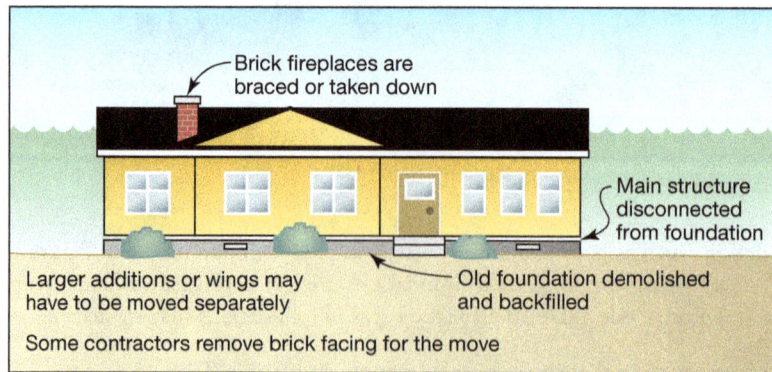

Table 15-2. Advantages and Disadvantages of Relocation

Advantages	Disadvantages
• Allows for substantially damaged or improved structure to be brought in to compliance with the NFIP • Significantly reduces flood risk to the structure and its contents • Uses established techniques • Requires qualified contractors who are often readily available • Can eliminate the need to purchase flood insurance or reduce the premium because the house is no longer in the floodplain • Reduces the physical, financial, and emotional strain that accompanies flood events	• May be cost-prohibitive • Requires locating a new site • Requires addressing disposition of the flood-prone site • May require additional costs to bring the structure up to current building codes for plumbing, electrical, and energy systems

SOURCE: FEMA 259
NFIP = National Flood Insurance Program

15.3.3   Dry Floodproofing

In dry floodproofing, the portion of a structure that is below the chosen flood protection level (walls and other exterior components) is sealed to make it watertight and impermeable to floodwaters. The objective is to make the walls and other exterior components impermeable to floodwaters. Watertight, impervious membrane sealant systems include wall coatings, waterproofing compounds, impermeable sheeting, and supplemental impermeable wall systems, such as cast-in-place concrete. Doors, windows, sewer and water lines, and vents are closed with permanent or removable shields or valves. Figure 15-11 is a schematic of a dry floodproofed home. Non-residential techniques are also applicable in residential situations. See Table 15-3 for the advantages and disadvantages of dry floodproofing.

WARNING

Dry floodproofing is not allowed under the NFIP for new and substantially damaged or improved residential structures in an SFHA. For additional information on dry floodproofing, see FEMA FIA-TB-3, *Non-Residential Floodproofing – Requirements and Certification for Buildings Located in Special Flood Hazard Areas in Accordance with the NFIP* (FEMA 1993a) and the *Substantial Improvement/Substantial Damage Desk Reference* (FEMA 2010b).

The expected duration of flooding is critical when deciding which sealant system to use because seepage can increase over time, rendering the floodproofing ineffective. Waterproofing compounds, sheeting, and sheathing may deteriorate or fail if exposed to floodwaters for extended periods. Sealant systems are also subject to damage (puncture) in areas that experience water flow of significant velocity, ice, or debris flow.

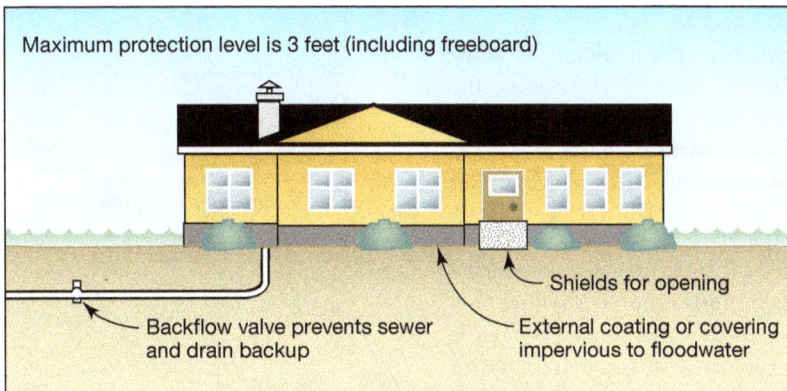

Figure 15-11. Dry floodproofed structure

Table 15-3. Advantages and Disadvantages of Dry Floodproofing

Advantages	Disadvantages
• Reduces the flood risk to the structure and contents even when the DFE is not exceeded • May be less costly than other retrofitting measures • Does not require the extra land that may be needed for floodwalls or reduced levees • Reduces the physical, financial, and emotional strain that accompanies flood events • Retains the structure in its present environment and may avoid significant changes in appearance	• Does not satisfy the NFIP requirement for bringing substantially damaged or improved residential structures into compliance • Requires ongoing maintenance • Does not reduce flood insurance premiums for residential structures unless community-wide basement exception is granted • Usually requires human intervention and adequate warning time for installation of protective measures • May provide no protection if measures fail or are exceeded during large floods • May result in more damage than flooding if design loads are exceeded, walls collapse, floors buckle, or the building floats • Prohibits the building from being occupied during a flood • May adversely affect the appearance of the building if shields are not aesthetically pleasing • May not reduce damage to the exterior of the building and other property • May lead to damage of the building and its contents if the sealant system leaks

SOURCE: FEMA 259

NFIP = National Flood Insurance Program
DFE = design flood elevation

15.3.4 Wet Floodproofing

Wet floodproofing involves modifying a building to allow floodwaters to enter it in such a way that damage to the structure and its contents is minimized. A schematic of a home that has been wet floodproofed is shown in Figure 15-12. See Table 15-4 for a list of the advantages and disadvantages of wet floodproofing.

Wet floodproofing is often used for structures with basements and crawlspaces when other mitigation techniques are technically infeasible or too costly. Wet floodproofing is generally appropriate if a structure has space available to temporarily store damageable items during the flood event. Utilities and furnaces situated below the DFE should be relocated to higher ground while remaining sub-DFE materials vulnerable to flood damage should be replaced with flood damage-resistant building materials. FEMA TB-2, *Flood Damage-Resistant Materials Requirements* (FEMA 2008b), provides guidance concerning the use of flood damage-resistant building components.

WARNING

Wet floodproofing is not allowed under the NFIP for new and substantially damaged or improved structures located in an SFHA. Refer to FEMA FIA-TB-7, *Wet Floodproofing Requirements for Structures Located in Special Flood Hazard Areas in Accordance with the NFIP* (FEMA 1993b).

CROSS REFERENCE

For additional information about wet floodproofing, see FEMA P-348, *Protecting Building Utilities From Flood Damage: Principles and Practices for the Design and Construction of Flood Resistant Building Utility Systems* (FEMA 1999).

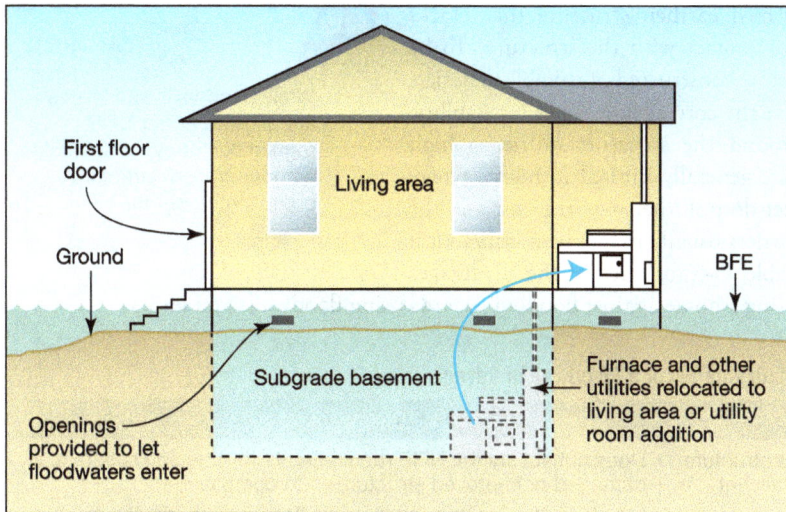

Figure 15-12.
Wet floodproofed
structure

Table 15-4. Advantages and Disadvantages of Wet Floodproofing

Advantages	Disadvantages
• Reduces the risk of flood damage to a building and its contents, even with minor mitigation • Greatly reduces loads on walls and floors due to equalized hydrostatic pressure • May be eligible for flood insurance coverage of cost of relocating or storing contents, except basement contents, after a flood warning is issued • Costs less than other measures • Does not require extra land • Reduces the physical, financial, and emotional strain that accompanies flood events	• Does not satisfy the NFIP requirement for bringing substantially damaged or improved structures into compliance • Usually requires a flood warning to prepare the building and contents for flooding • Requires human intervention to evacuate contents from the flood-prone area • Results in a structure that is wet on the inside and possibly contaminated by sewage, chemicals, and other materials borne by floodwaters and may require extensive cleanup • Prohibits the building from being occupied during a flood • May make the structure uninhabitable for some period after flooding • Limits the use of the floodable area • May require ongoing maintenance • May require additional costs to bring the structure up to current building codes for plumbing, electrical, and energy systems • Requires care when pumping out basements to avoid foundation wall collapse

SOURCE: FEMA 259
NFIP = National Flood Insurance Program

15.3.5   Floodwalls and Levees

Another retrofitting approach is to construct a barrier between the structure and source of flooding. The two basic types of barriers are floodwalls and levees. Small levees that protect a single home can be built to any height but are usually limited to 6 feet due to cost, aesthetics, access, water pressure, and space. The height of floodwalls is usually limited to 4 feet. Local zoning and building codes may also restrict use, size, and location.

A levee is typically a compacted earthen structure that blocks floodwaters from coming into contact with the structure. To be effective over time, levees must be constructed of suitable materials (i.e., impervious soils) and have the correct side slopes for stability. Levees may completely surround the structure or tie to high ground at each end. Levees are generally limited to homes where floodwaters are less than 5 feet deep. Otherwise, the cost and the land area required for such barriers usually make them impractical for the average owner. See Table 15-5 for a list of the advantages and disadvantages for retrofitting a home against flooding hazards using floodwalls and levees.

WARNING

While floodwalls and levees are allowed under NFIP regulations, they do not make a noncompliant structure compliant under the NFIP.

Table 15-5. Advantages and Disadvantages of a Floodwall or Levee

Advantages	Disadvantages
• Protects the area around the structure from inundation without significant changes to the structure • Eliminates pressure from floodwaters that would cause structural damage to the home or other structures in the protected area • Costs less to build than elevating or relocating the structure • Allows the structure to be occupied during construction • Reduces flood risk to the structure and its contents • Reduces the physical, financial, and emotional strain that accompanies flood events	• Does not satisfy the NFIP requirements for bringing substantially damaged or improved structures into compliance • May fail or be overtopped by large floods or floods of long duration • May be expensive • Requires periodic maintenance • Requires interior drainage • May affect local drainage, possibly resulting in water problems for others • Does not reduce flood insurance premiums • May restrict access to structure • Requires considerable land (levees only) • Does not eliminate the need to evacuate during floods • May require warning and human intervention for closures • May violate applicable codes or regulations

SOURCE: FEMA 259

NFIP = National Flood Insurance Program

Floodwalls are engineered barriers designed to keep floodwaters from coming into contact with the structure. Floodwalls can be constructed in a wide variety of shapes and sizes but are typically built of reinforced concrete and/or masonry materials.

See Figure 15-13 for an example of a home protected by both a floodwall and a levee.

15.3.6 Multihazard Mitigation

The architect, engineer, or code official must recognize that retrofitting a residential structure for flooding may affect how the structure will react to hazards other than flooding. Non-flood-related hazards such as earthquake and wind forces should also be considered when retrofitting for flood-related hazards such as water-borne ice and debris-impact forces, erosion forces, and mudslide impacts. Retrofitting a structure to withstand only floodwater forces may impair the structure's resistance to the multiple hazards mentioned above. Thus, it is important to approach retrofitting with a multi-hazard perspective.

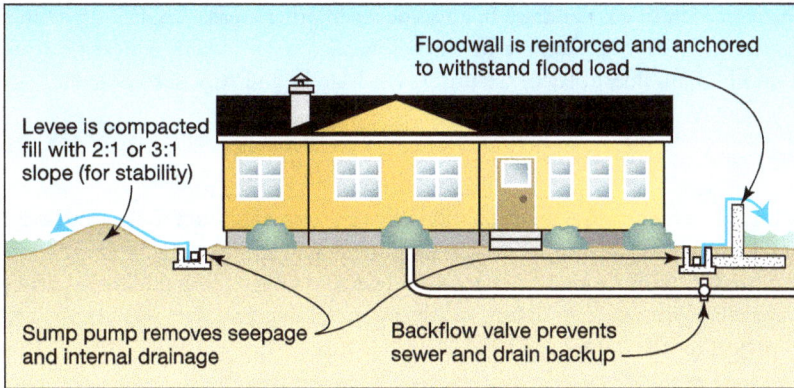

Figure 15-13.
Home protected by a
floodwall and a levee

15.4 High-Wind Mitigation

The high-wind natural hazards that affect the hurricane-prone regions of the United States are hurricanes, tropical storms, typhoons, nor'easters, and tornadoes. This section addresses protecting existing residential structures from hurricane damage. The evaluation process and implementation methods for wind retrofit projects discussed in this section are described more fully in FEMA P-804, *Wind Retrofit Guide for Residential Building* (FEMA 2010c) (see Figure 15-14).

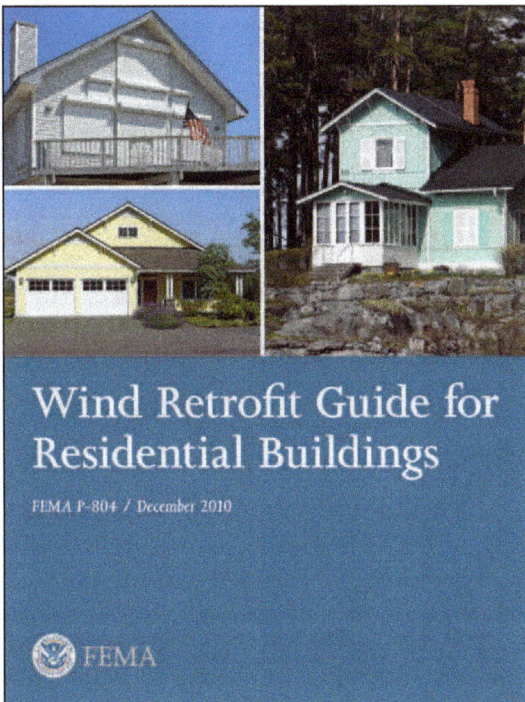

NOTE

Unless otherwise stated, all wind speeds in FEMA P-804 are ASCE 7-05 3-second gust wind speeds and correspond to design requirements set forth in ASCE 7-05 and 2006 IRC and 2009 IRC. Because of the changes in the ASCE 7-10 wind speed map, it is not appropriate to use the ASCE 7-10 wind speed map in combination with the provisions of ASCE 7-05 and the older codes.

Figure 15-14.
FEMA P-804, *Wind Retrofit Guide for Residential Buildings*

Hurricane-force winds are most common in coastal areas but also occur in other areas. ASCE 7-05 defines the hurricane-prone regions as the U.S. Atlantic Ocean and Gulf of Mexico coasts where the design wind speed is greater than 90 mph, and Hawaii, Puerto Rico, Guam, Virgin Islands, and American Samoa.

15.4.1 Evaluating Existing Homes

Executing a successful retrofit on any home requires an evaluation of its existing condition to determine age and condition; overall structural integrity; any weaknesses in the building envelope, structure, or foundation; whether the home can be retrofitted to improve resistance to wind-related damage; how the home can be retrofit for the Mitigation Packages (see Section 15.4.2); how much the Mitigation Packages would cost; and the most cost-effective retrofit project for the home.

A qualified individual should evaluate the home and provide recommendations to the homeowner. Qualified professionals may include building science professionals such as registered architects and engineers, building officials, and evaluators who are certified through other acceptable wind retrofit programs such as the FORTIFIED *for Existing Homes* Program from the Insurance Institute for Business & Home Safety (IBHS 2010).

The purposes of the evaluation are to identify any repairs that are needed before a wind retrofit project can be undertaken, the feasibility of the retrofit project, whether prescriptive retrofits can be performed on the home or whether an engineering solution should be developed, and whether the home is a good candidate for any of the wind retrofit Mitigation Packages described in Section 15.4.2. The purpose of the evaluation is ***not*** to determine whether the building meets the current building code.

15.4.2 Wind Retrofit Mitigation Packages

The wind retrofit projects described in this section, and more fully in FEMA P-804, are divided into the Basic Mitigation Package, Intermediate Mitigation Package, and Advanced Mitigation Package. Additional mitigation measures are presented at the end of this section. The packages should be implemented cumulatively, beginning with the Basic Mitigation Package. This means that for a home to successfully meet the criteria of the Advanced Mitigation Package, it must also meet the criteria of the Basic and Intermediate Mitigation Packages. The retrofits in each package are shown in Figure 15-15.

> **NOTE**
>
> In wind retrofitting, the most cost-effective techniques normally involve strengthening the weakest structural links and improving the water penetration resistance of the building envelope. To identify the weakest links, the designer should start at the top of the building and work down the load path.

The wind mitigation retrofits for each package, if implemented correctly, will improve the performance of residential buildings when subjected to high winds. Although the information in this section can be helpful to homeowners, it is intended primarily for evaluators, contractors, and design professionals. The retrofits described for each Mitigation Package and throughout this section are not necessarily listed in the order in which they should be performed. The order in which retrofits should be performed depends on the configuration of the home and should be determined once the desired Mitigation Package is chosen. For example, when the Advanced Mitigation Package is selected, the homeowner should consider retrofitting the roof-to-wall connections when retrofitting the soffits (part of the Basic Mitigation Package).

Basic Mitigation Package Retrofits

Option 1: Improvements with roof covering replacement

OR

Option 2: Improvements without roof covering replacement

Additional Required Retrofits
- Strengthening vents and soffits
- Strengthening overhangs at gable end walls
- Protecting openings per the Intermediate Package, if located in the windborne debris region

Intermediate Mitigation Package Retrofits
- Protecting windows and entry doors from windborne debris
- Protecting garage doors from wind pressure and garage door glazing from windborne debris
- Bracing gable end walls
- Strengthening connections of attached structures

Advanced Mitigation Package Retrofits
- Developing a continuous load path
- Protecting windows, entry doors, and garage doors from windborne debris and wind pressure

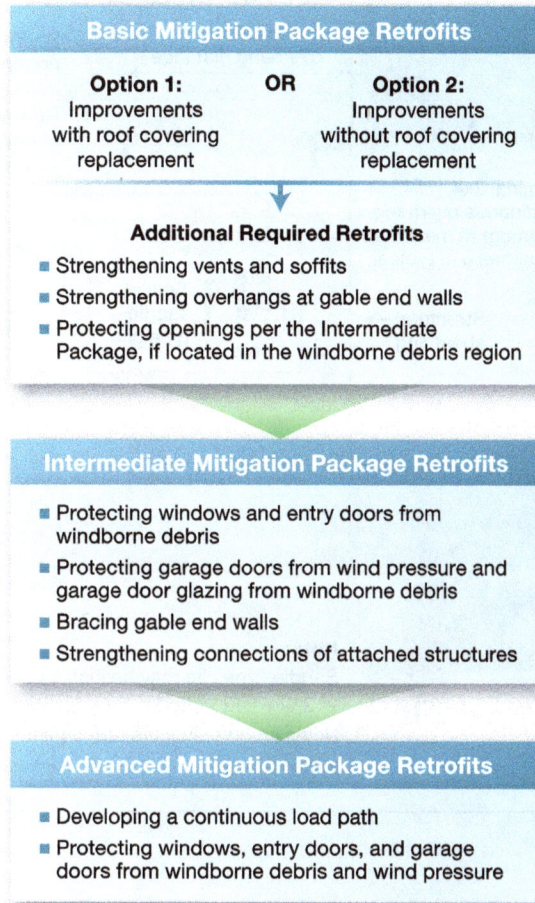

Figure 15-15.
Wind Retrofit Mitigation Packages
SOURCE: FEMA P-804

15.4.2.1 Basic Mitigation Package

The Basic Mitigation Package focuses on securing the roof system and improving the water intrusion resistance of the home. Figures 15-16 and 15-17 show two retrofits that fall into the Basic Mitigation Package. One of the first decisions to make when implementing the Basic Mitigation Package is whether to use Option 1 or Option 2. The evaluation will identify whether the roof covering needs to be replaced (see Section 3.1.1 of FEMA P-804 for more information).

If the home is located in a wind-borne debris region, the opening protection measures described in the Intermediate Mitigation Package should be performed for the Basic Mitigation Package in addition to the other retrofits. The opening protection measures include installing an approved impact-resistant covering or component at each exterior window, skylight, entry door, and garage door opening.

FEMA P-804 includes procedures, material specifications, and fastening schedules (when applicable) to facilitate implementation of the Basic Mitigation Package. Alternative methods and materials are also discussed to facilitate installation for a variety of as-built conditions.

Figure 15-16.
Bracing gable end
overhangs

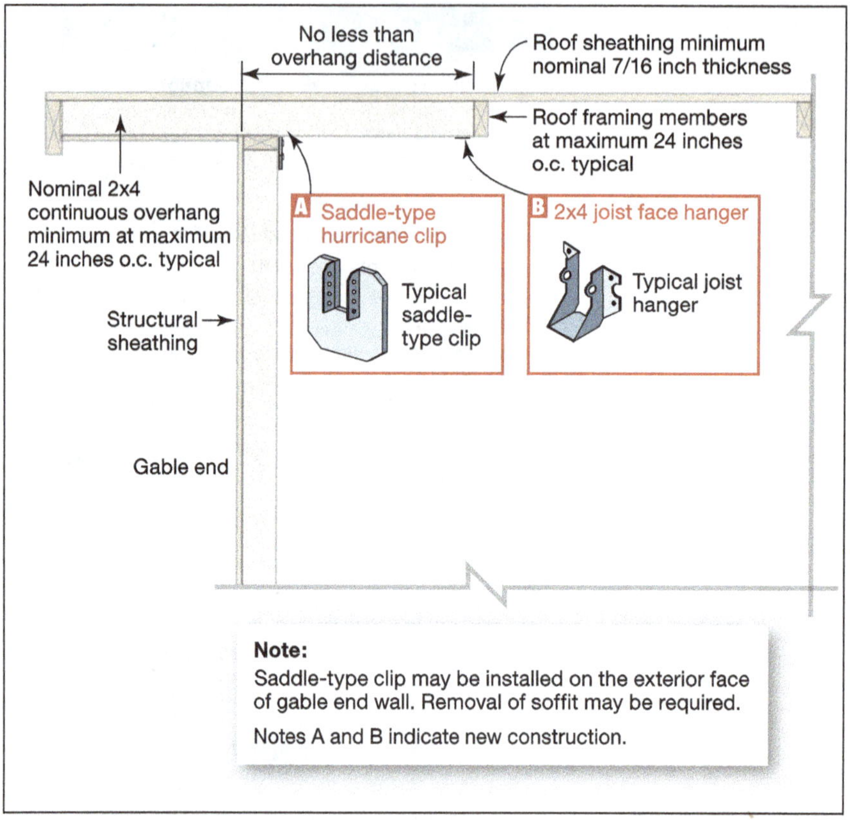

No less than
overhang distance

Roof sheathing minimum
nominal 7/16 inch thickness

Roof framing members
at maximum 24 inches
o.c. typical

Nominal 2x4
continuous overhang
minimum at maximum
24 inches o.c. typical

Structural
sheathing

Gable end

A Saddle-type
hurricane clip

Typical
saddle-
type clip

B 2x4 joist face hanger

Typical joist
hanger

Note:
Saddle-type clip may be installed on the exterior face
of gable end wall. Removal of soffit may be required.

Notes A and B indicate new construction.

Figure 15-17.
Sprayed polyurethane
foam adhesive to secure
roof deck panels

15.4.2.2 Intermediate Mitigation Package

For the Intermediate Mitigation Package to be effective, the measures in the Basic Mitigation Package must first be successfully completed. The Intermediate Mitigation Package includes protecting windows and entry doors from wind-borne debris, protecting garage doors from wind pressure and garage door glazing from wind-borne debris, bracing gable end walls over 4 feet tall, and strengthening the connections of attached structures such as porches and carports.

15.4.2.3 Advanced Mitigation Package

The Advanced Mitigation Package is the most comprehensive package of retrofits. This package can be effective only if the Basic Mitigation Package (with or without replacing the roof covering) and Intermediate Mitigation Package are also implemented. The Advanced Mitigation Package requires a more invasive inspection than the other two packages. Homes that are undergoing substantial renovation or are being rebuilt after a disaster are typically the best candidates for the Advanced Mitigation Package. The Advanced Mitigation Package requires the homeowner to provide a continuous load path as shown in Figure 15-18 and further protect openings.

15.4.2.4 Additional Mitigation Measures

The wind retrofit Mitigation Packages include important retrofits that reduce the risk of wind-related damage, but the risk cannot be eliminated entirely. By maintaining an awareness of vulnerabilities of and around a home, the homeowner can reduce the risk of wind-related damage even further. Although the mitigation measures prescribed to address these vulnerabilities are important to understand, they are not a part of the Mitigation Packages and are not eligible for HMA program funding. These additional measures, described in greater detail in FEMA P-804, include securing the exterior wall covering, implementing tree fall prevention measures, and protecting exterior equipment.

15.4.3 FEMA Wind Retrofit Grant Programs

Despite the significant damage experienced by all types of buildings during high-wind events, grant applications for wind retrofit projects have focused more on non-residential and commercial buildings than on residential buildings. FEMA developed FEMA P-804 to encourage wind mitigation of existing residential buildings.

FEMA administers two HMA grant programs that fund wind retrofit projects: HMGP and the PDM Program. Hazard mitigation is defined as any sustained action taken to reduce or eliminate long-term risk to people and property from natural hazards and their effects. The HMA process has five stages, starting with mitigation planning and ending with successful execution of a project (see Figure 15-19).

Through FEMA's HMA grant programs, applications for an individual home or groups of homes undergoing wind retrofit projects can be submitted for approval. If applications are approved, Federal funding is provided for 75 percent of the total project cost, significantly reducing the homeowner's expenses for the project. The remaining 25 percent of eligible project costs can be paid for directly or covered by donated labor, time, and materials. Refer to current HMA guidance for more details on cost-sharing (FEMA 2010a). More information on Federal assistance through HMA programs is also available in Chapter 5 of FEMA P-804.

Figure 15-18.
Continuous load path
for wind-uplift of a
residential, wood-frame
building

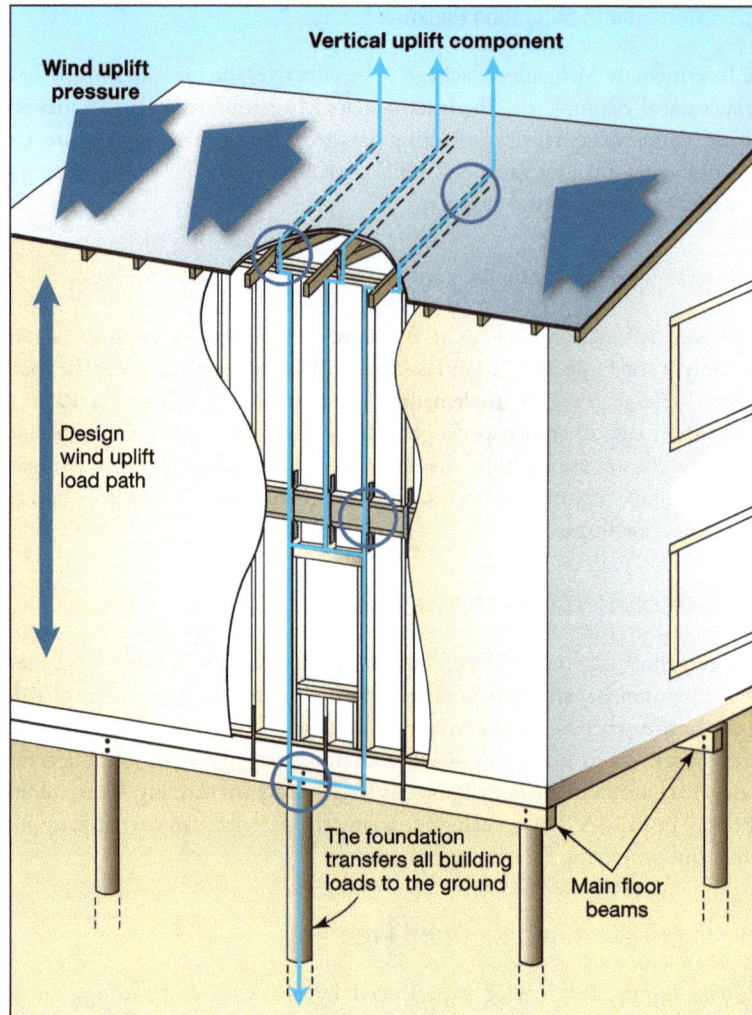

Homeowners should consider both qualitative and quantitative benefits and costs when deciding on a wind retrofit project. Applying for Federal assistance through HMA programs (as described in Chapter 5 of FEMA P-804) requires an analysis or comparison of the benefits to society compared to the cost of the project. Benefits such as reduced insurance premiums are not considered because they are an individual benefit. To assist with calculating the quantitative benefits and costs of implementing a project, FEMA developed Benefit-Cost Analysis (BCA) software, Version 4.5.5 (FEMA 2009). See Appendix C of FEMA P-804 for additional information on using the BCA software. Communities are encouraged to use the software regardless of whether they will apply for Federal funding. The software can be used to calculate project benefits such as avoided damage to the home, avoided displacement costs, and avoided loss of building contents. The evaluation discussed in Section 15.4.1 should identify all of the necessary input data needed for using the BCA software. Appendix C of FEMA P-804 provides a step-by-step guide to using the software to evaluate the cost-effectiveness of a wind retrofit project.

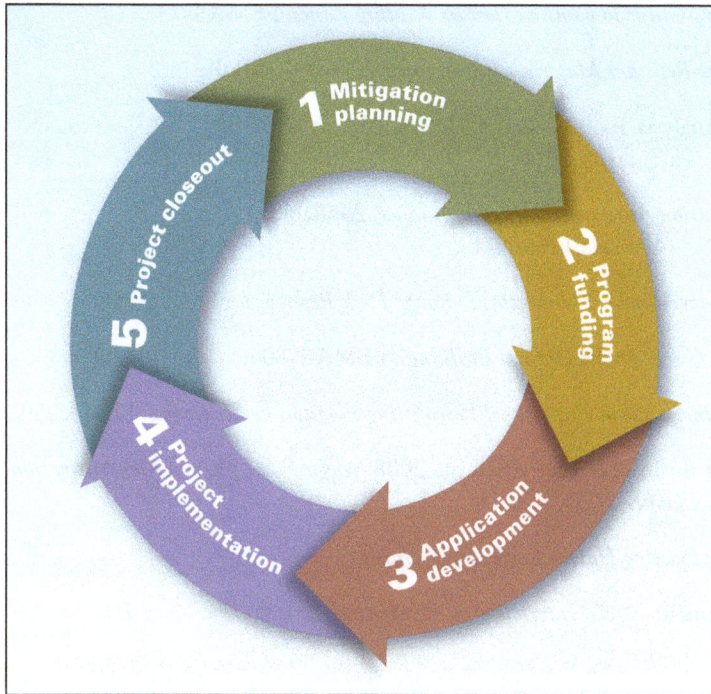

Figure 15-19.
HMA grant process
SOURCE: FEMA P-804

15.5 References

ASCE (American Society of Civil Engineers). 2005. *Minimum Design Loads for Buildings and Other Structures.* ASCE 7-05.

ASCE. *Seismic Evaluation of Existing Buildings.* ASCE 31

ASCE. *Seismic Rehabilitation of Existing Buildings.* ASCE 41.

FEMA (Federal Emergency Management Agency). 1993a. *Non-Residential Floodproofing – Requirements and Certification for Buildings Located in Special Flood Hazard Areas in Accordance with the National Flood Insurance Program.* FIA-TB-3.

FEMA. 1993b. *Wet Floodproofing Requirements for Structures Located in Special Flood Hazard Areas in Accordance with the National Flood Insurance Program.* FIA-TB-7.

FEMA. 1999. *Protecting Building Utilities from Flood Damage.* FEMA P-348.

FEMA. 2005. *Earthquake Safety Guide for Homeowners.* FEMA 530.

FEMA. 2006. *Homebuilders' Guide to Earthquake-Resistant Design and Construction.* FEMA 232.

FEMA. 2008a. *Home Builder's Guide to Construction in Wildfire Zones*. FEMA P-737.

FEMA. 2008b. *Flood Damage-Resistant Materials Requirements*. Technical Bulletin 2.

FEMA. 2009. Benefit-Cost Analysis Tool, Version 4.5.5. Available at http://www.bchelpline.com/Download.aspx. Accessed January 2011.

FEMA. 2010a. *Hazard Mitigation Assistance Unified Guidance*. Available at http://www.fema.gov/library/viewRecord.do?id=4225. Accessed June 2011.

FEMA. 2010b. *Substantial Improvement/Substantial Damage Desk Reference*. FEMA P-758.

FEMA. 2010c. *Wind Retrofit Guide for Residential Buildings*. FEMA P-804.

FEMA. 2011. *Engineering Principles and Practices of Retrofitting Floodprone Structures*. FEMA 259.

IBHS (Insurance Institute for Business & Home Safety). 2008. *Mega Fires: The Case for Mitigation – The Witch Creek Wildfire, October 21-31, 2007*.

IBHS. 2010. *FORTIFIED for Existing Homes Engineering Guide*.

ICC (International Code Council). 2006. *International Residential Code for One- and Two-Family Dwellings*. 2006 IRC.

ICC. 2009a. *International Residential Code for One- and Two-Family Dwellings*. 2009 IRC.

ICC. *International Wildland-Urban Interface Code* (IWUIC).

NFPA (National Fire Protection Association). *Code for Fire Protection of Historical Structures*.

NFPA. *Standard for Fire Protection Infrastructure for Land Development in Suburban and Rural Areas*. NFPA 1141.

NFPA. *Standard for Fire-Retardant Treated Wood and Fire-Retardant Coatings for Building Materials*. NFPA 703.

NFPA. *Standard for Reducing Structure Ignition Hazards from Wildland Fire*. NFPA 1144.

Acronyms

A

AAMA	American Architectural Manufacturers Association
ACI	American Concrete Institute
AF&PA	American Forest & Paper Association
AHJ	Authority Having Jurisdiction
AISI	American Iron and Steel Institute
ANSI	American National Standards Institute
ASCE	American Society of Civil Engineers
ASD	Allowable Stress Design
ASTM	American Society for Testing and Materials
AWPA	American Wood Protection Association

B

BCA	Benefit-Cost Analysis
BCEGS	Building Code Effectiveness Grading Schedule
BFE	base flood elevation
BUR	built-up roof

C

C&C	components and cladding

CBRA	Coastal Barrier Resources Act
CBRS	Coastal Barrier Resource System
CCM	Coastal Construction Manual
CEA	California Earthquake Authority
CMU	concrete masonry unit
CRS	Community Rating System

D

DASMA	Door & Access Systems Manufacturers Association
DFE	design flood elevation

E

EIFS	exterior insulating finishing system
ELF	Equivalent Lateral Force

F

FBC	Florida Building Code
FEMA	Federal Emergency Management Agency
FIRM	Flood Insurance Rate Map
FIS	Flood Insurance Study
FM	Factory Mutual
FRP	fiber-reinforced polymer
FS	factor of safety

G

GSA	General Services Administration

H

HMA	Hazard Mitigation Assistance
HMGP	Hazard Mitigation Grant Program

I

IBC	International Building Code
IBHS	Institute for Business and Home Safety
ICC	International Code Council
IRC	International Residential Code
ISO	Insurance Services Office

L

lb	pound(s)
LEED	Leadership in Energy and Environmental Design
LiMWA	Limit of Moderate Wave Action
LPS	lightning protection system
LRFD	Load and Resistance Factor Design

M

MEPS	molded expanded polystyrene
mph	miles per hour
MWFRS	main wind force-resisting system

N

NAHB	National Association of Home Builders

NAVD	North American Vertical Datum
NDS	National Design Specification
NFIP	National Flood Insurance Program
NFPA	National Fire Protection Association
NGVD	National Geodetic Vertical Datum
NRCA	National Roofing Contractors Association
NRCS	Natural Resources Conservation Service

O

o.c.	on center
OH	overhang
OSB	oriented strand board

P

PDM	Pre-Disaster Mitigation (Program)
plf	pound(s) per linear foot
psf	pound(s) per square foot
psi	pound(s) per square inch

S

SBC	Standard Building Code
SBS	styrene-butadiene-styrene
S-DRY	surface-dry lumber with <=19 percent moisture content
SFHA	Special Flood Hazard Area
SFIP	Standard Flood Insurance Policy
SPRI	Single-Ply Roofing Institute

T

TMS	The Masonry Society

U

UBC	Uniform Building Code
UL	Underwriters Laboratories
USACE	U.S. Army Corps of Engineers
USDN	U.S. Department of the Navy
USGBC	U.S. Green Buildings Council
USGS	U.S. Geological Survey
UV	ultraviolet

W

WFCM	Wood Frame Construction Manual
Wind-MAP	Windstorm Market Assistance Program (New Jersey)
WPPC	Wood Products Promotion Council

Y

yr	year(s)

Glossary

0-9

100-year flood – See *Base flood*.

500-year flood – Flood that has as 0.2-percent probability of being equaled or exceeded in any given year.

A

Acceptable level of risk – The level of risk judged by the building owner and designer to be appropriate for a particular building.

Adjacent grade – Elevation of the natural or graded ground surface, or structural fill, abutting the walls of a building. See also *Highest adjacent grade* and *Lowest adjacent grade*.

Angle of internal friction (soil) – A measure of the soil's ability to resist shear forces without failure.

Appurtenant structure – Under the National Flood Insurance Program, an "appurtenant structure" is "a structure which is on the same parcel of property as the principal structure to be insured and the use of which is incidental to the use of the principal structure."

B

Barrier island – A long, narrow sand island parallel to the mainland that protects the coast from erosion.

Base flood – Flood that has as 1-percent probability of being equaled or exceeded in any given year. Also known as the 100-year flood.

Base Flood Elevation (BFE) – The water surface elevation resulting from a flood that has a 1 percent chance of equaling or exceeding that level in any given year. Elevation of the base flood in relation to a specified datum, such as the National Geodetic Vertical Datum or the North American Vertical Datum. The Base Flood Elevation is the basis of the insurance and floodplain management requirements of the National Flood Insurance Program.

Basement – Under the National Flood Insurance Program, any area of a building having its floor subgrade on all sides. (Note: What is typically referred to as a "walkout basement," which has a floor that is at or above grade on at least one side, is not considered a basement under the National Flood Insurance Program.)

Beach nourishment – A project type that typically involve dredging or excavating hundreds of thousands to millions of cubic yards of sediment, and placing it along the shoreline.

Bearing capacity (soils) – A measure of the ability of soil to support gravity loads without soil failure or excessive settlement.

Berm – Horizontal portion of the backshore beach formed by sediments deposited by waves.

Best Practices – Techniques that exceed the minimum requirements of model building codes; design and construction standards; or Federal, State, and local regulations.

Breakaway wall – Under the National Flood Insurance Program, a wall that is not part of the structural support of the building and is intended through its design and construction to collapse under specific lateral loading forces without causing damage to the elevated portion of the building or supporting foundation system. Breakaway walls are required by the National Flood Insurance Program regulations for any enclosures constructed below the Base Flood Elevation beneath elevated buildings in Coastal High Hazard Areas (also referred to as Zone V). In addition, breakaway walls are recommended in areas where flood waters flow at high velocities or contain ice or other debris.

Building code – Regulations adopted by local governments that establish standards for construction, modification, and repair of buildings and other structures.

Building use – What occupants will do in the building. The intended use of the building will affect its layout, form, and function.

Building envelope – Cladding, roofing, exterior walls, glazing, door assemblies, window assemblies, skylight assemblies, and other components enclosing the building.

Building systems – Exposed structural, window, or roof systems.

Built-up roof covering – Two or more layers of felt cemented together and surfaced with a cap sheet, mineral aggregate, smooth coating, or similar surfacing material.

Bulkhead – Wall or other structure, often of wood, steel, stone, or concrete, designed to retain or prevent sliding or erosion of the land. Occasionally, bulkheads are used to protect against wave action.

C

Cladding – Exterior surface of the building envelope that is directly loaded by the wind.

Closed foundation – A foundation that does not allow water to pass easily through the foundation elements below an elevated building. Examples of closed foundations include crawlspace foundations and stem wall foundations, which are usually filled with compacted soil, slab-on-grade foundations, and continuous perimeter foundation walls.

Coastal A Zone – The portion of the coastal SFHA referenced by building codes and standards, where base flood wave heights are between 1.5 and 3 feet, and where wave characteristics are deemed sufficient to damage many NFIP-compliant structures on shallow or solid wall foundations.

Coastal barrier – Depositional geologic feature such as a bay barrier, tombolo, barrier spit, or barrier island that consists of unconsolidated sedimentary materials; is subject to wave, tidal, and wind energies; and protects landward aquatic habitats from direct wave attack.

Coastal Barrier Resources Act of 1982 (CBRA) – Act (Public Law 97-348) that established the Coastal Barrier Resources System (CBRS). The act prohibits the provision of new flood insurance coverage on or after October 1, 1983, for any new construction or substantial improvements of structures located on any designated undeveloped coastal barrier within the CBRS. The CBRS was expanded by the Coastal Barrier Improvement Act of 1991. The date on which an area is added to the CBRS is the date of CBRS designation for that area.

Coastal flood hazard area – An area subject to inundation by storm surge and, in some instances, wave action caused by storms or seismic forces. Usually along an open coast, bay, or inlet.

Coastal geology – The origin, structure, and characteristics of the rocks and sediments that make up the coastal region.

Coastal High Hazard Area – Under the National Flood Insurance Program, an area of special flood hazard extending from offshore to the inland limit of a primary frontal dune along an open coast and any other area subject to high-velocity wave action from storms or seismic sources. On a Flood Insurance Rate Map, the Coastal High Hazard Area is designated Zone V, VE, or V1-V30. These zones designate areas subject to inundation by the base flood, where wave heights or wave runup depths are 3.0 feet or higher.

Coastal processes – The physical processes that act upon and shape the coastline. These processes, which influence the configuration, orientation, and movement of the coast, include tides and fluctuating water levels, waves, currents, and winds.

Coastal sediment budget – The quantification of the amounts and rates of sediment transport, erosion, and deposition within a defined region.

Coastal Special Flood Hazard Area – The portion of the Special Flood Hazard Area where the source of flooding is coastal surge or inundation. It includes Zone VE and Coastal A Zone.

Code official – Officer or other designated authority charged with the administration and enforcement of the code, or a duly authorized representative, such as a building, zoning, planning, or floodplain management official.

Column foundation – Foundation consisting of vertical support members with a height-to-least-lateral-dimension ratio greater than three. Columns are set in holes and backfilled with compacted material. They are usually made of concrete or masonry and often must be braced. Columns are sometimes known as posts, particularly if they are made of wood.

Components and Cladding (C&C) – American Society of Civil Engineers (ASCE) 7-10 defines C&C as "... elements of the building envelope that do not qualify as part of the MWFRS [Main Wind Force Resisting System]." These elements include roof sheathing, roof coverings, exterior siding, windows, doors, soffits, fascia, and chimneys.

Conditions Greater than Design Conditions – Design loads and conditions are based on some probability of exceedance, and it is always possible that design loads and conditions can be exceeded. Designers can anticipate this and modify their initial design to better accommodate higher forces and more extreme conditions. The benefits of doing so often exceed the costs of building higher and stronger.

Connector – Mechanical device for securing two or more pieces, parts, or members together, including anchors, wall ties, and fasteners.

Consequence – Both the short- and long-term effects of an event for the building. See *Risk*.

Constructability – Ultimately, designs will only be successful if they can be implemented by contractors. Complex designs with many custom details may be difficult to construct and could lead to a variety of problems, both during construction and once the building is occupied.

Continuous load paths – The structural condition required to resist loads acting on a building. The continuous load path starts at the point or surface where loads are applied, moves through the building, continues through the foundation, and terminates where the loads are transferred to the soils that support the building.

Corrosion-resistant metal – Any nonferrous metal or any metal having an unbroken surfacing of nonferrous metal, or steel with not less than 10 percent chromium or with not less than 0.20 percent copper.

D

Dead load – Weight of all materials of construction incorporated into the building, including but not limited to walls, floors, roofs, ceilings, stairways, built-in partitions, finishes, cladding, and other similarly incorporated architectural and structural items and fixed service equipment. See also *Loads*.

Debris – Solid objects or masses carried by or floating on the surface of moving water.

Debris impact loads – Loads imposed on a structure by the impact of floodborne debris. These loads are often sudden and large. Though difficult to predict, debris impact loads must be considered when structures are designed and constructed. See also *Loads*.

Deck – Exterior floor supported on at least two opposing sides by an adjacent structure and/or posts, piers, or other independent supports.

Design event – The minimum code-required event (for natural hazards, such as flood, wind, and earthquake) and associated loads that the structure must be designed to resist.

Design flood – The greater of either (1) the base flood or (2) the flood associated with the flood hazard area depicted on a community's flood hazard map, or otherwise legally designated.

Design Flood Elevation (DFE) – Elevation of the design flood, or the flood protection elevation required by a community, including wave effects, relative to the National Geodetic Vertical Datum, North American Vertical Datum, or other datum. The DFE is the locally adopted regulatory flood elevation. If a community regulates to minimum National Flood Insurance Program (NFIP) requirements, the

DFE is identical to the Base Flood Elevation (BFE). If a community chooses to exceed minimum NFIP requirements, the DFE exceeds the BFE.

Design flood protection depth – Vertical distance between the eroded ground elevation and the Design Flood Elevation.

Design stillwater flood depth – Vertical distance between the eroded ground elevation and the design stillwater flood elevation.

Design stillwater flood elevation – Stillwater elevation associated with the design flood, excluding wave effects, relative to the National Geodetic Vertical Datum, North American Vertical Datum, or other datum.

Development – Under the National Flood Insurance Program, any manmade change to improved or unimproved real estate, including but not limited to buildings or other structures, mining, dredging, filling, grading, paving, excavation, or drilling operations or storage of equipment or materials.

Dry floodproofing – A flood retrofitting technique in which the portion of a structure below the flood protection level (walls and other exterior components) is sealed to be impermeable to the passage of floodwaters.

Dune – See *Frontal dune* and *Primary frontal dune*.

Dune toe – Junction of the gentle slope seaward of the dune and the dune face, which is marked by a slope of 1 on 10 or steeper.

Effective Flood Insurance Rate Map – See *Flood Insurance Rate Map*.

Elevation – Raising a structure to prevent floodwaters from reaching damageable portions.

Enclosure – The portion of an elevated building below the lowest floor that is partially or fully shut in by rigid walls.

Encroachment – The placement of an object in a floodplain that hinders the passage of water or otherwise affects the flood flows.

Erodible soil – Soil subject to wearing away and movement due to the effects of wind, water, or other geological processes during a flood or storm or over a period of years.

Erosion – Under the National Flood Insurance Program, the process of the gradual wearing away of land masses.

Erosion analysis – Analysis of the short- and long-term erosion potential of soil or strata, including the effects of flooding or storm surge, moving water, wave action, and the interaction of water and structural components.

Exterior-mounted mechanical equipment – Includes, but is not limited to, exhaust fans, vent hoods, air conditioning units, duct work, pool motors, and well pumps.

F

Federal Emergency Management Agency (FEMA) – Independent agency created in 1979 to provide a single point of accountability for all Federal activities related to disaster mitigation and emergency preparedness, response, and recovery. FEMA administers the National Flood Insurance Program.

Federal Insurance and Mitigation Administration (FIMA) – The component of the Federal Emergency Management Agency directly responsible for administering the flood insurance aspects of the National Flood Insurance Program as well as a range of programs designed to reduce future losses to homes, businesses, schools, public buildings, and critical facilities from floods, earthquakes, tornadoes, and other natural disasters.

Fill – Material such as soil, gravel, or crushed stone placed in an area to increase ground elevations or change soil properties. See also *Structural fill*.

Flood – Under the National Flood Insurance Program, either a general and temporary condition or partial or complete inundation of normally dry land areas from:

(1) the overflow of inland or tidal waters;

(2) the unusual and rapid accumulation or runoff of surface waters from any source;

(3) mudslides (i.e., mudflows) that are proximately caused by flooding as defined in (2) and are akin to a river of liquid and flowing mud on the surfaces of normally dry land areas, as when the earth is carried by a current of water and deposited along the path of the current; or

(4) the collapse or subsidence of land along the shore of a lake or other body of water as a result of erosion or undermining caused by waves or currents of water exceeding anticipated cyclical levels or suddenly caused by an unusually high water level in a natural body of water, accompanied by a severe storm, or by an unanticipated force of nature, such as flash flood or abnormal tidal surge, or by some similarly unusual and unforeseeable event which results in flooding as defined in (1), above.

Flood-damage-resistant material – Any construction material capable of withstanding direct and prolonged contact (i.e., at least 72 hours) with flood waters without suffering significant damage (i.e., damage that requires more than cleanup or low-cost cosmetic repair, such as painting).

Flood elevation – Height of the water surface above an established elevation datum such as the National Geodetic Vertical Datum, North American Vertical Datum, or mean sea level.

Flood hazard area – The greater of the following: (1) the area of special flood hazard, as defined under the National Flood Insurance Program, or (2) the area designated as a flood hazard area on a community's legally adopted flood hazard map, or otherwise legally designated.

Flood insurance – Insurance coverage provided under the National Flood Insurance Program.

Flood Insurance Rate Map (FIRM) – Under the National Flood Insurance Program, an official map of a community, on which the Federal Emergency Management Agency has delineated both the special hazard areas and the risk premium zones applicable to the community. (Note: The latest FIRM issued for a community is referred to as the "effective FIRM" for that community.)

Flood Insurance Study (FIS) – Under the National Flood Insurance Program, an examination, evaluation, and determination of flood hazards and, if appropriate, corresponding water surface elevations, or an examination, evaluation, and determination of mudslide (i.e., mudflow) and flood-related erosion hazards in a community or communities. (Note: The National Flood Insurance Program regulations refer to Flood Insurance Studies as "flood elevation studies.")

Flood-related erosion area or flood-related erosion prone area – A land area adjoining the shore of a lake or other body of water, which due to the composition of the shoreline or bank and high water levels or wind-driven currents, is likely to suffer flood-related erosion.

Flooding – See *Flood*.

Floodplain – Under the National Flood Insurance Program, any land area susceptible to being inundated by water from any source. See also *Flood*.

Floodplain management – Operation of an overall program of corrective and preventive measures for reducing flood damage, including but not limited to emergency preparedness plans, flood control works, and floodplain management regulations.

Floodplain management regulations – Under the National Flood Insurance Program, zoning ordinances, subdivision regulations, building codes, health regulations, special purpose ordinances (such as floodplain ordinance, grading ordinance, and erosion control ordinance), and other applications of police power. The term describes State or local regulations, in any combination thereof, that promulgate standards for the purpose of flood damage prevention and reduction.

Floodwall – A flood retrofitting technique that consists of engineered barriers designed to keep floodwaters from coming into contact with the structure.

Footing – Enlarged base of a foundation wall, pier, post, or column designed to spread the load of the structure so that it does not exceed the soil bearing capacity.

Footprint – Land area occupied by a structure.

Freeboard – Under the National Flood Insurance Program, a factor of safety, usually expressed in feet above a flood level, for the purposes of floodplain management. Freeboard is intended to compensate for the many unknown factors that could contribute to flood heights greater than the heights calculated for a selected size flood and floodway conditions, such as the hydrological effect of urbanization of the watershed. Freeboard is additional height incorporated into the Design Flood Elevation, and may be required by State or local regulations or be desired by a property owner.

Frontal dune – Ridge or mound of unconsolidated sandy soil extending continuously alongshore landward of the sand beach and defined by relatively steep slopes abutting markedly flatter and lower regions on each side.

Frontal dune reservoir – Dune cross-section above 100-year stillwater level and seaward of dune peak.

G

Gabion – Rock-filled cage made of wire or metal that is placed on slopes or embankments to protect them from erosion caused by flowing or fast-moving water.

Geomorphology – The origin, structure, and characteristics of the rocks and sediments that make up the coastal region.

Glazing – Glass or transparent or translucent plastic sheet in windows, doors, skylights, and shutters.

Grade beam – Section of a concrete slab that is thicker than the slab and acts as a footing to provide stability, often under load-bearing or critical structural walls. Grade beams are occasionally installed to provide lateral support for vertical foundation members where they enter the ground.

H

High-velocity wave action – Condition in which wave heights or wave runup depths are 3.0 feet or higher.

Highest adjacent grade – Elevation of the highest natural or regraded ground surface, or structural fill, that abuts the walls of a building.

Hurricane – Tropical cyclone, formed in the atmosphere over warm ocean areas, in which wind speeds reach 74 miles per hour or more and blow in a large spiral around a relatively calm center or "eye." Hurricane circulation is counter-clockwise in the northern hemisphere and clockwise in the southern hemisphere.

Hurricane clip or strap – Structural connector, usually metal, used to tie roof, wall, floor, and foundation members together so that they resist wind forces.

Hurricane-prone region – In the United States and its territories, hurricane-prone regions are defined by The American Society of Civil Engineers (ASCE) 7-10 as: (1) The U.S. Atlantic Ocean and Gulf of Mexico coasts where the basic wind speed for Risk Category II buildings is greater than 115 mph, and (2) Hawaii, Puerto Rico, Guam, the Virgin Islands, and American Samoa.

Hydrodynamic loads – Loads imposed on an object, such as a building, by water flowing against and around it. Among these loads are positive frontal pressure against the structure, drag effect along the sides, and negative pressure on the downstream side.

Hydrostatic loads – Loads imposed on a surface, such as a wall or floor slab, by a standing mass of water. The water pressure increases with the square of the water depth.

I

Initial costs – Include property evaluation, acquisition, permitting, design, and construction.

Interior mechanical equipment – Includes, but is not limited to, furnaces, boilers, water heaters, and distribution ductwork.

J

Jetting (of piles) – Use of a high-pressure stream of water to embed a pile in sandy soil. See also *Pile foundation*.

Jetty – Wall built from the shore out into the water to restrain currents or protect a structure.

Joist – Any of the parallel structural members of a floor system that support, and are usually immediately beneath, the floor.

L

Lacustrine flood hazard area – Area subject to inundation from lakes.

Landslide – Occurs when slopes become unstable and loose material slides or flows under the influence of gravity. Often, landslides are triggered by other events such as erosion at the toe of a steep slope, earthquakes, floods, or heavy rains, but can be worsened by human actions such as destruction of vegetation or uncontrolled pedestrian access on steep slopes.

Levee – Typically a compacted earthen structure that blocks floodwaters from coming into contact with the structure, a levee is a manmade structure built parallel to a waterway to contain, control, or divert the flow of water. A levee system may also include concrete or steel floodwalls, fixed or operable floodgates and other closure structures, pump stations for rainwater drainage, and other elements, all of which must perform as designed to prevent failure.

Limit of Moderate Wave Action (LiMWA) – A line indicating the limit of the 1.5-foot wave height during the base flood. FEMA requires new flood studies in coastal areas to delineate the LiMWA.

Littoral drift – Movement of sand by littoral (longshore) currents in a direction parallel to the beach along the shore.

Live loads – Loads produced by the use and occupancy of the building or other structure. Live loads do not include construction or environmental loads such as wind load, snow load, rain load, earthquake load, flood load, or dead load. See also *Loads*.

Load-bearing wall – Wall that supports any vertical load in addition to its own weight. See also *Non-load-bearing wall*.

Loads – Forces or other actions that result from the weight of all building materials, occupants and their possessions, environmental effects, differential movement, and restrained dimensional changes. Loads can be either permanent or variable. Permanent loads rarely vary over time or are of small magnitude. All other loads are variable loads.

Location – The location of the building determines the nature and intensity of hazards to which the building will be exposed, loads and conditions that the building must withstand, and building regulations that must be satisfied. See also *Siting*.

Long-term costs – Include preventive maintenance and repair and replacement of deteriorated or damaged building components. A hazard-resistant design can result in lower long-term costs by preventing or reducing losses from natural hazards events.

Lowest adjacent grade (LAG) – Elevation of the lowest natural or regraded ground surface, or structural fill, that abuts the walls of a building. See also *Highest adjacent grade*.

Lowest floor – Under the National Flood Insurance Program (NFIP), "lowest floor" of a building includes the floor of a basement. The NFIP regulations define a basement as "... any area of a building having its floor subgrade (below ground level) on all sides." For insurance rating purposes, this definition applies even when the subgrade floor is not enclosed by full-height walls.

Lowest horizontal structural member – In an elevated building, the lowest beam, joist, or other horizontal member that supports the building. Grade beams installed to support vertical foundation members where they enter the ground are not considered lowest horizontal structural members.

M

Main Wind Force Resisting System (MWFRS) – Consists of the foundation; floor supports (e.g., joists, beams); columns; roof raters or trusses; and bracing, walls, and diaphragms that assist in transferring loads. The American Society of Civil Engineers (ASCE) 7-10 defines the MWFRS as "... an assemblage of structural elements assigned to provide support and stability for the overall structure."

Manufactured home – Under the National Flood Insurance Program, a structure, transportable in one or more sections, built on a permanent chassis and designed for use with or without a permanent foundation when attached to the required utilities. Does not include recreational vehicles.

Marsh – Wetland dominated by herbaceous or non-woody plants often developing in shallow ponds or depressions, river margins, tidal areas, and estuaries.

Masonry – Built-up construction of building units made of clay, shale, concrete, glass, gypsum, stone, or other approved units bonded together with or without mortar or grout or other accepted methods of joining.

Mean return period – The average time (in years) between landfall or nearby passage of a tropical storm or hurricane.

Mean water elevation – The surface across which waves propagate. The mean water elevation is calculated as the stillwater elevation plus the wave setup.

Mean sea level (MSL) – Average height of the sea for all stages of the tide, usually determined from hourly height observations over a 19-year period on an open coast or in adjacent waters having free access to the sea. See also *National Geodetic Vertical Datum*.

Metal roof panel – Interlocking metal sheet having a minimum installed weather exposure of 3 square feet per sheet.

Minimal Wave Action area (MiWA) – The portion of the coastal Special Flood Hazard Area where base flood wave heights are less than 1.5 feet.

Mitigation – Any action taken to reduce or permanently eliminate the long-term risk to life and property from natural hazards.

Mitigation Directorate – Component of the Federal Emergency Management Agency directly responsible for administering the flood hazard identification and floodplain management aspects of the National Flood Insurance Program.

Moderate Wave Action area (MoWA) – See *Coastal A Zone*.

N

National Flood Insurance Program (NFIP) – Federal program created by Congress in 1968 that makes flood insurance available in communities that enact and enforce satisfactory floodplain management regulations.

National Geodetic Vertical Datum (NGVD) – Datum established in 1929 and used as a basis for measuring flood, ground, and structural elevations, previously referred to as Sea Level Datum or Mean Sea Level. The Base Flood Elevations shown on most of the Flood Insurance Rate Maps issued by the Federal Emergency Management Agency are referenced to NGVD or, more recently, to the *North American Vertical Datum*.

Naturally decay-resistant wood – Wood whose composition provides it with some measure of resistance to decay and attack by insects, without preservative treatment (e.g., heartwood of cedar, black locust, black walnut, and redwood).

New construction – *For the purpose of determining flood insurance rates* under the National Flood Insurance Program, structures for which the start of construction commenced on or after the effective date of the initial Flood Insurance Rate Map or after December 31, 1974, whichever is later, including any subsequent improvements to such structures. (See also *Post-FIRM structure*.) *For floodplain management purposes*, new construction means structures for which the start of construction commenced on or after the effective date of a floodplain management regulation adopted by a community and includes any subsequent improvements to such structures.

Non-load-bearing wall – Wall that does not support vertical loads other than its own weight. See also *Load-bearing wall*.

Nor'easter – A type of storm that occurs along the East Coast of the United States where the wind comes from the northeast. Nor'easters can cause coastal flooding, coastal erosion, hurricane-force winds, and heavy snow.

North American Vertical Datum (NAVD) – Datum established in 1988 and used as a basis for measuring flood, ground, and structural elevations. NAVD is used in many recent Flood Insurance Studies rather than the National Geodetic Vertical Datum.

O

Open foundation – A foundation that allows water to pass through the foundation of an elevated building, which reduces the lateral flood loads the foundation must resist. Examples of open foundations are pile, pier, and column foundations.

Operational costs – Costs associated with the use of the building, such as the cost of utilities and insurance. Optimizing energy efficiency may result in a higher initial cost but save in operational costs.

Oriented strand board (OSB) – Mat-formed wood structural panel product composed of thin rectangular wood strands or wafers arranged in oriented layers and bonded with waterproof adhesive.

Overwash – Occurs when low-lying coastal lands are overtopped and eroded by storm surge and waves such that the eroded sediments are carried landward by floodwaters, burying uplands, roads, and at-grade structures.

P

Pier foundation – Foundation consisting of isolated masonry or cast-in-place concrete structural elements extending into firm materials. Piers are relatively short in comparison to their width, which is usually greater than or equal to 12 times their vertical dimension. Piers derive their load-carrying capacity through skin friction, end bearing, or a combination of both.

Pile foundation – Foundation consisting of concrete, wood, or steel structural elements driven or jetted into the ground or cast-in-place. Piles are relatively slender in comparison to their length, which usually exceeds 12 times their horizontal dimension. Piles derive their load-carrying capacity through skin friction, end bearing, or a combination of both.

Platform framing – A floor assembly consisting of beams, joists, and a subfloor that creates a platform that supports the exterior and interior walls.

Plywood – Wood structural panel composed of plies of wood veneer arranged in cross-aligned layers. The plies are bonded with an adhesive that cures when heat and pressure are applied.

Post-FIRM structure – For purposes of determining insurance rates under the National Flood Insurance Program, structures for which the start of construction commenced on or after the effective date of an initial Flood Insurance Rate Map or after December 31, 1974, whichever is later, including any subsequent improvements to such structures. This term should not be confused with the term new construction as it is used in floodplain management.

Post foundation – Foundation consisting of vertical support members set in holes and backfilled with compacted material. Posts are usually made of wood and usually must be braced. Posts are also known as columns, but columns are usually made of concrete or masonry.

Precast concrete – Structural concrete element cast elsewhere than its final position in the structure. See also *Cast-in-place concrete*.

Pressure-treated wood – Wood impregnated under pressure with compounds that reduce the susceptibility of the wood to flame spread or to deterioration caused by fungi, insects, or marine borers.

Premium – Amount of insurance coverage.

Primary frontal dune – Under the National Flood Insurance Program, a continuous or nearly continuous mound or ridge of sand with relatively steep seaward and landward slopes immediately landward and adjacent to the beach and subject to erosion and overtopping from high tides and waves during major coastal storms. The inland limit of the primary frontal dune occurs at the point where there is a distinct change from a relatively steep slope to a relatively mild slope.

R

Rating factor (insurance) – A factor used to determine the amount to be charged for a certain amount of insurance coverage (premium).

Recurrence interval – The frequency of occurrence of a natural hazard as referred to in most design codes and standards.

Reinforced concrete – Structural concrete reinforced with steel bars.

Relocation – The moving of a structure to a location that is less prone to flooding and flood-related hazards such as erosion.

Residual risk – The level of risk that is not offset by hazard-resistant design or insurance, and that must be accepted by the property owner.

Retrofit – Any change or combination of adjustments made to an existing structure intended to reduce or eliminate damage to that structure from flooding, erosion, high winds, earthquakes, or other hazards.

Revetment – Facing of stone, cement, sandbags, or other materials placed on an earthen wall or embankment to protect it from erosion or scour caused by flood waters or wave action.

Riprap – Broken stone, cut stone blocks, or rubble that is placed on slopes to protect them from erosion or scour caused by flood waters or wave action.

Risk – Potential losses associated with a hazard, defined in terms of expected probability and frequency, exposure, and consequences. Risk is associated with three factors: threat, vulnerability, and consequence.

Risk assessment – Process of quantifying the total risk to a coastal building (i.e., the risk associated with all the significant natural hazards that may impact the building).

Risk category – As defined in American Society of Civil Engineers (ASCE) 7-10 and the 2012 International Building Code, a building's risk category is based on the risk to human life, health, and welfare associated with potential damage or failure of the building. These risk categories dictate which design event is used when calculating performance expectations of the building, specifically the loads the building is expected to resist.

Risk reduction – The process of reducing or offsetting risks. Risk reduction is comprised of two aspects: physical risk reduction and risk management through insurance.

Risk tolerance – Some owners are willing and able to assume a high degree of financial and other risks, while other owners are very conservative and seek to minimize potential building damage and future costs.

Riverine SFHA – The portion of the Special Flood Hazard Area mapped as Zone AE and where the source of flooding is riverine, not coastal.

Roof deck – Flat or sloped roof surface not including its supporting members or vertical supports.

Sand dunes – Under the National Flood Insurance Program, natural or artificial ridges or mounds of sand landward of the beach.

Scour – Removal of soil or fill material by the flow of flood waters. Flow moving past a fixed object accelerates, often forming eddies or vortices and scouring loose sediment from the immediate vicinity of the object. The term is frequently used to describe storm-induced, localized conical erosion around pilings and other foundation supports, where the obstruction of flow increases turbulence. See also *Erosion*.

Seawall – Solid barricade built at the water's edge to protect the shore and prevent inland flooding.

Setback – For the purpose of this Manual, a State or local requirement that prohibits new construction and certain improvements and repairs to existing coastal buildings in areas expected to be lost to shoreline retreat.

Shearwall – Load-bearing wall or non-load-bearing wall that transfers in-plane lateral forces from lateral loads acting on a structure to its foundation.

Shoreline retreat – Progressive movement of the shoreline in a landward direction; caused by the composite effect of all storms over decades and centuries and expressed as an annual average erosion rate. Shoreline retreat is essentially the horizontal component of erosion and is relevant to long-term land use decisions and the siting of buildings.

Single-ply membrane – Roofing membrane that is field-applied with one layer of membrane material (either homogeneous or composite) rather than multiple layers. The four primary types of single-ply membranes are chlorosulfonated polyethylene (CSPE) (Hypalon), ethylene propylene diene monomer (EPDM), polyvinyl chloride (PVC), and thermoplastic polyolefin (TPO).

Siting – Choosing the location for the development or redevelopment of a structure.

Special Flood Hazard Area (SFHA) – Under the National Flood Insurance Program, an area having special flood, mudslide (i.e., mudflow), or flood-related erosion hazards, and shown on a Flood Hazard Boundary Map or Flood Insurance Rate Map as Zone A, AO, A1-A30, AE, A99, AH, V, V1-V30, VE, M, or E. The area has a 1 percent chance, or greater, of flooding in any given year.

Start of construction (for other than new construction or substantial improvements under the Coastal Barrier Resources Act) – Under the National Flood Insurance Program, date the building permit was issued, provided the actual start of construction, repair, reconstruction, rehabilitation, addition placement, or other improvement was within 180 days of the permit date. The actual start means either the first placement of permanent construction of a structure on a site such as the pouring of slab or footings, the installation of piles, the construction of columns, or any work beyond the stage of excavation; or the placement of a manufactured home on a foundation. Permanent construction does not include land preparation, such as clearing, grading, and filling; nor the installation of streets or walkways; excavation for a basement, footings, piers, or foundations or the erection of temporary forms; or the installation on the property of accessory buildings, such as garages or sheds not occupied as dwelling units or not part of the main structure. For a substantial improvement, the actual start of construction means the first alteration of any wall, ceiling, floor, or other structural part of a building, whether or not that alteration affects the external dimensions of the building.

State Coordinating Agency – Under the National Flood Insurance Program, the agency of the State government, or other office designated by the Governor of the State or by State statute to assist in the implementation of the National Flood Insurance Program in that State.

Stillwater elevation – The elevations of the water surface resulting solely from storm surge (i.e., the rise in the surface of the ocean due to the action of wind and the drop in atmospheric pressure association with hurricanes and other storms).

Storm surge – Water pushed toward the shore by the force of the winds swirling around a storm. It is the greatest cause of loss of life due to hurricanes.

Storm tide – Combined effect of storm surge, existing astronomical tide conditions, and breaking wave setup.

Structural concrete – All concrete used for structural purposes, including plain concrete and reinforced concrete.

Structural fill – Fill compacted to a specified density to provide structural support or protection to a structure. See also *Fill*.

Structure – *For floodplain management purposes* under the National Flood Insurance Program (NFIP), a walled and roofed building, gas or liquid storage tank, or manufactured home that is principally above ground. *For insurance coverage purposes* under the NFIP, structure means a walled and roofed building, other than a gas or liquid storage tank, that is principally above ground and affixed to a permanent site, as well as a manufactured home on a permanent foundation. For the latter purpose, the term includes a building undergoing construction, alteration, or repair, but does not include building materials or supplies intended for use in such construction, alteration, or repair, unless such materials or supplies are within an enclosed building on the premises.

Substantial damage – Under the National Flood Insurance Program, damage to a building (regardless of the cause) is considered substantial damage if the cost of restoring the building to its before-damage condition would equal or exceed 50 percent of the market value of the structure before the damage occurred.

Substantial improvement – Under the National Flood Insurance Program, improvement of a building (such as reconstruction, rehabilitation, or addition) is considered a substantial improvement if its cost equals or exceeds 50 percent of the market value of the building before the start of construction of the improvement. This term includes structures that have incurred substantial damage, regardless of the actual repair work performed. The term does not, however, include either (1) any project for improvement of a structure to correct existing violations of State or local health, sanitary, or safety code specifications which have been identified by the local code enforcement official and which are the minimum necessary to ensure safe living conditions, or (2) any alteration of a "historic structure," provided that the alteration will not preclude the structure's continued designation as a "historic structure."

Super typhoons – Storms with sustained winds equal to or greater than 150 mph.

T

Threat – The probability that an even of a given recurrence interval will affect the building within a specified period. See *Risk*.

Tornado – A rapidly rotating vortex or funnel of air extending groundward from a cumulonimbus cloud

Tributary area – The area of the floor, wall, roof, or other surface that is supported by the element. The tributary area is generally a rectangle formed by one-half the distance to the adjacent element in each applicable direction.

Tropical cyclone – A low-pressure system that generally forms in the tropics, and is often accompanied by thunderstorms.

Tropical depression – Tropical cyclone with some rotary circulation at the water surface. With maximum sustained wind speeds of up to 39 miles per hour, it is the second phase in the development of a hurricane.

Tropical disturbance – Tropical cyclone that maintains its identity for at least 24 hours and is marked by moving thunderstorms and with slight or no rotary circulation at the water surface. Winds are not strong. It is a common phenomenon in the tropics and is the first discernable stage in the development of a hurricane.

Tropical storm – Tropical cyclone that has 1-minute sustained wind speeds averaging 39 to 74 miles per hour (mph).

Tsunami – Long-period water waves generated by undersea shallow-focus earthquakes, undersea crustal displacements (subduction of tectonic plates), landslides, or volcanic activity.

Typhoon – Name given to a hurricane in the area of the western Pacific Ocean west of 180 degrees longitude.

U

Underlayment – One or more layers of felt, sheathing paper, non-bituminous saturated felt, or other approved material over which a steep-sloped roof covering is applied.

Undermining – Process whereby the vertical component of erosion or scour exceeds the depth of the base of a building foundation or the level below which the bearing strength of the foundation is compromised.

Uplift – Hydrostatic pressure caused by water under a building. It can be strong enough lift a building off its foundation, especially when the building is not properly anchored to its foundation.

V

Variance – Under the National Flood Insurance Program, grant of relief by a community from the terms of a floodplain management regulation.

Violation – Under the National Flood Insurance Program (NFIP), the failure of a structure or other development to be fully compliant with the community's floodplain management regulations. A structure or other development without the elevation certificate, other certifications, or other evidence of compliance required in Sections 60.3(b)(5), (c)(4), (c)(10), (d)(3), (e)(2), (e)(4), or (e)(5) of the NFIP regulations is presumed to be in violation until such time as that documentation is provided.

Vulnerability – Weaknesses in the building or site location that may result in damage. See *Risk*.

W

Water surface elevation – Under the National Flood Insurance Program, the height, in relation to the National Geodetic Vertical Datum of 1929 (or other datum, where specified), of floods of various magnitudes and frequencies in the floodplains of coastal or riverine areas.

Wave – Ridge, deformation, or undulation of the water surface.

Wave height – Vertical distance between the wave crest and wave trough. Wave crest elevation is the elevation of the crest of a wave, referenced to the National Geodetic Vertical Datum, North American Vertical Datum, or other datum.

Wave overtopping – Occurs when waves run up and over a dune or barrier.

Wave runup – Is the rush of water up a slope or structure. Wave runup occurs as waves break and run up beaches, sloping surfaces, and vertical surfaces.

Wave runup depth – At any point is equal to the maximum wave runup elevation minus the lowest eroded ground elevation at that point.

Wave runup elevation – Is the elevation reached by wave runup, referenced to the National Geodetic Vertical Datum or other datum.

Wave setup – Increase in the stillwater surface near the shoreline due to the presence of breaking waves. Wave setup typically adds 1.5 to 2.5 feet to the 100-year stillwater flood elevation and should be discussed in the Flood Insurance Study.

Wave slam – The action of wave crests striking the elevated portion of a structure.

Wet floodproofing – A flood retrofitting technique that involves modifying a structure to allow floodwaters to enter it in such a way that damage to a structure and its contents is minimized.

Z

Zone A – Under the National Flood Insurance Program, area subject to inundation by the 100-year flood where wave action does not occur or where waves are less than 3 feet high, designated Zone A, AE, A1-A30, A0, AH, or AR on a Flood Insurance Rate Map.

Zone AE – The portion of the Special Flood Hazard Area (SFHA) not mapped as Zone VE. It includes the Moderate Wave Action area, the Minimal Wave Action area, and the riverine SFHA.

Zone B – Areas subject to inundation by the flood that has a 0.2-percent chance of being equaled or exceeded during any given year, often referred to the as 500-year flood. Zone B is provided on older flood maps, on newer maps this is referred to as "shaded Zone X."

Zone C – Designates areas where the annual probability of flooding is less than 0.2 percent. Zone C is provided on older flood maps, on newer maps this is referred to as "unshaded Zone X."

Zone V – See *Coastal High Hazard Area.*

Zone VE – The portion of the coastal Special Flood Hazard Area where base flood wave heights are 3 feet or greater, or where other damaging base flood wave effects have been identified, or where the primary frontal dune has been identified.

Zone X – Under the National Flood Insurance Program, areas where the flood hazard is lower than that in the Special Flood Hazard Area. Shaded Zone X shown on recent Flood Insurance Rate Maps (Zone B on older maps) designate areas subject to inundation by the 500-year flood. Unshaded Zone X (Zone C on older Flood Insurance Rate Maps) designate areas where the annual probability of flooding is less than 0.2 percent.

Zone X (Shaded) – Areas subject to inundation by the flood that has a 0.2-percent chance of being equaled or exceeded during any given year, often referred to the as 500-year flood.

Zone X (Unshaded) – Designates areas where the annual probability of flooding is less than 0.2 percent.

Index, Volume II

Bold text indicates chapter titles or major headings. Italicized page numbers indicates a figure or table.

Z

www.ingramcontent.com/pod-product-compliance
Lightning Source LLC
Chambersburg PA
CBHW080225270326
41926CB00020B/4147